T0342311

Modeling Infectious Diseases

IN HUMANS AND ANIMALS

Modeling Infectious Diseases

IN HUMANS AND ANIMALS

Matt J. Keeling and Pejman Rohani

PRINCETON UNIVERSITY PRESS

PRINCETON AND OXFORD

Published by Princeton University Press, 41 William Street, Princeton, New Jersey 08540
In the United Kingdom: Princeton University Press, 3 Market Place, Woodstock, Oxfordshire OX20 1SY
All Rights Reserved

Library of Congress Control Number: 2006939548
ISBN: 978-0-691-11617-4

British Library Cataloging-in-Publication Data is available

This book has been composed in Times Roman

press.princeton.edu

10 9 8 7 6 5 4 3 2 1

Dedicated to Alison and Erin,
for always making me laugh and smile.
Lots of love
—Matt Keeling

I dedicate this book to my wife Rebecca,
and my daughters Mae and Sophia
I love and cherish you all
—Pej Rohani

Contents

Acknowledgments xiii

Chapter 1 Introduction 1

1.1 Types of Disease 1
1.2 Characterization of Diseases 3
1.3 Control of Infectious Diseases 5
1.4 What Are Mathematical Models? 7
1.5 What Models Can Do 8
1.6 What Models Cannot Do 10
1.7 What Is a Good Model? 10
1.8 Layout of This Book 11
1.9 What Else Should You Know? 13

Chapter 2 Introduction to Simple Epidemic Models 15

2.1 Formulating the Deterministic *SIR* Model 16
 2.1.1 The *SIR* Model Without Demography 19
  2.1.1.1 The Threshold Phenomenon 19
  2.1.1.2 Epidemic Burnout 21
  2.1.1.3 Worked Example: Influenza in a Boarding School 26
 2.1.2 The *SIR* Model With Demography 26
  2.1.2.1 The Equilibrium State 28
  2.1.2.2 Stability Properties 29
  2.1.2.3 Oscillatory Dynamics 30
  2.1.2.4 Mean Age at Infection 31
2.2 Infection-Induced Mortality and *SI* Models 34
 2.2.1 Mortality Throughout Infection 34
  2.2.1.1 Density-Dependent Transmission 35
  2.2.1.2 Frequency Dependent Transmission 36
 2.2.2 Mortality Late in Infection 37
 2.2.3 Fatal Infections 38
2.3 Without Immunity: The *SIS* Model 39
2.4 Waning Immunity: The *SIRS* Model 40
2.5 Adding a Latent Period: The *SEIR* Model 41
2.6 Infections with a Carrier State 44
2.7 Discrete-Time Models 46
2.8 Parameterization 48
 2.8.1 Estimating R_0 from Reported Cases 50
 2.8.2 Estimating R_0 from Seroprevalence Data 51
 2.8.3 Estimating Parameters in General 52
2.9 Summary 52

Chapter 3   Host Heterogeneities   54

3.1 Risk-Structure: Sexually Transmitted Infections   55
    3.1.1 Modeling Risk Structure   57
        3.1.1.1 High-Risk and Low-Risk Groups   57
        3.1.1.2 Initial Dynamics   59
        3.1.1.3 Equilibrium Prevalence   62
        3.1.1.4 Targeted Control   63
        3.1.1.5 Generalizing the Model   64
        3.1.1.6 Parameterization   64
    3.1.2 Two Applications of Risk Structure   69
        3.1.2.1 Early Dynamics of HIV   71
        3.1.2.2 Chlamydia Infections in Koalas   74
    3.1.3 Other Types of Risk Structure   76
3.2 Age-Structure: Childhood Infections   77
    3.2.1 Basic Methodology   78
        3.2.1.1 Initial Dynamics   80
        3.2.1.2 Equilibrium Prevalence   80
        3.2.1.3 Control by Vaccination   81
        3.2.1.3 Parameterization   82
    3.2.2 Applications of Age Structure   84
        3.2.2.1 Dynamics of Measles   84
        3.2.2.2 Spread and Control of BSE   89
3.3 Dependence on Time Since Infection   93
    3.3.1 $SEIR$ and Multi-Compartment Models   94
    3.3.2 Models with Memory   98
    3.3.3 Application: SARS   100
3.4 Future Directions   102
3.5 Summary   103

Chapter 4   Multi-Pathogen/Multi-Host Models   105

4.1 Multiple Pathogens   106
    4.1.1 Complete Cross-Immunity   107
        4.1.1.1 Evolutionary Implications   109
    4.1.2 No Cross-Immunity   112
        4.1.2.1 Application: The Interaction of Measles and
                   Whooping Cough   112
        4.1.2.2 Application: Multiple Malaria Strains   115
    4.1.3 Enhanced Susceptibility   116
    4.1.4 Partial Cross-Immunity   118
        4.1.4.1 Evolutionary Implications   120
        4.1.4.2 Oscillations Driven by Cross-Immunity   122
    4.1.5 A General Framework   125
4.2 Multiple Hosts   128
    4.2.1 Shared Hosts   130
        4.2.1.1 Application: Transmission of Foot-and-Mouth Disease   131
        4.2.1.2 Application: Parapoxvirus and the Decline of the
                   Red Squirrel   133

4.2.2 Vectored Transmission   135
    4.2.2.1 Mosquito Vectors   136
    4.2.2.2 Sessile Vectors   141
4.2.3 Zoonoses   143
    4.2.3.1 Directly Transmitted Zoonoses   144
    4.2.3.2 Vector-Borne Zoonoses: West Nile Virus   148
4.3 Future Directions   151
4.4 Summary   153

Chapter 5   Temporally Forced Models   155

5.1 Historical Background   155
  5.1.1 Seasonality in Other Systems   158
5.2 Modeling Forcing in Childhood Infectious Diseases: Measles   159
  5.2.1 Dynamical Consequences of Seasonality: Harmonic and
       Subharmonic Resonance   160
  5.2.2 Mechanisms of Multi-Annual Cycles   163
  5.2.3 Bifurcation Diagrams   164
  5.2.4 Multiple Attractors and Their Basins   167
  5.2.5 Which Forcing Function?   171
  5.2.6 Dynamical Trasitions in Seasonally Forced Systems   178
5.3 Seasonality in Other Diseases   181
  5.3.1 Other Childhood Infections   181
  5.3.2 Seasonality in Wildlife Populations   183
    5.3.2.1 Seasonal Births   183
    5.3.2.2 Application: Rabbit Hemorrhagic Disease   185
5.4 Summary   187

Chapter 6   Stochastic Dynamics   190

6.1 Observational Noise   193
6.2 Process Noise   193
  6.2.1 Constant Noise   195
  6.2.2 Scaled Noise   197
  6.2.3 Random Parameters   198
  6.2.4 Summary   199
    6.2.4.1 Contrasting Types of Noise   199
    6.2.4.2 Advantages and Disadvantages   200
6.3 Event-Driven Approaches   200
  6.3.1 Basic Methodology   201
    6.3.1.1 The SIS Model   202
  6.3.2 The General Approach   203
    6.3.2.1 Simulation Time   203
  6.3.3 Stochastic Extinctions and The Critical Community Size   205
    6.3.3.1 The Importance of Imports   209
    6.3.3.2 Measures of Persistence   212
    6.3.3.3 Vaccination in a Stochastic Environment   213
  6.3.4 Application: Porcine Reproductive and Respiratory Syndrome 214
  6.3.5 Individual-Based Models   217

6.4 Parameterization of Stochastic Models   219
6.5 Interaction of Noise with Heterogeneities   219
    6.5.1 Temporal Forcing   219
    6.5.2 Risk Structure   220
    6.5.3 Spatial Structure   221
6.6 Analytical Methods   222
    6.6.1 Fokker-Plank Equations   222
    6.6.2 Master Equations   223
    6.6.3 Moment Equations   227
6.7 Future Directions   230
6.8 Summary   230

Chapter 7   Spatial Models   232

7.1 Concepts   233
    7.1.1 Heterogeneity   233
    7.1.2 Interaction   235
    7.1.3 Isolation   236
    7.1.4 Localized Extinction   236
    7.1.5 Scale   236
7.2 Metapopulations   237
    7.2.1 Types of Interaction   240
        7.2.1.1 Plants   240
        7.2.1.2 Animals   241
        7.2.1.3 Humans   242
        7.2.1.4 Commuter Approximations   243
    7.2.2 Coupling and Synchrony   245
    7.2.3 Extinction and Rescue Effects   246
    7.2.4 Levins-Type Metapopulations   250
    7.2.5 Application to the Spread of Wildlife Infections   251
        7.2.5.1 Phocine Distemper Virus   252
        7.2.5.2 Rabies in Raccoons   252
7.3 Lattice-Based Models   255
    7.3.1 Coupled Lattice Models   255
    7.3.2 Cellular Automata   257
        7.3.2.1 The Contact Process   258
        7.3.2.2 The Forest-Fire Model   259
        7.3.2.3 Application: Power laws in Childhood Epidemic Data 260
7.4 Continuous-Space Continuous-Population Models   262
    7.4.1 Reaction-Diffusion Equations   262
    7.4.2 Integro-Differential Equations   265
7.5 Individual-Based Models   268
    7.5.1 Application: Spatial Spread of Citrus Tristeza Virus   269
    7.5.2 Applilcation: Spread of Foot-and-mouth Disease in the
           United Kingdom   274
7.6 Networks   276
    7.6.1 Network Types   277
        7.6.1.1 Random Networks   277
        7.6.1.2 Lattices   277

7.6.1.3 Small World Networks 279
7.6.1.4 Spatial Networks 279
7.6.1.5 Scale-Free Networks 279
7.6.2 Simulation of Epidemics on Networks 280
7.7 Which Model to Use? 282
7.8 Approximations 283
7.8.1 Pair-Wise Models for Networks 283
7.8.2 Pair-Wise Models for Spatial Processes 286
7.9 Future Directions 287
7.10 Summary 288

Chapter 8 Controlling Infectious Diseases 291

8.1 Vaccination 292
8.1.1 Pediatric Vaccination 292
8.1.2 Wildlife Vaccination 296
8.1.3 Random Mass Vaccination 297
8.1.4 Imperfect Vaccines and Boosting 298
8.1.5 Pulse Vaccination 301
8.1.6 Age-Structured Vaccination 303
8.1.6.1 Application: Rubella Vaccination 304
8.1.7 Targeted Vaccination 306
8.2 Contact Tracing and Isolation 308
8.2.1 Simple Isolation 309
8.2.2 Contact Tracing to Find Infection 312
8.3 Case Study: Smallpox, Contact Tracing, and Isolation 313
8.4 Case Study: Foot-and-Mouth Disease, Spatial Spread, and Local Control 321
8.5 Case Study: Swine Fever Virus, Seasonal Dynamics, and Pulsed Control 327
8.5.1 Equilibrium Properties 329
8.5.2 Dynamical Properties 331
8.6 Future Directions 333
8.7 Summary 334

References 337

Index 361

Parameter Glossary 367

Acknowledgments

The writing of this book has truly benefited from discussions over the years with our long-term collaborators, especially Bryan Grenfell, David Earn, Aaron King, Ottar Bjørnstad, Neil Ferguson, Julia Gog, and Graham Medley. We are also grateful to the various reviewers who were very generous with advice and suggestions, especially Andy Dobson, Hans Heesterbeek, and Marc Lipsitch. In addition, we would like to sincerely thank various group members who gave thoughtful and detailed comments on various drafts of this book: Matt Bonds, Marc Choisy, Ken Eames, Jim MacDonald, Jon Read, Mike Tildesley, and Helen Wearing.

Modeling Infectious Diseases

IN HUMANS AND ANIMALS

Chapter One

―――――――――――――――

Introduction

This book is designed as an introduction to the modeling of infectious diseases. We start with the simplest of mathematical models and show how the inclusion of appropriate elements of biological complexity leads to improved understanding of disease dynamics and control. Throughout, our emphasis is on the development of models, and their use either as predictive tools or as a means of understanding fundamental epidemiological processes. Although many theoretical results can be proved analytically for very simple models, we have generally focused on results obtained by computer simulation, providing analytical results only where they lead to a more generic interpretation of model behavior. Where practical, we have illustrated the general modeling principles with applied examples from the recent literature. We hope this book motivates readers to develop their own models for diseases of interest, expanding on the model frameworks given here.

1.1. TYPES OF DISEASE

The Oxford English Dictionary defines a disease as "a condition of the body, or of some part or organ of the body, in which its functions are disturbed or deranged; a morbid physical condition; a departure from the state of health, especially when caused by structural change." This definition encompasses a wide range of ailments from AIDS to arthritis, from the common cold to cancer. The fine-scale classification of diseases varies drastically between different scientific disciplines. Medical doctors and veterinary clinicians, for example, are primarily interested in treating human patients or animals and, as such, are most concerned about the infection's pathophysiology (affecting, for example, the central nervous system) or clinical symptoms (for example, secretory diarrhea). Microbiologists, on the other hand, focus on the natural history of the causative organism: What is the etiological agent (a virus, bacterium, protozoan, fungus, or prion)? and what are the ideal conditions for its growth? Finally, epidemiologists are most interested in features that determine patterns of disease and its transmission.

In general terms, we may organize diseases according to several overlapping classifications (Figure 1.1). Diseases can be either infectious or noninfectious. Infectious diseases (such as influenza) can be passed between individuals, whereas noninfectious diseases (such as arthritis) develop over an individual's lifespan. The epidemiology of noninfectious diseases is primarily a study of risk factors associated with the chance of developing the disease (for example, the increased risk of lung cancer attributable to smoking). In contrast, the primary risk factor for catching an infectious disease is the presence of infectious cases in the local population—this tenet is reflected in all the mathematical models presented in this book. These two categories, infectious and noninfectious, are not necessarily mutually exclusive. Infection with the human papillomavirus (HPV), for example, is firmly associated with (although not necessary for developing) cervical cancer,

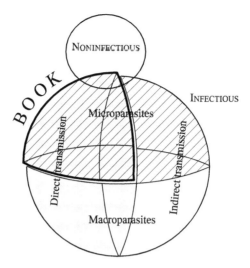

Figure 1.1. A representation of the various types of diseases. The two main groups of infectious and noninfectious diseases are represented by the two circles. The infectious group can be further subdivided into directly transmitted (gray) and indirectly transmitted infections as well as into diseases due to microparasites (hatched) and macroparasites. The focus of this book are the diseases subsumed within the intersection of the hatched and gray areas, which is offset and bounded by a thick black line.

thus bridging the two fields. This book focuses on infectious diseases, where models have great predictive power at the population scale and over relatively short time scales.

Infectious diseases can be further subdivided (Figure 1.1). The infecting pathogen can be either a microparasite (hatched in diagram) or a macroparasite. Microparasites, as the same suggests, are small (usually single-cell organisms) and are either viruses, bacteria, protoza, or prions; macroparasites are any larger form of pathogen and include helminths and flukes. Although the biological distinction between these two groups of organisms is clear, from a modeling perspective the boundaries are less well defined. In general, microparasitic infections develop rapidly from a small number of infecting particles so the internal dynamics of the pathogen within the host can often be safely ignored. As a result, we are not interested in the precise abundance of pathogens within the host; instead we focus on the host's infection status. In contrast, macroparasites such as helminths have a *complex life cycle* within the host which often needs to be modeled explicitly. In addition, the worm burden, or the number of parasites within the host, represents an important contributing factor to pathogenicity and disease transmission. We focus in this book on microparasites, where extensive long-term data and a good mechanistic understanding of the transmission dynamics have led to a wealth of well-parameterized models.

Infectious diseases (both macro- and microparasitic) can also be subdivided into two further categories (Figure 1.1), depending on whether transmission of infection is direct (shaded gray) or indirect. Direct transmission is when infection is caught by close contact with an infectious individual. The great majority of microparasitic diseases, such as influenza, measles, and HIV, are directly transmitted, although there are exceptions such as cholera, which is waterborne. Generally, directly transmitted pathogens do not survive for long outside the host organism. In contrast, indirectly transmitted parasites are passed between hosts via the environment; most macroparasitic diseases, such as those caused by helminths and schistosomes, are indirectly transmitted, spending part of their life cycle outside of their hosts. In addition, there is a class of diseases where transmission is via a secondary host or vector, usually insects such as mosquitoes, tsetse flies, or ticks. However, this transmission route can be considered as two sequential direct transmission events, from the primary host to the insect and then from the insect to another primary host.

The models and diseases of this book are focused toward the study of directly transmitted, microparasitic infectious diseases. As such, this subset represents only a fraction of the whole field of epidemiological modeling and analysis, but one in which major advances have occurred over recent decades.

Worldwide there are about 1,415 known human pathogens of which 217 (15%) are viruses or prions and 518 (38%) are bacteria or rickettsia; hence around 53% are microparasites (Cleaveland et al. 2001). Of these pathogens, 868 (61%) are zoonotic and can therefore be transmitted from animals to humans. Around 616 pathogens of domestic livestock are known, of which around 18% are viral and 25% bacterial. However, if we restrict our attention to the 70 pathogens listed by the Office International des Epizooties (which contain the most prominent and infectious livestock diseases), we find that 77% are microparasites (Cleaveland et al. 2001). The lower number of known livestock pathogens compared to human pathogens probably reflects to some degree our natural anthropocentric bias. Similarly, very few infectious diseases of wildlife are known or studied in any detail, and yet wildlife reservoirs may be important sources of novel emerging human infections. It is therefore clear that the study of microparasitic infectious diseases encompasses a huge variety of hosts and diseases.

1.2. CHARACTERIZATION OF DISEASES

The progress of an infectious microparasitic disease is defined qualitatively in terms of the level of pathogen within the host, which in turn is determined by the growth rate of the pathogen and the interaction between the pathogen and the host's immune response. Figure 1.2 shows a much simplified infection profile. Initially, the host is *susceptible* to infection: No pathogen is present; just a low-level nonspecific immunity within the host. At time 0, the host encounters an infectious individual and becomes infected with a microparasite; the abundance of the parasite grows over time. During this early phase the individual may exhibit no obvious signs of infection and the abundance of pathogen may be too low to allow further transmission—individuals in this phase are said to be in the *exposed* class. Once the level of parasite is sufficiently large within the host, the potential exists to transmit the infection to other susceptible individuals; the host is *infectious*. Finally, once the individual's immune system has cleared the parasite and the host is therefore no longer infectious, they are referred to as *recovered*.

This fundamental classification (as susceptible, exposed, infectious, or recovered) solely depends on the host's ability to transmit the pathogen. This has two implications. First, the disease status of the host is irrelevant—it is not important whether the individual is showing symptoms; an individual who feels perfectly healthy can be excreting large amounts of pathogen (Figure 1.2). Second, the boundaries between exposed and infectious (and infectious and recovered) are somewhat fuzzy because the ability to transmit does not simply switch on and off. This uncertainty is further complicated by the variability in responses between different individuals and the variability in pathogen levels over the infectious period; it is only with the recent advances in molecular techniques that these

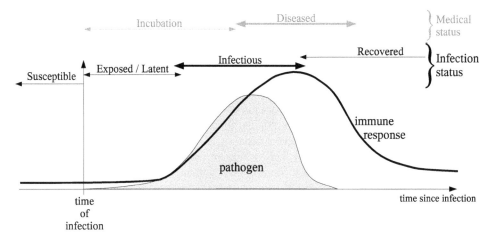

Figure 1.2. A caracature of the time-line of infection, showing the dynamics of the pathogen (gray area) and the host immune response (black line) as well as labeling the various infection classes: susceptible, exposed, infectious, and recovered. Note that the diseased period, when symptoms are experienced, is not necessarily correlated with any particular infection class.

within-host individual-level details are beginning to emerge. Our classification of hosts as susceptible, exposed, infectious, or recovered can therefore be compared to the ecological concept of a metapopulation (Levins 1969; Hanski and Gilpin 1991), in which the within-host density of the pathogen is ignored and each host is simply classified as being in one of a limited number of categories.

Although Figure 1.2 shows an example of a disease profile that might be modeled as $SEIR$ (susceptible-exposed-infectious-recovered), other within-host profiles are also common. Often, it is mathematically simpler and justifiable at the population scale to ignore the exposed class, reducing the number of equations by one and leading to SIR dynamics. Some infections, especially of plants, are more appropriately described by the SI (susceptible-infectious) paradigm; for such diseases, the host is infectious soon after it is infected, such that the exposed period can be safely ignored, and remains infectious until its death. Other infectious diseases, in particular sexually transmitted infections (such as gonorrhoea), are better described by an SIS (susceptible-infectious-susceptible) framework, because once recovered (or following treatment) the host is once again susceptible to infection. In the majority of cases this renewed susceptibility is due to the vast antigenic variation associated with sexually transmitted diseases. Finally, many diseases have profiles that are individualistic and require specific model formulation. Smallpox has a definite short prodromal period before the symptoms emerge when the infected individual is mobile and can widely disseminate the virus but infectiousness has not reached its peak. Hepatitis B has a carrier state such that some infected individuals do not fully recover but transmit at a low level for the rest of their lives. Chlamydia (and many other sexually transmitted diseases) may be asymptomatic, such that some infected individuals do not suffer from the disease even though they are able to transmit infection. Similarly, infections such as meningitis or MRSA (methicillin resistant streptococus aurius) are widespread in the general population and usually benign, with only occasional symptomatic outbreaks. All of these more complex epidemiological behaviors require greater subdivision of the population and therefore models that deal explicitly with these extra classes.

Although such qualitative descriptions of disease dynamics allow us to understand the behavior of infection within an individual and may even shed some light on potential transmission, if we are to extrapolate from the individual-level dynamics to the population-scale epidemic, numerical values are required for many of the key parameters. Two fundamental quantities govern the population-level epidemic dynamics: the basic reproductive ratio, R_0, and the timescale of infection, which is measured by the infectious period for SIS and SIR infections or by a mixture of exposed and infectious periods in diseases with $SEIR$ dynamics (for details, see Chapter 2). The basic reproductive ratio is one of the most critical epidemiological parameters because it defines the average number of secondary cases an average primary case produces in a totally susceptible population. Among other things, this single parameter allows us to determine whether a disease can successfully invade or not, the threshold level of vaccination required for eradication, and the long-term proportion of susceptible individuals when the infection is endemic.

One of the key features of epidemiological modeling is the huge variability in infection profiles, parameter values, and timescales. Many childhood infectious diseases (such as measles, rubella, or chickenpox) follow the classic $SEIR$ profile, have high basic reproductive ratios ($R_0 \approx 17$ for both measles and whooping cough in England and Wales from 1945 to 1965), and short infected periods (of less than one month). In contrast, diseases such as HIV have a much more complex infection profile with transmission rates varying as a function of time since infection, R_0 is crucially dependent on sexual behavior ($R_0 \approx 4$ for the homosexual population in the United Kingdom, whereas $R_0 \approx 11$ for female prostitutes in Kenya), and infection is lifelong. Between these two extremes lies a vast array of other infectious diseases, with their own particular characteristics and parameters.

1.3. CONTROL OF INFECTIOUS DISEASES

One of the primary reasons for studying infectious diseases is to improve control and ultimately to eradicate the infection from the population. Models can be a powerful tool in this approach, allowing us to optimize the use of limited resources or simply to target control measures more efficiently. Several forms of control measure exist; all operate by reducing the average amount of transmission between infectious and susceptible individuals. Which control strategy (or mixture of strategies) is used will depend on the disease, the hosts, and the scale of the epidemic.

The practice of vaccination began with Edward Jenner in 1796 who developed a vaccine against smallpox—which remains the only disease to date that has been eradicated world-wide. Vaccination acts by stimulating a host immune response, such that immunized individuals are protected against infection. Vaccination is generally applied prophylactically to a large proportion of the population, so as to greatly reduce the number of susceptible individuals. Such prophylactic vaccination campaigns have successfully reduced the incidence of many childhood infections in the developed world by vaccinating the vast majority of young children and infants. In 1988, the World Heath Organization (WHO) resolved to use similar campaigns to eradicate polio worldwide by 2005—this is still ongoing work although much progress has been made to date.

Although vaccination offers a very powerful method of disease control, there are many associated difficulties. Generally, vaccines are not 100% effective, and therefore only a proportion of vaccinated individuals are protected. Some vaccines can have adverse side effects; the vaccine against smallpox can be harmful (sometimes fatal) to those with eczema, asthma, or are immuno-suppressed, and may even cause cases of smallpox. Some vaccines provide only limited immunity, whether this is due to the natural waning of immunity in the host or to antigenic variation in the pathogen. Finally, in the face of a novel (or unexpected) epidemic, reactive vaccination may prove to be too slow to prevent a large outbreak. Therefore, in many situations, alternative control measures are necessary.

Vaccination operates by reducing the number of susceptible individuals in the population.

Quarantine, or the isolation of known or suspected infectious individuals, is one of the oldest known forms of disease control. During the fifteenth and sisteenth century, Venice, Italy, practiced a policy of quarantine against all ships arriving from areas infected with plague, and in 1665 the village of Eyam in Derbyshire, UK, famously quarantined themselves in an effort to prevent the plague spreading to neighboring villages. Today quarantining is still a powerful control measure; was used to combat SARS in 2003, and it is a rapid first response against many invading pathogens. Quarantining essentially operates by preventing infectious individuals from mixing with susceptible individuals, hence stopping transmission. The primary advantage of quarantining is that it is simple and generic; quarantining is effective even when the causative agent is unknown. However, quarantining can be applied only once an infectious individual is identified, by which time the individual may have been transmitting infection for many days. In addition, unless the number of cases is small, quarantining can be a prohibitive drain on resources.

Quarantining operates by reducing the number of infected individuals freely mixing in the population.

Culling acts by depleting the host population by killing hosts. From recent years, there are three clear examples of culling as a means of control. During the 2001 foot-and-mouth epidemic in the United Kingdom, culling was used as a fast and effective control measure. Ring culling, removing all citrus trees within, 1900 feet of identified infected trees, is currently being used to control Citrus Canker disease (caused by the *Xanthomonas axonopodis* pv. *citri* bacterium) in Florida. For this disease, proximity was judged to be the main risk factor, with around 95% of inoculum dispersal being within 1,900 feet. Finally, large-scale field trials in the United Kingdom and Ireland have examined the effect of culling badger populations on the spread of bovine tuberculosis in cattle—based on the assumption that badgers act as a reservoir of infection.

Obviously culling is applicable only to animal and plant diseases, and even then it is used only against harmful, rapidly spreading, pathogens when other control measures are ineffective. Culling is usually indiscriminate, killing both infected and susceptible hosts and thereby reducing transmission in two distinct ways. However, culling is often locally targeted such that this severe action is limited to regions of high risk. It is vitally important that culling measures are highly targeted and tightly controlled—there is a clear trade-off between sufficient culling to control the epidemic and excess culling that could be more detrimental than an uncontrolled epidemic. Models can be extremely powerful tools of discrimination in such situations.

Culling operates by reducing both the number of infected and susceptible individuals in the population.

Contact tracing, although not a control measure in itself, is an important tool in efficiently targeting other control measures and therefore limiting disease spread. Contact tracing operates by questioning infected individuals about their behavior, identifying potential transmission contacts, and therefore finding individuals who are likely to be infected but are not yet symptomatic. The individuals identified by contact tracing can then be vaccinated, quarantined, or hospitalized, depending on the nature of the infection.

Contact tracing operates by refining the targeting of other control measures.

1.4. WHAT ARE MATHEMATICAL MODELS?

Recent years have seen an increasing trend in the number of publications, both in high-profile journals and more generally, that utilize mathematical models (Figure 1.3). This is associated with an increased understanding of what models can offer in terms of prediction and insight. Any model can be typically thought of as a conceptual tool that explains how an object (or system of objects) will behave. A mathematical model uses the language of mathematics to produce a more refined and precise description of the system. In epidemiology, models allow us to translate between behavior at various scales, or extrapolate from a known set of conditions to another. As such, models allow us to predict the population-level epidemic dynamics from an individual-level knowledge of epidemiological factors, the long-term behavior from the early invasion dynamics, or the impact of vaccination on the spread of infection.

Models come in a variety of forms—from highly complex models that (like jet aircraft) need a range of experts to create and maintain them, to simple "toy" models that (like bicycles) can be easily understood, modified, and adapted. The decision whether to travel by bike or aircraft depends on several factors, such as time, distance, and cost. Similarly, which sort of model is the most appropriate depends on the precision or generality required, the available data, and the time frame in which results are needed. By definition, all models are "wrong," in the sense that even the most complex will make some simplifying assumptions. It is, therefore, difficult to express definitively which model is "right," though naturally we are interested in developing models that capture the essential features of a system. Ultimately, we are faced with a rather subjective measure of the *usefulness* of any model.

Formulating a model for a particular problem is a trade-off between three important and often conflicting elements: accuracy, transparency, and flexibility. *Accuracy*, the ability to reproduce the observed data and reliably predict future dynamics, is clearly vital, but whether a qualitative or quantitative fit is necessary depends on the details of the problem. A qualitative fit may be sufficient to gain insights into the dynamics of an infectious disease, but a good quantitative fit is generally necessary if the model is used to advise on future control policies. Accuracy generally improves with increasing model complexity and the inclusion of more heterogeneities and relevant biological detail. Clearly, the feasibility of model complexity is compromised by computational power, the mechanistic understanding of disease natural history, and the availability of necessary parameters. Consequently, the accuracy of any model is always limited. *Transparency* comes from being able to understand (either analytically or more often numerically) how the various

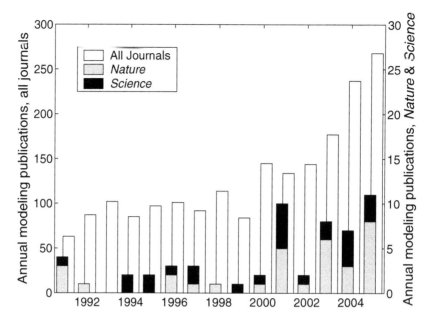

Figure 1.3. An indication of the increasing importance and use of mathematical models in the epidemiological literature. White bars show the approximate number of publications in the entire scientific literature that utilize models of infectious diseases. (Data are obtained from ISI Web of Science, and include all publications that contain in their title or abstract the words "epidemic," and "infect*," and either "model*" or "simulat*.") The gray and black bars show the number of these publications to be found in *Nature* and *Science* respectively, providing some indication of the high impact of such work. (These papers were identified from their title and abstract.) Note the different scales for general papers and those in *Nature* or *Science*.

model components influence the dynamics and interact. This is usually achieved by successively adding or removing components and building upon general intuitions from simpler models. As the number of model components increases, it becomes more difficult to assess the role of each component and its interactions with the whole. Transparency is, therefore, often in direct opposition to accuracy. *Flexibility* measures the ease with which the model can be adapted to new situations; this is vital if the model is to evaluate control policies or predict future disease levels in an ever-changing environment. Most mechanistic models (such as those within this book) are based on well-understood disease transmission principles and are therefore highly flexible, whereas "black-box" time-series tools (such as neural nets) that may be able to accurately reproduce a given time series of reported cases are less amenable to modification.

1.5. WHAT MODELS CAN DO

Models have two distinct roles, *prediction* and *understanding*, which are related to the model properties of accuracy and transparency, and therefore can often be in conflict. We usually require a high degree of accuracy from any predictive model, whereas transparency is a more important quality of models used to improve our understanding.

Prediction is the most obvious use of models. It requires that the model is as accurate as possible and therefore includes all of the known complexities and population-level heterogeneities. Predictive models can have great power in specific situations, guiding difficult policy decisions where a trade-off between two (or more) alternative control strategies exists. It is interesting to contrast how models were used during the 2001 UK foot-and-mouth epidemic to how models were used to investigate the control of potential bio-terrorist releases of smallpox. Both of these scenarios called for detailed, accurate predictive modeling. Chapter 8 provides a comprehensive description of the types of models that could be used to tackle these two problems.

During the UK foot-and-mouth epidemic in 2001, two primary questions were addressed using models: first, was the epidemic "under control" and second, whether additional targeted culling would lead to a reduction in the total loss of livestock. Three distinct models were used, based on different judgements of the known dynamical complexity. Each of these models had their own advantages and problems, but fortunately—due to the robustness of the problem—all three models provided similar advice: A large-scale epidemic was predicted and additional, locally targeted culling would reduce the overall loss of livestock by dramatically reducing the number of cases (Keeling 2005).

Smallpox is a potential bio-terrorist weapon given the high mortality rate and the large number of susceptible individuals in the population. Here the main question focused on the best method of control, mass-vaccination or targeted measures. Mass-vaccination is obviously most effective against a large-scale outbreak, but due to side effects of the vaccine, a large-scale vaccination campaign could cause more health problems than a small-scale epidemic. Again a variety of models were used, ranging from the very simple (Meltzer et al. 2001) to highly complex (Halloran et al. 2002). However, these models provided conflicting advice, in part due to uncertainties in the epidemiological parameters and to the different underlying assumptions. It is still an open problem to determine under what conditions it is optimal to mass-vaccinate against smallpox.

Accurate predictive models can have an additional use as a statistical tool. The failure to accurately predict epidemic behavior in a particular area can act as a diagnostic warning that underlying parameters and behavior may be different from the norm. For example, whereas isolated cases of meningitis may be the norm, several clustered cases can signify the start of a localized epidemic; accurate models should be able to predict a threshold number of cases above which prompt action is required (Stollenwerk and Jansen 2003). Similarly, during an eradication campaign, regions that do not respond as rapidly as the models predict could be detected and targeted for more intensive control measures. Finally, detailed modeling and robust statistical analysis of reported and hospitalized cases may be able to identify the early emergence of an epidemic.

Models can also be used to understand how an infectious disease spreads in the real world, and how various complexities affect the dynamics. In essence, models provide epidemiologists with a ideal world in which individual factors can be examined in isolation and where every facet of the disease spread is recorded in perfect detail. With such tools we can examine, in a fairly robust and generic framework, a range of issues such as the effects of variable numbers of partners on the spread of sexually transmitted diseases, the effects of increased transmission between children during school terms, or the effects of localized spread of infection.

It might seem that such modeling approaches are driven purely by scientific curiosity with little relevance to practical matters or particular infections. However, the insights gained from such modeling are often robust and generic, and therefore can be applied to a wide variety of particular problems. Moreover, the understanding gained can help us to develop more sophisticated predictive models and gather more relevant epidemiological data, allowing us to decide which elements are important and which can be neglected. Finally, it is only by developing an intuition for infection patterns, building from simple models to more complex ones, that we can begin to understand all the rich complexities and dynamics that are observed in the real world.

Although some of the model examples given in this book are predictive in nature, because they are accurate characterizations of reality, the majority of this book is devoted to obtaining a deeper understanding of epidemiological patterns. However, the techniques can be utilized to build more complex predictive models.

1.6. WHAT MODELS CANNOT DO

Models also have their limitations. It is impossible to build a fully accurate model; there will always be some element of the host behavior or quirk of the disease that is unknown or even unknowable. Consider trying to make an accurate model for a human airborne infection (say influenza); such a model would need to account for variations in transmission with temperature and climate, capture the day-to-day movement and interaction of individuals, and encompass the variability in susceptibility due to genetic factors or past infections. Even if such a model could be built, the chance nature of transmission would still prevent perfect prediction. We will never be able to predict the precise course of an epidemic, or which people will be infected. The best that we can hope for is models that provide confidence intervals on the epidemic behavior and determine the risk of infection for various groups of hosts.

1.7. WHAT IS A GOOD MODEL?

It is clear from what has already been said that no model is perfect, and no model can accurately predict the detailed outcome of an infection process. However, two key points define a good model. First, a model should be *suited to its purpose*—that is, it should be as simple as possible, but no simpler—having an appropriate balance of accuracy, transparency, and flexibility. A model that is designed to help us understand the behavior of an infectious disease should concentrate on the characteristics that are of interest while simplifying all others. A model built for accurate prediction should provide a comprehensive picture of the full dynamics, and include all the relevant features of the disease and host, although determining which factors are relevant and which may be safely ignored is a complex and skilled process. Second, the model should be *parameterizable* (where necessary) *from available data*. Thus, although a predictive model requires the inclusion of many features, it is important that they can all be parameterized from available data. Hence, in many situations—such as at the start of an emerging (novel) epidemic—it may be impossible to produce a good predictive model. In contrast, if we are interested only in understanding an epidemic pattern, there is far less need for a model to accurately represent a particular scenario, and so parameterization and availability of data are less important. Therefore, it is clear that what constitutes a good model is context dependent. Throughout this book we have attempted to use only examples of good modeling practice.

1.8. LAYOUT OF THIS BOOK

This book is divided into seven major chapters (plus this introduction), which deal with different characteristic patterns of epidemics and the models that can be used to understand and capture their behavior. Each chapter can be further subdivided into methods and applications or case studies: Methods explain the underlying principles of models, and applications show how this type of model has been used in understanding specific disease dynamics in human and animal populations. Finally, to help with a rapid understanding of each chapter, crucial synopsis of the main points are highlighted throughout the chapter as follows:

This sentence is an important summary of this section.

The chapters are as follows:

1. Introduction

This chapter introduces the basic concepts and ideas of modeling, as well as providing a brief overview of epidemiological characteristics and behavior.

2. Introduction to Simple Epidemic

This first true chapter reviews the basic building blocks of most epidemiological models: the compartmental SIS and SIR models. For these models it is possible to develop some analytical results, which are useful in the understanding of simple epidemics and in our interpretation of more complex scenarios. Therefore, although much of the analytical detail of this chapter has been considered elsewhere in far greater depth, this work is included to provide a firm foundation to further developments. In addition, we discuss the dynamics of other compartmental models, such as those with exposed or carrier classes.

3. Host Heterogeneities

Almost all populations (with the exception of large clonal agricultures) can be sub-divided into different groups, depending upon characteristics that may influence the risk of catching and transmitting an infection. For example, an individual's pattern of sexual behavior clearly determines the likelihood of catching sexually transmitted diseases. Models that include such heterogeneities, therefore, are a better representation of reality in such cases. Other important population-level heterogeneities include age, gender, behavior, and even generic susceptibility, although this may be difficult to ascertain. Understanding how such heterogeneities influence transmission allows us to determine which individuals in a population are most at risk and the most effective means of targeting control.

4. Multi-Pathogen Multi-Host Models

Many diseases can be caught and transmitted by numerous hosts (e.g., most livestock species are susceptible to foot-and-mouth disease); other diseases require an obligatory second host species to complete the transmission cycle (e.g., vector-borne diseases such as malaria). These are all examples of multi-host single-pathogen systems. The converse situation, single-host multi-pathogen, occurs if we are interested in the competition between two strains that are to some degree cross-reactive. The dynamics of many diseases can be fully explained only as the interaction of many cross-reactive strains. The prediction of future worldwide influenza strains and the possibility of pandemics are based on such models.

5. Temporally Forced Models

Many diseases undergo periodic forcing from some external environmental factor. Examples include the opening and closing of school terms for childhood diseases, climatic variations affecting the transmission of diseases by arthropod vectors, or the annual planting and harvesting of agricultural crops. Such simple periodic perturbations to the basic models can have dramatic consequences, driving regular multi-year epidemic cycles or even complex/chaotic dynamics. Much of this dynamical behavior is illustrated for measles infection, where there is a rich history in studying the seasonal behavior observed in major cities in Europe and the United States. The concept of a bifurcation diagram is introduced, which provides a powerful and intuitive visualization of epidemic patterns because a key parameter is varied.

6. Stochastic Dynamics

All diseases are subject to stochasticity in terms of the chance nature of transmission, and so, in principle, a stochastic model is always more realistic than a deterministic one. However, the relative magnitude of stochastic fluctuations reduces as the number of cases increases; therefore, in large populations, with a high level of disease incidence, a deterministic model may be a good approximation. However, when the population is small or the disease is rare (for example, due to vaccination or early during an epidemic), stochasticity can have a major impact. In particular, stochasticity can have three major effects: It pushes the system away from the deterministic attractor such that transients may play a significant role, it can cause chance extinctions of the disease, and finally it introduces variances and co-variances that can influence the deterministic behavior. Hence, if we are interested in eradication of a disease, or if irregular epidemics are observed, stochastic modeling is generally necessary.

7. Spatial Models

Spatial heterogeneities occur at a range of scales, although it is the two extremes that are most commonly studied. At the local scale, strong correlations emerge between the infectious status of interacting individuals, such that infected hosts are spatially aggregated and patches of susceptibles exist. It is with such individual-based models that we can capture the wavelike spread of invading diseases through populations. At the other extreme, there are the heterogeneities between distinct populations, such as different towns and cities, or different geographic regions. Models for such scenarios act at a much larger scale, and are generally concerned with the correlation between the populations and the effects of the transmission between them. This chapter provides a comprehensive review of a large range of model types that are used to capture the spatial spread of infection.

8. Controling Infectious Diseases

The final chapter deals with the applied issue of control, and as such focuses on issues that are of public health and veterinary importance. In particular it discusses how the models and understanding gained in the previous chapters can be used to optimally target control measures so as to minimize the impact of infection. Vaccination, quarantining, culling, and contact tracing are all together, as well as detailed studies of smallpox, foot-and-mouth disease, and swine fever virus.

1.8.1. Accompanying Software

Given that much of the focus in this book is on understanding epidemiological problems that are analytically intractable, a variety of programs are available at the Web page:

http://press.princeton.edu/titles/8459.html

The purpose of these programs is to help those interested in further exploring the models presented and to aid in the development of new models. For each model, four different versions of the program are included: a *Java* program that allows parameters to be adjusted and the dynamics to be inspected graphically, and programs in *C, Fortran* and *Matlab* that can be freely adapted and tailored to suit a given situation. These programs serve to ease and facilitate the use of models in epidemiological problems—they are not intended as a programming guide and are not necessarily the most refined or efficient programming approach.

1.9. WHAT ELSE SHOULD YOU KNOW?

An epidemiological modeler requires a wide arsenal of tools and techniques in addition to an understanding of disease behavior and the ability to construct models (which is the focus of this book). Here we give a brief summary of potentially useful additional techniques and disciplines, and point the reader to general texts that provide an introduction to the subject matter.

In much of this book, we go about modeling by first considering the underlying assumptions about the processes involved and how these scale the number of infected and susceptible individuals. We then proceed to express these assumptions in terms of mathematical equations, which are then analyzed. The actual process of analysis usually involves computing the solution numerically, because the models are often analytically intractable. A variety of off-the-shelf software may be used to solve ordinary differential equations. Examples include *ModelMaker* and *Stella*. Numerous scientific computing packages also permit quite sophisticated modeling, such as *Mathematica, Maple, R, Mathcad,* and *Matlab*. Our personal preference is to carry out (nearly) all model analyses in code written in a low-level computing language, such as *C* or *Fortran*. Almost all the figures in this book were generated by first simulating the equations using *C*-code, followed by analyses of these computer-generated data in *Matlab*, which has the additional advantage of superb graphics capabilities.

A sound knowledge of *statistics* is obviously an essential asset, allowing us to link models with available data, and providing a framework for analyzing the results of model simulations. Statistics therefore has three main purposes in epidemiological modeling. First, it allows us to analyze any data that are available and to use this information to derive parameters (and associated confidence intervals) for our model. Second, statistics provides a powerful set of tools to compare model output with available data, with the general aim of showing that the model is a good fit. Finally, statistics can be used to compare the results of multiple model simulations, and so elucidate the differences between them—this is particularly important if the models are stochastic. Statistics itself is a very diverse subject area and no single publication could possibly cover the entire disciple. However, *Statistics in Theory and Practise* by Robert Lupton (1993) or *Introductory Statistics* by Ronald and Thomas Wonnacott (1990) provide a good introduction to the basics, whereas

Markov Chain Monte Carlo by Dani Gamerman (1997) is a suitable primer for the rapidly expanding field of MCMC and Bayesian analysis.

When we are lucky enough to have long-term case reporting data for a particular infection (as is the case with measles in England and Wales), then techniques from *time series analysis* become important. Although technically part of the discipline of statistics, time series analysis is a subject area in its own right with many powerful tools that can be used to examine longitudinal data and to extract meaningful patterns, such as periodic oscillations or density dependence. The two books, *Time Series Analysis* by James Hamilton (1994) and *Non-linear Time Series, A Dynamical System Approach* by Howell Tong (1990) review the range of techniques and possible applications.

From a more model-based perspective, both *dynamical systems* and *numerical techniques* play important roles. Dynamical systems provides an understanding of model behavior, in terms of fixed points, attractors, and stability. Some basic knowledge of this subject area has been assumed throughout the book (generally in terms of equilibria and their stability), but readers seeking a more detailed knowledge could consult *Introduction to the Modern Theory of Dynamical Systems* by Anatole Katok and Boris Hasselblatt (1996) or *Dynamical Systems: Differential Equations, Maps and Chaotic Behavior* by D. K. Arrowsmith and C. M. Place (1992).

Numerical techniques are often important; the models that are developed are unlikely to be analytically tractable; therefore, it is imperative that we can compute the solutions accurately (and rapidly). The *Numerical Recipes* series of books give a wide range of techniques and sample routines, in a variety of programming languages, that can be readily used. Throughout this book, the vast majority of differential equations models have been integrated using the fourth-order Runge-Kutta scheme, which provides a good balance between accuracy and simplicity; stochastic models and partial-differential equation models have their own particular methods of simulation that are discussed at the relevant places in the text.

Many other subject areas impinge upon epidemiological modeling. There are clear parallels between *ecology* and epidemiology, with many ecological concepts possessing a counterpoint in epidemiological theory. In addition, if we are interested in wildlife diseases, then a solid ecological knowledge and model of the host dynamics in the absence of infection is a necessary starting point. Over the past few years, molecular and genetic techniques have provided the epidemiologist with unparallelled insights into the dynamics of infection within the host organism and how the pathogen operates at a fundamental level. Nowadays it is common to study infections where the genome of the causative agent has been fully sequenced and annotated. An understanding of *immunology* and *genetics* is vital if epidemiological modelers are to take advantage of this growing body of knowledge that underpins the basic infection behavior. Finally, we are often interested in how infectious diseases relate to humans and human activities. In such circumstances, it is often impossible to separate disease control from issues of *economics* and *social sciences*.

Chapter Two

Introduction to Simple Epidemic Models

The process of modeling in epidemiology has, at its heart, the same underlying philosophy and goals as ecological modeling. Both endeavors share the ultimate aim of attempting to understand the prevalence and distribution of a species, together with the factors that determine incidence, spread, and persistence (Anderson and May 1979; May and Anderson 1979; Earn et al. 1998; Shea 1998; Bascompte and Rodriguez-Trelles 1998). Whereas in ecology the precise abundance of a species is often of great interest, establishing or predicting the exact number of, for example, virus particles in a population (or even within an individual) is both daunting and infeasible.[1] Instead, modelers concentrate on the simpler task of categorising individuals in the "host" population according to their infection status. As such, these epidemiological models can be compared to the metapopulation models used in ecology (Levins 1969; Hanski and Gilpin 1997), where each individual host is considered as a patch of resource for the pathogen, with transmission and recovery analogous to dispersal and extinction (Nee 1994; Rohani et al. 2002).

There is a long and distinguished history of mathematical modeling in epidemiology, going back to the eighteenth century (Bernoulli 1760). However, it was not until the early 1900s that the increasingly popular dynamical systems approaches were applied to epidemiology. Since then, theoretical epidemiology has witnessed numerous significant conceptual and technical developments. Although these historical advances are both interesting and important, we will side-step a detailed account of these progressions and instead refer interested readers to the lucid texts by Bailey (1975), Anderson and May (1991), Grenfell and Dobson (1995), Daley and Gani (1999), and Hethcote (2000).

In this chapter, we start with the simplest epidemiological models and consider both infections that are strongly immunizing as well as those that do not give rise to immunity. In either case, the underlying philosophy is to assume individuals are either susceptible to infection, currently infectious, or recovered (previously infected and consequently immune). Although the progress between these classes could be presented as a verbal argument, to make quantitative predictions we must translate them into formal mathematical terms. This chapter presents the mathematical equations describing these models, together with the kinds of model analyses that have proved useful to epidemiologists. These approaches encompass both deterministic and probabilistic frameworks. The preliminary models will, of necessity, be somewhat primitive and ignore a number of well-known and important heterogeneities, such as differential susceptibility to infection, contact networks, variation in the immunological response, and transmissibility. Many of these complexities are addressed in subsequent chapters.

---

[1] A clear exception to this is the study of macroparasitic infections (such as helminthic worms) where the worm burden is of great interest because it can substantially affect both host and parasite demography. In this book, we do not deal with such systems and refer interested readers to Anderson and May's (1991) thorough treatment of the subject.

2.1. FORMULATING THE DETERMINISTIC SIR MODEL

In order to develop a model, we first need to discuss terminology. Infectious diseases are typically categorized as either acute or chronic. The term acute refers to "fast" infections, where relatively rapid immune response removes pathogens after a short period of time (days or weeks). Examples of acute infections include influenza, distemper, rabies, chickenpox, and rubella. Chronic infections, on the other hand, last for much longer periods (months or years) and examples include herpes and chlamydia. We start the development of models by focusing on acute infections, assuming the pathogen causes illness for a period of time followed by (typically lifelong) immunity. This scenario is mathematically best described by the so-called S-I-R models (Dietz 1967). This formalism, which was initially studied in depth by Kermack and McKendrick (1927), categorizes hosts within a population as **S**usceptible (if previously unexposed to the pathogen), **I**nfected (if currently colonized by the pathogen), and **R**ecovered (if they have successfully cleared the infection).

Now that we know how many categories there are and how these categories are defined, the question becomes how individuals move from one to the other. In the simplest case (ignoring population demography—births, deaths, and migration), we have only the transitions $S \to I$ and $I \to R$. The second of these is easier, so we deal with it first. Those infected can move to the recovered class only once they have fought off the infection. For acute infections, it is generally observed that the amount of time spent in the infectious class (the "infectious period") is distributed around some mean value, which can be often estimated accurately from clinical data. From a modeling perspective, this translates into the probability of an individual moving from I to R being dependent on how long they have been in the I class. However, modelers often make the simplifying assumption that the recovery rate γ (which is the inverse of the infectious period) is constant; this leads to far more straightforward equations and exponentially distributed infectious periods. In Section 3.3 we deal with the dynamical consequences of alternative, more realistic formulations of the infectious period.

The progression from S to I clearly involves disease transmission, which is determined by three distinct factors: the prevalence of infecteds, the underlying population contact structure, and the probability of transmission given contact. For a directly transmitted pathogen, there has to be contact between susceptible and infected individuals and the probability of this happening is determined by the respective levels of S and I, as well as the inherent contact structure of the host population. Finally, we need to take into account the likelihood that a contact between a susceptible and an infectious person results in transmission.

These conceptual descriptions of the model can be represented by a flow diagram. The flow diagram for the SIR model uses black arrows to show the movement between the S and I classes and the I and R classes. The fact that the level of the infectious disease influences the rate at which a susceptible individual moves into the infected class is shown by the dotted gray arrow. This book uses such flow diagrams to illustrate the essential epidemiological characteristics. In general, demography will be ignored in these diagrams to reduce the number of arrows and hence improve their clarity.

Flow diagrams provide a useful graphical method of illustrating the main epidemiological assumptions underlying a model.

The previous paragraphs make the derivation of the transmission term seem relatively straightforward. Unfortunately, the precise structure of the transmission term is plagued by controversy and conflicting nomenclature. To explain some of these issues, we start by defining the *force of infection*, λ, which is defined as the *per capita* rate at which susceptible individuals contract the infection. Thus, the rate at which new infecteds are produced is λX, where X is the *number* of individuals in class S. This force of infection is intuitively proportional to the number of infectious individuals. For directly transmitted pathogens, where transmission requires contact between infecteds and susceptibles, two general possibilities exist depending on how we expect the contact structure to change with population size: $\lambda = \beta Y / N$ and $\lambda = \beta Y$ (where Y is the number of infectious individuals, N is the total population size, and β is the product of the contact rates and transmission probability). The first of these formulations will be referred to as frequency dependent (or mass action) transmission and the second as density dependent (or pseudo mass action) transmission. (We note, however, that Hamer (1906) refers to $\lambda = \beta Y$ as mass-action; this duality of nomenclature causes much confusion). A mechanistic derivation of the transmission term is provided in Box 2.1.

It is important to distinguish between these two basic assumptions in terms of the underlying structure of contacts within the population. Frequency-dependent transmission reflects the situation where the number of contacts is independent of the population size. At least as far as directly transmitted diseases are concerned, this agrees with our natural intuition about human populations. We would not expect someone living in, for example, London (population 7 million), or New York (population 8 million), to transmit an infectious disease over 50 times more than someone living in Cambridge, United Kingdom (population 130,000) or Cambridge, Massachusetts (population 100,000). The number of close contacts that are likely to result in disease transmission will be determined by social constraints, resulting in similar patterns of transmission in any large town or city. Indeed, estimates of measles transmission rates in England and Wales demonstrate no relationship with population size (Bjørnstad et al. 2002). In contrast, density-dependent transmission assumes that as the population size (or more accurately, as the density of individuals) increases, so does the contact rate. The rationale is that if more individuals are crowded into a given area (and individuals effectively move at random), then the contact rate will be greatly increased. As a rule of thumb, frequency-dependent (mass action) transmission is considered appropriate for vector-borne pathogens and those with heterogeneous contact structure. Density-dependent (psuedo mass action) transmission, however, is generally considered to be more applicable to plant and animal diseases, although care must be taken in the distinction between number and density of organisms (for further discussion, we refer interested readers to McCallum et al. 2001; Begon et al. 2002).

The distinction between these two transmission mechanisms becomes pronounced when host population size varies, otherwise the $1/N$ term can be absorbed into the parameterization of β in the mass-action term. As a simplification to our notation, it is convenient to let $S(= X/N)$ and $I(= Y/N)$ to be the *proportion* of the population that are susceptible or infectious, respectively. In this new notation our mass-action (frequency-dependent) assumption becomes βSI, which informs about the rate at which new infectious individuals (as a proportion of the total population size) are infected.

In some instances, such as when we need to employ integer-valued stochastic models (Chapter 6), variables need to reflect numbers rather than proportions. To distinguish between these different approaches, this book will consistently use X, Y, and Z to represent the *numbers* in each class and S, I, and R to represent *proportions* (see Parameter Glossary).

Box 2.1 The Transmission Term

Here, we derive from first principles the frequency-dependent (mass action) transmission term (also called proportionate mixing; Anderson and May 1992), which is commonly used in epidemic models. It assumes homogenous mixing in the population, which means everyone interacts with equal probability with everyone else; it discards possible heterogeneities arising from age, space, or behavioral aspects (see Chapters 3 and 7).

Consider a susceptible individual with an average κ contacts per unit of time. Of these, a fraction $I = Y/N$ are contacts with infected individuals (where Y is the *number* of infectives and N is the total population size). Thus, during a small time interval (from t to $t + \delta t$), the number of contacts with infecteds is $(\kappa Y/N) \times (\delta t)$. If we define c as the probability of successful disease transmission following a contact, then $1 - c$ is the probability that transmission does not take place. Then, by independence of contacts, the probability (denoted by $1 - \delta q$) that a susceptible individual escapes infection following $(\kappa Y/N \times \delta t)$ contacts is

$$1 - \delta q = (1 - c)^{(\kappa Y/N)\delta t}.$$

Hence, the probability that the individual is infected following any of these contacts is simply δq.

We now define $\beta = -\kappa \log(1 - c)$ and substitute into the expression for $1 - \delta q$, which allows us to rewrite the probability of transmission in a small time interval δt as

$$\delta q = 1 - e^{-\beta Y \delta t/N}.$$

To translate this probability into the *rate* at which transmission occurs, first we expand the exponential term (recalling that $e^x = 1 + x + \frac{x^2}{2!} + \frac{x^3}{3!} \ldots$), divide both sides by δt, and take the limit of $\delta q/\delta t$ as $\delta t \to 0$. This gives:

$$\frac{dq}{dt} = \beta Y/N,$$

which is the transmission rate per susceptible individual. In fact, this quantity is often represented by λ and referred to as the "force of infection"—it measures the per capita probability of acquiring the infection (Anderson and May 1991). Then, by extention, the total rate of transmission to the entire susceptible population is given by

$$\frac{dX}{dt} = -\lambda X = -\beta XY/N,$$

where X is defined as the *number* of susceptibles in the population. If we rescale the variables (by substituting $S = X/N$ and $I = Y/N$) so that we are dealing with *fractions* (or densities), the above equation becomes

$$\frac{dS}{dt} = -\beta IS.$$

The transmission term is generally described by frequency dependence $\beta XY/N$ (or βSI), or by density dependence βXY.

The differences between frequency- and density-dependent transmission become important if the population size changes or we are trying to parameterize disease models across a range of population sizes.

2.1.1. The SIR Model Without Demography

To introduce the model equations, it is easiest to consider a "closed population" without demographics (no births, deaths, or migration). The scenario we have in mind is a large naive population into which a low level of infectious agent is introduced and where the resulting epidemic occurs sufficiently quickly that demographic processes are not influential. We also assume homogeneous mixing, whereby intricacies affecting the pattern of contacts are discarded, yielding βSI as the transmission term. Given the premise that underlying epidemiological probabilities are constant, we get the following SIR equations:

$$\frac{dS}{dt} = -\beta SI, \tag{2.1}$$

This is online program 2.1

$$\frac{dI}{dt} = \beta SI - \gamma I, \tag{2.2}$$

$$\frac{dR}{dt} = \gamma I. \tag{2.3}$$

The parameter γ is called the removal or recovery rate, though often we are more interested in its reciprocal $(1/\gamma)$, which determines the average infectious period. For most diseases, the infectious period can be estimated relatively precisely from epidemiological data. Note that epidemiologists typically do not write the equation for the R class because we know that $S + I + R = 1$, hence knowing S and I will allow us to calculate R. These equations have the initial conditions $S(0) > 0$, $I(0) > 0$, and $R(0) = 0$. An example of the epidemic progression generated from these equations is presented in Figure 2.1; the conversion of susceptible to infectious to recovered individuals is clear.

Despite its extreme simplicity, this model (equations (2.1) to (2.3)) cannot be solved explicitly. That is, we cannot obtain an exact analytical expression for the dynamics of S and I though time; instead the model has to be solved numerically. Nevertheless, the model has been invaluable for highlighting at least two very important qualitative epidemiological principles.

2.1.1.1. The Threshold Phenomenon

First, let us consider the initial stages after $I(0)$ infectives are introduced into a population consisting of $S(0)$ susceptibles. What factors will determine whether an epidemic will occur or if the infection will fail to invade? To answer this, we start by rewriting equation (2.2) in the form

$$\frac{dI}{dt} = I(\beta S - \gamma). \tag{2.4}$$

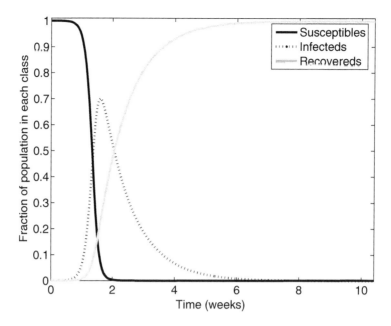

Figure 2.1. The time-evolution of model variables, with an initially entirely susceptible population and a single infectious individual. The figure is plotted assuming $\beta = 520$ per year (or 1.428 per day) and $1/\gamma = 7$ days, giving $R_0 = 10$. (See the following pages for a definition of the crucial parameter R_0)

If the initial fraction of susceptibles ($S(0)$) is less than γ/β, then $\frac{dI}{dt} < 0$ and the infection "dies out." This is a famous result due to Kermack and McKendrick (1927) and is referred to as the "threshold phenomenon" because initially the proportion of susceptibles in the population must exceed this critical threshold for an infection to invade (see also Ransom 1880, 1881; Hamer 1897). Alternatively, we can interpret this result as requiring γ/β, the relative removal rate, to be small enough to permit the disease to spread. The inverse of the relative removal rate is called the **basic reproductive ratio** (universally represented by the symbol R_0) and is one of the most important quantities in epidemiology. It is defined as:

> *the average number of secondary cases arising from an average primary case in an entirely susceptible population*

and essentially measures the maximum reproductive potential for an infectious disease (Diekman and Heesterbeek 2000). We can use R_0 to re-express the threshold phenomenon; assuming everyone in the population is initially susceptible ($S(0) = 1$), a pathogen can invade only if $R_0 > 1$. This makes very good sense because any infection that, on average, cannot successfully transmit to more than one new host is not going to spread (Lloyd-Smith et al. 2005). Some example diseases with their estimated R_0s are presented in Table 2.1; due to differences in demographic rates, rural-urban gradients, and contact structure, different human populations may be associated with different values of R_0 for the same disease (Anderson and May 1982). The value of R_0 depends on both the disease and the host population. Mathematically, we can calculate R_0 as the rate at which new cases are produced by an infectious individual (when the entire population is susceptible)

TABLE 2.1.
Some Estimated Basic Reproductive Ratios.

Infectious Disease	Host	Estimated R_0	Reference
FIV	Domestic Cats	1.1–1.5	Smith (2001)
Rabies	Dogs (Kenya)	2.44	Kitala et al. (2002)
Phocine Distemper	Seals	2–3	Swinton et al. (1998)
Tuberculosis	Cattle	2.6	Goodchild and Clifton-Hadley (2001)
Influenza	Humans	3–4	Murray (1989)
Foot-and-Mouth Disease	Livestock farms (UK)	3.5–4.5	Ferguson et al. (2001b)
Smallpox	Humans	3.5–6	Gani and Leach (2001)
Rubella	Humans (UK)	6–7	Anderson and May (1991)
Chickenpox	Humans (UK)	10–12	Anderson and May (1991)
Measles	Humans (UK)	16–18	Anderson and May (1982)
Whooping Cough	Humans (UK)	16–18	Anderson and May (1982)

multiplied by the average infectious period:

For an infectious disease with an average infectious period given by $1/\gamma$ and a transmission rate β, its basic reproductive ratio R_0 is determined by β/γ.

In a closed population, an infectious disease with a specified R_0 can invade only if there is a threshold fraction of susceptibles greater than $1/R_0$.

Vaccination can be used to reduce the proportion of susceptibles below $1/R_0$ and hence eradicate the disease; full details are given in Chapter 8.

2.1.1.2. Epidemic Burnout

The above observations are informative about the initial stages, after an infectious agent has been introduced. Another important lesson to be learned from this simple SIR model concerns the long-term (or "asymptotic") state. Let us first divide equation (2.1) by equation (2.3):

$$\frac{dS}{dR} = -\frac{\beta S}{\gamma},$$
$$= -R_0 S. \tag{2.5}$$

Upon integrating with respect to R, we obtain

$$S(t) = S(0)e^{-R(t)R_0}, \tag{2.6}$$

assuming $R(0) = 0$. So, as the epidemic develops, the number of susceptibles declines and therefore, with a delay to take the infectious period into account, the number of recovereds increases. We note that S always remains above zero because e^{-RR_0} is always positive; in fact given that $R \leq 1$, S must remain above e^{-R_0}. Therefore, there will always be some susceptibles in the population who escape infection. This leads to the another important

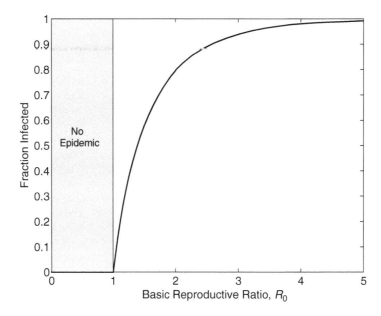

Figure 2.2. The total fraction of the population infected as a function of disease R_0. The curve is obtained by solving equation (2.7) using the Newton-Raphson method, and assuming that initially the entire population is susceptible, $S(0) = 1$, which generates the largest epidemic size.

and rather counter-intuitive conclusion that emerges from this simple model:

The chain of transmission eventually breaks due to the decline in infectives, *not* due to a complete lack of susceptibles.

This approach to model analysis can also shed some light on the fraction of the population who eventually contract an infection (Kermack and McKendrick 1927, Waltmann 1974). As shown in the steps leading to equation (2.6), it is possible to remove the variable I from the system by the division of equation (2.1) by equation (2.3), which (after integrating) gave an expression for S in terms of R. Bearing in mind that by definition $S + I + R = 1$, and that the epidemic ends when $I = 0$, we can rewrite the long-term behavior of equation (2.6):

$$S(\infty) = 1 - R(\infty) = S(0)e^{-R(\infty)R_0}$$

$$\Rightarrow \quad 1 - R(\infty) - S(0)e^{-R(\infty)R_0} = 0. \tag{2.7}$$

where $R(\infty)$ is the final proportion of recovered individuals, which is equal to the total proportion of the population that gets infected.

This equation is transcendental and hence an exact solution is not possible. However, by noting that when $R(\infty) = 0$, equation (2.7) is positive, whereas if $R(\infty) = 1$ then the equation is negative, we know that at some point in between the value must be zero and a solution exists. Using standard methods, such as the Newton-Raphson (Press et al. 1988), or even by trial and error, it is possible to obtain an approximate numerical solution for equation (2.7); this is shown for the standard assumption of $S(0) = 1$ in Figure 2.2. This figure reinforces the message that if $R_0 < 1$, then no epidemic occurs. It also demonstrates the principle that whenever an infectious disease has a sufficiently large basic reproductive

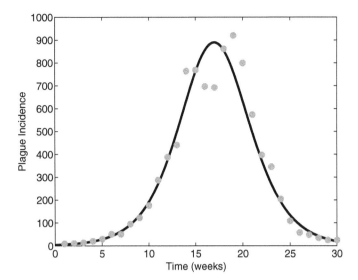

Figure 2.3. The epidemic curve. The filled circles represent weekly deaths from plague in Bombay from December 17, 1905 to July 21, 1906. The solid line is Kermack and McKendrick's approximate solution given by $dR/dt = 890 \operatorname{sech}^2(0.2t - 3.4)$.

ratio (R_0 is larger than approximately 5), essentially everyone (>99%) in a well-mixed population is likely to contract it.

Note that the expression derived in equation (2.7) is not specifically dependent on the structure of the SIR model. It can be alternatively derived from a probabilistic argument as follows: If a single individual has been infected, then assuming the rest of the population are susceptible, they will on average infect R_0 others. Therefore, the probability of a randomly selected individual escaping infection and remaining susceptible is $\exp(-R_0/N)$. Now, if Z individuals have been infected, then the probability of an individual escaping infection is $\exp(-ZR_0/N)$. If at the end of the epidemic a proportion $R(\infty) = Z/N$ have been infected, then the probability of remaining susceptible is clearly $S(\infty) = \exp(-R(\infty)R_0)$, which again must be equal to $1 - R(\infty)$. Therefore, we once again find that: $1 - R(\infty) = \exp(-R(\infty)R_0)$, which is independent of the exact structure of the model.

As we mentioned above, an exact solution of the SIR model (equations (2.1) to (2.3)) is not feasible due to the nonlinear transmission term, βSI. It is possible, however, to obtain an approximate solution for the "epidemic curve," which is defined as the number of new cases per time interval (Waltmann 1974; Hethcote 2000). A classic example of the epidemic curve is provided in Figure 2.3 which shows the number of deaths per week from the plague in Bombay during 1905–1906. Assuming new cases are identified once an individual exhibits the characteristic symptoms of the infection, we can get a handle on the epidemic curve by exploring the equation involving dR/dt (details provided in Box 2.2). This gives the following inelegant but useful approximation:

$$\frac{dR}{dt} = \frac{\gamma \alpha^2}{2S(0)R_0^2} \operatorname{sech}^2\left(\frac{1}{2}\alpha\gamma t - \phi\right). \tag{2.8}$$

The quantities α and ϕ depend in a complex way on the parameters and initial conditions, and are defined in Box 2.2. For any specific epidemic, the parameters in equation (2.8) can

Box 2.2 The Epidemic Curve

To obtain an expression for the epidemic curve, we start by considering equations (2.1)–(2.3). As shown in the steps leading to equation (2.6), it is possible to remove the variable I from the system by the division of equation (2.1) by equation (2.3), which (after integrating) gives an expression for S in terms of R. Bearing in mind that by definition $S + I + R = 1$, we can rewrite equation (2.3):

$$\frac{dR}{dt} = \gamma(1 - S - R).$$

After substituting for S from equation (2.6), this gives

$$\frac{dR}{dt} = \gamma(1 - S(0)e^{-R_0 R} - R). \tag{2.9}$$

As it stands, this equation is not solvable. If, however, we assume that $R_0 R$ is small, we can Taylor expand the exponential term to obtain:

$$\frac{dR}{dt} = \gamma\left(1 - S(0) + \left(S(0)R_0 - 1\right)R - \frac{S(0)R_0^2}{2}R^2\right). \tag{2.10}$$

It is messy but possible to solve this equation. Omitting the intermediate steps, we get

$$R(t) = \frac{1}{R_0^2 S(0)}\left(S(0)R_0 - 1 + \alpha \tanh\left(\frac{1}{2}\alpha\gamma t - \phi\right)\right), \tag{2.11}$$

where

$$\alpha = \left[(S(0)R_0 - 1)^2 + 2S(0)I(0)R_0^2\right]^{\frac{1}{2}},$$

and

$$\phi = \tanh^{-1}\left[\frac{1}{\alpha}(S(0)R_0 - 1)\right].$$

To obtain the epidemic curve as a function of time, we need to differentiate equation (2.11) with respect to time, giving

$$\text{reported cases per unit time} \sim \frac{dR}{dt} = \frac{\gamma\alpha^2}{2S(0)R_0^2}\operatorname{sech}^2\left(\frac{1}{2}\alpha\gamma t - \phi\right). \tag{2.12}$$

As usual, it is important to scrutinize the assumptions made while deriving this result. The key step was in going from equation (2.9) to equation (2.10) and it involved the assumption that $R_0 R$ is small. This condition is most likely to be met at the start of the epidemic (when $R \ll 1$) or if the infection has a very small R_0. Hence, the approximation will, in general, probably not be very accurate for highly infectious diseases such as measles, whooping cough, or rubella with estimated R_0 values of 10 or higher (see Table 2.1.). In addition, the ease with which these equations can be numerically integrated largely negates the need for such involved approximations on a regular basis.

be estimated from the data, as shown in Figure 2.3. Note that the derivation of this equation requires $R_0 R$ to be small, which is unlikely to be the case for many infectious diseases, especially near the end of the epidemic when R typically approaches 1 (see Figure 2.2), but is a good approximation during the early stages often up to the peak of the epidemic. For most practical purposes, however, we need to numerically solve the SIR equations to calculate how variables such as X and Y change through time. The basic issues involved in such an endeavor are discussed in Box 2.3 and the associated figure.

Box 2.3 Numerical Integration

Given the nonlinearity in epidemiological models, it is typically not possible to derive an exact equation that predicts the evolution of model variables through time. In such cases, it is necessary to resort to numerical integration methods. By far the simplest algorithm is called Euler's method, due to the eighteenth century Swiss mathematician. It involves the translation of differential equations into discrete-time analogues. Simply, if we are interested in integrating the SIR equations, we consider the change in each variable during a very small time interval, δt. This is approximately given by the rate of change of that variable at time t multiplied by δt. For example, the equation describing the dynamics of the fraction of infectives is given by

$$I(t + \delta t) \approx I(t) + \delta t \times \frac{dI}{dt}$$
$$\approx I(t) + \delta t \times (\beta S(t)I(t) - (\mu + \gamma)I(t)).$$

Using this scheme, we can simulate the dynamics of a system of ordinary differential equations (ODEs) through time. The problem with the method, however, is that it is rather crude and possesses low accuracy (the error in predicted trajectories scales with δt rather than a higher power). In extreme cases (with δt too large), the method has been known to generate spurious dynamics such as cyclic trajectories when it is possible to demonstrate analytically that all solutions converge to a stable equilibrium (see figure below).

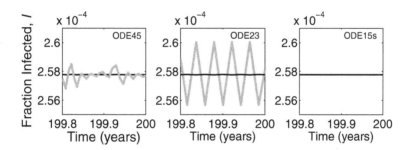

This figure demonstrates $SEIR$ (see Section 2.5) dynamics predicted by different numerical integration schemes. The gray line depicts model predictions using untransformed variables, and the black line represents log-transformed variables (see below). The three different integration schemes demonstrated are the Runge-Kutta Dormand-Prince ("ODE45" in Matlab; left), Runge-Kutta Bogacki and Shampine pairs ("ODE23"; center), and Gear's method ("ODE15s"; right) (see Shampine and Reichelt 1997 for more details of these methods). As shown by the fluctuations in the left and center, numerical integration schemes can generate spurious dynamics, highlighting the importance of a thorough examination of model dynamics using alternative methods, as well as potentially substantial gains from log-transforming variables. (Parameter values are $\beta = 1250$ per year, $\mu = 0.02$ per year, $1/\sigma = 8$ days, and $1/\gamma = 5$ days.)

Such numerical issues can be easily overcome using more sophisticated integration methods, where the deviation from the true solution scales with higher powers of δt (see Press et al. (1988) for a review of different methods, as well as algorithms for their implementation).

However, no numerical scheme will ever be exact due to computational rounding error and therefore it is optimal to reformulate the equations so that such errors are minimized. Dietz (1976) suggested the use of log-transform variables ($\hat{x} = \log(S)$ and $\hat{y} = \log(I)$). The transformed SIR equations become

$$e^{\hat{x}}\frac{d\hat{x}}{dt} = \mu - \beta e^{\hat{y}+\hat{x}} - \mu e^{\hat{x}}$$

$$e^{\hat{y}}\frac{d\hat{y}}{dt} = \beta e^{\hat{x}+\hat{y}} - (\mu + \gamma)e^{\hat{y}},$$

which after simplification become

$$\frac{d\hat{x}}{dt} = \mu e^{-\hat{x}} - \beta e^{\hat{y}} - \mu$$

$$\frac{d\hat{y}}{dt} = \beta e^{\hat{x}} - \mu - \gamma.$$

These new equations, although looking inherently more complex, are far less prone to numerical error. In the above figure, we show that although some integration schemes applied to the standard SIR equations can generate spurious fluctuating dynamics (gray lines), the use of log-transformation of variables (black lines) can overcome this problem—as can the use of more sophisticated integration methods.

2.1.1.3. Worked Example: Influenza in a Boarding School

An interesting example of an epidemic with no host demography comes from an outbreak of influenza in a British boarding school in early 1978 (Anon 1978; Murray 1989). Soon after the start of the Easter term, three boys were reported to the school infirmary with the typical symptoms of influenza. Over the next few days, a very large fraction of the 763 boys in the school had contracted the infection (represented by circles in Figure 2.4). Within two weeks, the infection had become extinguished, as predicted by the simple SIR model without host demography.

We can get an understanding into the epidemiology of this particular strain of influenza A virus (identified by laboratory tests to be A/USSR/90/77 (H1N1)) by estimating the parameters for the SIR model from these data. Using a simple least squares procedure (minimizing the difference between predicted and observed cases), we find the best fit parameters yield an estimated active infectious period ($1/\gamma$) of 2.2 days and a mean transmission rate (β) of 1.66 per day. Therefore, the estimated R_0 of this virus during this epidemic is $\beta/\gamma = 1.66 \times 2.2$, which is 3.65. As shown in Figure 2.4, model dynamics with these parameters is in good agreement with the data. Note, however, that as pointed out by Wearing et al. (2005), the precise value of R_0 estimated from these data is substantially affected by the assumed model structure (in Section 3.3 we deal with this issue in more detail).

2.1.2. The SIR Model With Demography

In the last section, we presented the basic framework for the SIR model given the assumption that the time scale of disease spread is sufficiently fast so as not to be affected by population births and deaths. Some important epidemiological lessons were learned from this model, but ultimately, the formalism ensured the eventual extinction of the pathogen. If we are interested in exploring the longer-term persistence and endemic dynamics of an infectious disease, then clearly demographic processes will be important.

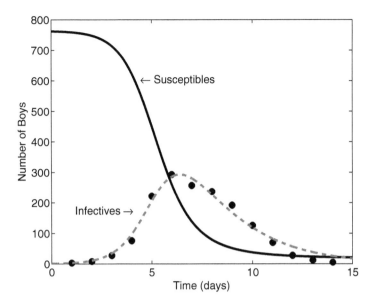

Figure 2.4. The *SIR* dynamics. The filled circles represent the number of boys with influenza in an English boarding school in 1978 (data from the March 4th edition of the *British Medical Journal*). The curves represent solutions from the *SIR* model fitted to the data using least squares. Estimated parameters are $\beta = 1.66$ per day and $1/\gamma = 2.2$ days, giving an R_0 of 3.65.

In particular, the most important ingredient necessary for endemicity in a population is the influx of new susceptibles through births.

The simplest and most common way of introducing demography into the *SIR* model is to assume there is a natural host "lifespan", $1/\mu$ years. Then, the *rate* at which individuals (in any epidemiological class) suffer natural mortality is given by μ. It is important to emphasize that this factor is independent of the disease and is not intended to reflect the pathogenicity of the infectious agent. Some authors have made the alternative assumption that mortality acts only on the recovered class (see Bailey 1975; Keeling et al. 2001a; Brauer 2002), which makes manipulation easier but is generally less popular among epidemiologists. Historically, it has been assumed that μ also represents the population's crude birth rate, thus ensuring that total population size does not change through time ($\frac{dS}{dt} + \frac{dI}{dt} + \frac{dR}{dt} = 0$). This framework is very much geared toward the study of human infections in developed nations—our approach would be different if the host population exhibited its own "interesting" dynamics (as is often the case with wildlife populations; see Chapter 5). Putting all these assumptions together, we get the generalized *SIR* model:

$$\frac{dS}{dt} = \mu - \beta SI - \mu S, \tag{2.13}$$

$$\frac{dI}{dt} = \beta SI - \gamma I - \mu I, \tag{2.14}$$

$$\frac{dR}{dt} = \gamma I - \mu R. \tag{2.15}$$

This is online program 2.2

Note that for many diseases, such as measles, newborns may have passive immunity derived via the placental transfer of maternal antibodies (if the mother had experienced infection or had been vaccinated). Given that the average age at which this immunity to measles and other childhood infections is lost (approximately 6 months) is considerably smaller than the typical age at infection (4–5 years in developed nations; Anderson & May 1982), the assumption that all newborns enter the susceptible class is not unreasonable. In cases when the mean age at infection is very small—in developing nations, for example— maternally derived immunity needs to be explicitly incorporated into models (see McLean and Anderson 1988a,b; and Hethcote 2000).

Before proceeding further, it is useful to establish the expression for R_0 in this model. Starting with the basic definition of R_0 (number of secondary infectives per index case in a naive population of susceptibles), we can look closely at equation (2.14) to work it out. The parameter β represents the transmission *rate* per infective, and the negative terms in the equation tells us that each infectious individual spends an average $\frac{1}{\gamma+\mu}$ time units in this class—the infectious period is effectively reduced due to some individuals dying while infectious. Therefore, if we assume the entire population is susceptible ($S = 1$), then the average number of new infections per infectious individual is determined by the transmission rate multiplied by the infectious period:

$$R_0 = \frac{\beta}{\gamma + \mu}. \tag{2.16}$$

This value is generally similar to, but always smaller than, R_0 for a closed population because the natural mortality rate reduces the average time an individual is infectious. Having established the expression for the R_0, we can now explore some of the properties of the system. This model has proved very useful, primarily for (1) establishing disease prevalence at equilibrium, (2) determining the conditions necessary for endemic equilibrium stability, (3) identifying the underlying oscillatory dynamics, and (4) predicting the threshold level of vaccination necessary for eradication. We will successively explore these features and discuss some of the relevant mathematical underpinnings.

2.1.2.1. The Equilibrium State

The inclusion of host demographic dynamics may permit a disease to persist in a population in the long term. One of the most useful ways of thinking about what may happen *eventually* is to explore when the system is at equilibrium, with $\frac{dS}{dt} = \frac{dI}{dt} = \frac{dR}{dt} = 0$. We therefore set each equation in the system (equations (2.13)–(2.15)) to zero and work out the values of the variables (now denoted by S^*, I^*, and R^*) that satisfy this condition.

Without needing to do much work, the *disease-free equilibrium* is self-evident. This is the scenario where the pathogen has suffered extinction and, in the long run, everyone in the population is susceptible. Mathematically, this state is expressed as $(S^*, I^*, R^*) = (1, 0, 0)$. Below, we discuss the likelihood of observing this state in the system.

Establishing the *endemic equilibrium* requires slightly more work. Perhaps counterintuitively, we start by setting the equation for the infectives (equation (2.14)) to zero:

$$\beta S I - (\gamma + \mu)I = 0. \tag{2.17}$$

After factoring for I, we have

$$I(\beta S - (\gamma + \mu)) = 0, \tag{2.18}$$

which is satisfied whenever $I^* = 0$ or $S^* = \frac{(\gamma+\mu)}{\beta}$. The first condition is simply the disease-free equilibrium, so we concentrate on the second. The quantity on the right-hand side of the equality should look familiar: It is the inverse of the R_0. This leads to an important result:

In the SIR model with births and deaths, the endemic equilibrium is characterized by the fraction of susceptibles in the population being the inverse of R_0.

Having established that $S^* = \frac{1}{R_0}$, we substitute this into equation (2.13) and solve for I^*. Missing out a few lines of algebra, we eventually arrive at

$$I^* = \frac{\mu}{\gamma}\left(1 - \frac{1}{R_0}\right) = \frac{\mu}{\beta}(R_0 - 1). \tag{2.19}$$

One universal condition on population variables is that they cannot be negative. Hence the endemic equilibrium is biologically feasible only if $R_0 > 1$, which agrees with our earlier ideas about when an epidemic is possible. Now, utilizing $S^* + I^* + R^* = 1$, we can obtain an expression for R^*. The endemic equilibrium is, therefore, given by:

$$(S^*, I^*, R^*) = \left(\frac{1}{R_0}, \frac{\mu}{\beta}(R_0 - 1), 1 - \frac{1}{R_0} - \frac{\mu}{\beta}(R_0 - 1)\right).$$

2.1.2.2. Stability Properties

So far, we have derived expressions for the disease-free and endemic equilibrium points of the SIR system, and the restrictions on parameter values for these equilibria to be biologically meaningful. We now would like to know how likely we are to observe them. In mathematical terms, this calls for a "stability analysis" of each equilibrium point, which would provide conditions on the parameter values necessary for the equilibrium to be (asymptotically) stable to small perturbations. The basic idea behind stability analysis is explained in Box 2.4. When this technique is applied to our two equilibrium states, we find that for the endemic equilibrium to be stable, R_0 must be greater than one, otherwise the disease-free equilibrium is stable. This makes good sense because the pathogen cannot invade if each infected host passes on the infection to fewer than one other host (i.e. $R_0 < 1$). However, if successively larger numbers are infected ($R_0 > 1$), then the "topping up" of the susceptible pool by reproduction ensures disease persistence in the long term.

In the SIR model with births and deaths, the endemic equilibrium is stable if $R_0 > 1$, otherwise the disease-free equilibrium state is stable.

2.1.2.3. Oscillatory Dynamics

An important issue for any dynamical system concerns the *manner* in which a stable equilibrium is eventually approached. Do trajectories undergo oscillations as they approach the equilibrium state or do they tend to reach the steady state smoothly? The SIR system is an excellent example of a "damped oscillator," which means the inherent dynamics contain a strong oscillatory component, but the amplitude of these fluctuations declines over time as the system equilibrates (Figure 2.5).

Figure 2.6, shows how the period of oscillations (as determined by equation (2.23) in Box 2.4) changes with the transmission rate (β) and the infectious period ($1/\gamma$). We note that (relative to the infectious period) the period of oscillations becomes longer as the reproductive ratio approaches one; this is also associated with a slower convergence toward the equilibrium.

Box 2.4 Equilibrium Analysis

Here, we outline the basic ideas and methods of determining the stability of equilibrium points. Although we are specifically concerned with the SIR model here, the description of the concepts will be quite general and could be applied to a range of models discussed in this book. Assume we have n variables of interest, N_i ($i = 1, 2, \ldots n$). The dynamics of the system are governed by n coupled Ordinary Differential Equations (ODEs; in the SIR equations, n is three):

$$\frac{dN_i}{dt} = f_i(N_1, N_2, \ldots, N_n), \qquad\qquad i = 1, 2, \ldots, n \qquad\qquad (2.20)$$

To explore the equilibrium dynamics, we must first determine the system's equilibrium state (or states). This is done by setting equations (2.20) to zero and solving for the solutions $N_1^*, N_2^*, \ldots, N_n^*$—noting that multiple solutions may exist. We know that if the system is at equilibrium, then it will remain at equilibrium (by definition). But what happens when the system is (inevitably) perturbed from this state? In mathematical jargon, we are interested in determining the consequences of small perturbations to the equilibrium state. This is achieved by looking at the rates of change of these variables when each variable is slightly shifted away from its equilibrium value. This is done by making the substitutions $N_i = N_i^* + \epsilon_i$ in equations (2.20) and exploring the growth or decline of the perturbation terms, ϵ_i, over time. In any specific case, we can carry out each of these steps, but there is a more generic methodology.

Mathematical results dating back some 200 years have established that for a series of equations such as those described by equations (2.20), the stability of an equilibrium point is determined by quantities known as *eigenvalues*, here represented by Λ_i. For a system of n ODEs, there will be n eigenvalues and stability is ensured if the real part of all eigenvalues are less than zero—these eigenvalues are often complex numbers. Having established the usefulness of eigenvalues, we need to explain how to calculate these terms. Before we can do that, a matrix, J, known as the *Jacobian* must be introduced. It is given by:

$$J = \begin{pmatrix} \dfrac{\partial f_1^*}{\partial N_1} & \dfrac{\partial f_1^*}{\partial N_2} & \cdots & \dfrac{\partial f_1^*}{\partial N_n} \\[2ex] \dfrac{\partial f_2^*}{\partial N_1} & \dfrac{\partial f_2^*}{\partial N_2} & \cdots & \dfrac{\partial f_2^*}{\partial N_n} \\[2ex] \vdots & & \ddots & \vdots \\[2ex] \dfrac{\partial f_n^*}{\partial N_1} & \dfrac{\partial f_n^*}{\partial N_2} & \cdots & \dfrac{\partial f_n^*}{\partial N_n} \end{pmatrix}.$$

The terms f_i^* refer to the functions $f_i(N_1, N_2, \ldots, N_n)$ calculated at equilibrium, i.e. $f_i(N_1^*, N_2^*, \ldots, N_n^*)$. The eigenvalues Λ_i ($i = 1, 2, \ldots, n$) are the solutions of $\det(J - \Lambda I) = 0$; where I is the identity matrix. That is, we subtract Λ from each diagonal element of the Jacobian and then work out the determinant of the matrix. This will give rise to a polynomial in Λ of order n. This is called the *characteristic polynomial*, which, when set to zero and solved, gives rise to the eigenvalues ($\Lambda_1, \Lambda_2, \ldots, \Lambda_n$).

Let us demonstrate these ideas by applying them to the SIR system of equations. Finding the two equilibrium states is described in the main text. So next, we work out the Jacobian:

$$J = \begin{pmatrix} -\beta I^* - \mu & -\beta S^* & 0 \\ \beta I^* & \beta S^* - (\mu + \gamma) & 0 \\ 0 & \gamma & -\mu \end{pmatrix}.$$

To obtain the characteristic polynomial, we subtract Λ from the diagonal elements and calculate the determinant. This gives:

$$(\beta I^* - \mu - \Lambda)(\beta S^* - (\mu + \gamma) - \Lambda)(-\mu - \Lambda) + (\beta I^*)(\beta S^*)(-\mu - \Lambda) = 0.$$

Notice that $(-\mu - \Lambda)$ can be factored immediately, giving one eigenvalue ($\Lambda_1 = -\mu$) that is negative. The remaining two solutions $\Lambda_{2,3}$ are found by solving the following quadratic equation:

$$(\beta I^* - \mu - \Lambda)(\beta S^* - (\mu + \gamma) - \Lambda) + \beta I^* \beta S^* = 0. \qquad (2.21)$$

Let us first consider the disease-free equilibrium. If we make the appropriate substitutions ($S^* = 1$ and $I^* = 0$), we have

$$(-\mu - \Lambda)(\beta - (\mu + \gamma) - \Lambda) = 0.$$

This clearly has two solutions, $\Lambda_2 = -\mu$ and $\Lambda_3 = \beta - (\mu + \gamma)$. For this equilibrium to be stable, we need to ensure all eigenvalues are negative, hence the stability criterion becomes $\beta < \mu + \gamma$, which translates into ensuring $R_0 < 1$.

To explore the endemic equilibrium, again we substitute the expressions for S^* and I^* into equation (2.21) and explore the condition required for the remaining two eigenvalues to be negative. After making some simplifications, we arrive at the following quadratic equation,

$$\Lambda^2 + \mu R_0 \Lambda + (\mu + \gamma)\mu(R_0 - 1) = 0, \qquad (2.22)$$

the solutions of which can be obtained by the standard formula, giving:

$$\Lambda_{2,3} = -\frac{\mu R_0}{2} \pm \frac{\sqrt{(\mu R_0)^2 - \frac{4}{AG}}}{2},$$

where the term $A = \frac{1}{\mu(R_0 - 1)}$ denotes the mean age at infection (see Section 2.1.2.4) and $G = \frac{1}{\mu + \gamma}$ determines the typical period of a host's infectivity.

To make further progress with this equation, we notice that often $(\mu R_0)^2$ is small enough to ignore and hence we can approximate the above solutions to

$$\Lambda_{2,3} \approx -\frac{\mu R_0}{2} \pm \frac{i}{\sqrt{AG}},$$

where $i = \sqrt{-1}$. Therefore, the endemic equilibrium is feasible only when R_0 is greater than one, but it is always stable. The fact that the largest ("dominant") eigenvalues are complex conjugates (they are of the form $\Lambda_{2,3} = x \pm iy$) tells us that the equilibrium is approached via oscillatory dynamics. The period of these damped oscillations, T, is determined by the inverse of the complex part of the eigenvalues multiplied by 2π:

$$T \sim 2\pi \sqrt{AG}. \qquad (2.23)$$

2.1.2.4. Mean Age at Infection

When dealing with infectious diseases in the real world, an important indicator of prevalence is the host's mean age at infection, A (if it possible to contract the infection

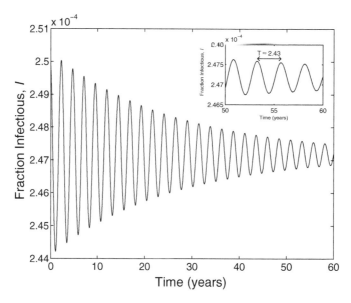

Figure 2.5. The SIR model's damped oscillations. The main figure shows how the fraction of infectives oscillates with decreasing amplitude as it settles toward the equilibrium. The inset shows a slice of the time-series with the period of fluctuations as predicted by equation (2.23). The figure is plotted assuming $1/\mu = 70$ years, $\beta = 520$ per year, and $1/\gamma = 7$ days, giving $R_0 = 10$. Initial conditions were $S(0) = 0.1$ and $I(0) = 2.5 \times 10^{-4}$.

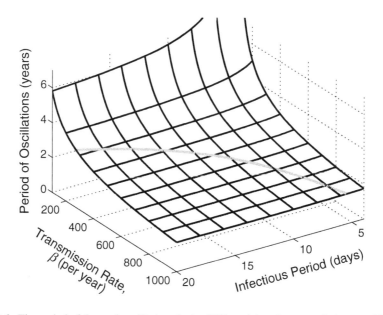

Figure 2.6. The period of damped oscillations in the SIR model as the transmission rate (β) and infectious period ($1/\gamma$) are varied. The gray line on the surface depicts the oscillatory period whenever the parameters combine to yield an R_0 of 15. The figure is plotted assuming $1/\mu = 70$ years.

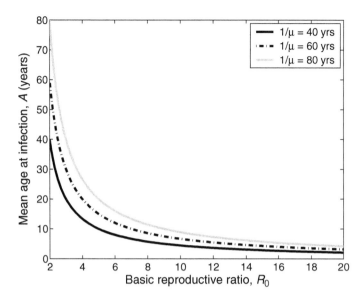

Figure 2.7. The mean age at infection for the SIR model as a function of the basic reproductive ratio (R_0) and the average life expectancy ($1/\mu$). Using equation (2.24), we see that as R_0 increases, there is a dramatic decline in the mean age at infection, whereas changes in μ are not as influential.

multiple times, then we would be interested in the mean age at first infection). This quantity can be measured in a straightforward way for many human or animal infections by analyzing age-specific serological data, which detect the presence of antibodies specific to pathogen antigens. The *rate* at which individuals seroconvert (from negative to positive) provides information on the force of infection (see Box 2.1). Age-stratified serological surveys, when conducted randomly through the population, can yield a relatively unbiased population level estimate of the mean age at infection. Typically, we may also have accurate independent estimates for some of the model parameters such as the host life expectancy (from demographic data) and the infectious period (from clinical epidemiology data). Thus, once we have an estimate of the mean age at infection from serological surveys, we can estimate the transmission rate β as long as we have an expression for A derived from the model (see Figure 2.7).

How do we calculate the average age at which susceptibles are infected, especially given a nonage-structured model? We can approach this question by taking equation (2.13) at equilibrium and calculating the mean time an individual remains susceptible—that is, the mean time from birth (or loss of maternally derived immunity) to infection. Ignoring the small, disease-independent mortality term, the average period spent in the susceptible class is approximated by the inverse of the force of infection, namely $\frac{1}{\beta I^*}$. Upon substituting for I^* from equation (2.19), we obtain an expression for the mean age at infection:

$$A \approx \frac{1}{\mu(R_0 - 1)}. \tag{2.24}$$

This equation can be rephrased as $R_0 - 1 \approx \frac{L}{A}$, where L is the host's life expectancy. The above step has proven historically very important in establishing a robust link between model parameters and population level quantities such as L and A. In their classic work,

Anderson and May (1982; 1991) used the estimates of A and L to calculate R_0 for numerous infections in different geographical regions and eras (see Table 2.1).

The mean age of (first) infection is equal to the average life expectancy of an individual divided by $R_0 - 1$.

2.2. INFECTION-INDUCED MORTALITY AND SI MODELS

The models described in the previous section have implicitly assumed that the infection is essentially benign. Transmission results in a period of illness, which is followed by recovery and lifelong immunity. This scenario is reasonable for largely harmless infections such as the common cold or chickenpox. However, numerous infectious diseases are associated with a substantial mortality risk. Examples include malaria, measles, whooping cough, SARS, and dengue fever, among others. How do we explore the consequences of infection-induced mortality? Specifically, how do we incorporate a mortality *probability* into the SIR equations? The obvious approach would be to add a term such as $-mI$ to equation (2.14), where m is a per capita disease-induced mortality *rate* for infected individuals. However, this may be tricky to interpret biologically or estimate from data. Instead, it is preferable to think about the probability, ρ, of an individual in the I class dying from the infection before either recovering or dying from natural causes. This is the quantity most likely estimated from clinical studies or case observations. Mathematically, this translates into the following equation:

$$\frac{dI}{dt} = \beta SI - (\gamma + \mu)I - \frac{\rho}{1 - \rho}(\gamma + \mu)I, \qquad (2.25)$$

where ρ represents the per capita *probability* of dying from the infection and takes values from zero to one. (The S equation remains as before.) Therefore, in order to convert this to a mortality rate, we should set $m = \frac{\rho}{1-\rho}(\gamma + \mu)$. This equation for the infection dynamics can be tidied up to give

$$\frac{dI}{dt} = \beta SI - \frac{(\gamma + \mu)}{1 - \rho}I. \qquad (2.26)$$

Note that as ρ approaches unity, new infectives die almost instantaneously and R_0 drops to zero; for such diseases it may be more appropriate to further subdivide the infectious period and allow mortality only in the later stages of infection (see Section 3.3).

2.2.1. Mortality Throughout Infection

Because infection actively removes individuals from the population, we can no longer implicitly assume that the population size is fixed—disease-induced mortality could lead to an ever-declining population size. One way around this is to incorporate a *fixed* birth rate (ν) into the susceptible equation (2.13), independent of the population size:

$$\frac{dS}{dt} = \nu - \beta SI - \mu S. \qquad (2.27)$$

However, the fact that the population size N can vary means that we need to consider the transmission term βSI in much more detail. Until now, it has made little difference whether density or frequency-dependent transmission was assumed, because the N term was constant and could be absorbed by rescaling the population size or rescaling the transmission rate β. However, because N is now variable, this is no longer possible and the choice of transmission mechanism can profoundly affect the dynamics. Given that this problem illustrates an important (and often confusing) issue in epidemiological modeling, we will deal with the two cases separately, paying particular attention to the underlying assumptions. To make matters explicit, we will deal with numbers rather than proportions throughout this section.

2.2.1.1. Density-Dependent Transmission

For density-dependent (pseudo mass action) transmission, we specifically consider the case where as the total population size N decreases, due to disease-induced mortality, there is reduced interaction between hosts. In this density-dependent formulation, the rate at which new cases are produced is βXY, which scales linearly with both the number of susceptibles and the number of infectious individuals. We start by considering the values of ν and μ and the dynamics of the disease-free state. In the absence of disease, we find that

$$\frac{dN}{dt} = \nu - \mu N \qquad \Rightarrow \qquad N \to \frac{\nu}{\mu} \tag{2.28}$$

This is online program 2.3

Hence, we can equate $\frac{\nu}{\mu}$ with the ecological concept of a carrying capacity for the population.

We can now repeat the kinds of analyses that were carried out on the generalized SIR model in Section 2.1.1.2. Once again, we find the system possesses two equilibrium points: one endemic (X^*, Y^*, Z^*) and one disease free ($\frac{\nu}{\mu}, 0, 0$). Missing out a few lines of algebra, we find the endemic equilibrium to be:

$$X^* = \frac{\mu + \gamma}{\beta(1 - \rho)} = \frac{\nu}{\mu R_0}, \tag{2.29}$$

$$Y^* = \frac{\mu}{\beta}(R_0 - 1), \tag{2.30}$$

$$Z^* = \frac{\gamma}{\beta}(R_0 - 1). \tag{2.31}$$

$$\Rightarrow N^* = \frac{\nu}{\mu R_0}[1 + (1 - \rho)(R_0 - 1)]. \tag{2.32}$$

In this case, $R_0 \left(= \frac{\beta(1-\rho)\nu}{(\mu+\gamma)\mu}\right)$ contains a correction term $(1 - \rho)$ that takes into account the reduced period of infectivity due to disease-induced mortality, as well as a term that takes into account the population size at the disease-free equilibrium. The condition necessary to ensure the feasibility of the endemic equilibrium (and hence the instability of the disease-free steady state) is found by ensuring $Y^* > 0$, which translates to $R_0 > 1$, as we have previously seen. This means that if $\rho > 0$, then the pathogen has to have a higher transmission rate per unit of infectious period ($\frac{\beta}{\mu+\gamma}$) in order to remain endemic, compared to a similar infectious disease that is benign—due to the fact that the effective infectious period is reduced by disease-induced mortality.

The stability properties of this system are similar to the basic SIR model. Indeed, stability analysis reveals the endemic equilibrium to be always locally asymptotically stable. The approach to the equilibrium is also via damped oscillations, the (natural) period of which are determined by the following equation:

$$T \sim 2\pi \sqrt{AG}.$$ (2.33)

This is identical to equation (2.23), though note that here the terms A (the average age of infection) and G (the average generation length of infection), respectively, both contain a correction term due to taking infection-induced mortality into account.

When disease-induced mortality is added to the SIR model with density-dependent transmission, the equilibrium and stability properties simply reflect a change in parameters.

2.2.1.2. Frequency-Dependent Transmission

For frequency-dependent (mass action) transmission, the calculation is somewhat more involved. Recall that if X and Y are the number of susceptible and infectious hosts, then the mass-action assumption means that the transmission rate is given by $\beta XY/N$. Previously, we rescaled the variables such that S and I represented fractions of the population, which removed the N term in the transmission rate, leaving us with more elegant-looking equations. However, in this scenario N is varying and that trick is no longer appropriate. Instead, we retain the notion that X and Y are numbers of hosts. Our equation for the number of infectious individuals is therefore:

$$\frac{dY}{dt} = \beta XY/N - \frac{(\gamma + \mu)}{1 - \rho} Y.$$ (2.34)

We note that the mass-action assumption means that even when the population size is reduced, each individual still has the same average number of contacts.

This is online program 2.4

It will not be a surprise that two equilibria exist: the endemic (X^*, Y^*, Z^*) and the disease-free $(\frac{v}{\mu}, 0, 0)$. The endemic equilibrium can again be found by setting the rates of change equal to zero:

$$X^* = \frac{v(1-\rho)(\gamma+\mu)}{\mu(\beta(1-\rho)-\mu\rho-\gamma\rho)} = \frac{N}{R_0}, \quad \Rightarrow \quad S^* = \frac{(\gamma+\mu)}{\beta(1-\rho)} = \frac{1}{R_0},$$

$$Y^* = \frac{v\beta(1-\rho)^2 - v(\mu+\gamma)(1-\rho)}{(\mu+\gamma)(\beta(1-\rho)-\mu\rho-\gamma\rho)}, \quad \Rightarrow \quad I^* = \frac{\mu}{\beta(1-\rho)}(R_0-1),$$ (2.35)

$$N^* = \frac{\beta v(1-\rho)^2}{\mu(\beta(1-\rho)-\mu\rho-\gamma\rho)} = \frac{v}{\mu}\left(\frac{R_0(1-\rho)}{(R_0-\rho)}\right).$$

As the expressions on the right-hand side demonstrate, equilibrium values are easier to understand intuitively if we deal with the proportion of the population in each state, rather than the absolute number. Again, we note that $R_0 > 1$ is necessary and sufficient for the disease to invade the population and for the equilibrium point to be feasible and stable.

For both the frequency- and density-dependent models, the population size can be reduced from its carrying capacity of $\frac{v}{\mu}$ by the impact of the disease. Not surprisingly, infectious diseases with the highest mortalities (ρ close to one) and largest R_0 have the greatest impact on the population (Figure 2.8). Although for low mortality levels both mixing assumptions lead to similar results, when the mortality is high the

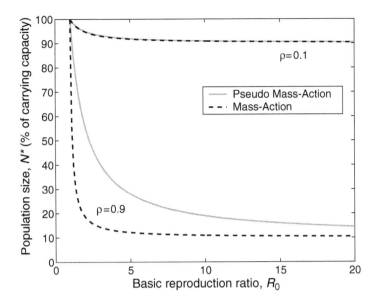

Figure 2.8. The population size, N^*, for diseases that are associated with mortality. Two different disease mortality probabilities are considered, $\rho = 0.1$ and $\rho = 0.9$, and the two mixing assumptions of density dependence (βXY, solid line) and frequency dependence ($\beta XY/N$, dashed line) are shown.

frequency-dependent (mass-action) assumption leads to the largest drop in the total population size. This is because pseudo mass-action mixing places a natural damping on transmission, such that as the population size decreases so does the contact rate between individuals, limiting disease spread and reducing disease-induced mortality. From this relatively simple example it is clear that when population sizes change, our assumption about mixing behavior can have profound effects on the dynamics.

When disease-induced mortality is added to the SIR model with frequency-dependent transmission, the equilibrium and stability properties can change substantially, especially if the probability of mortality is high.

2.2.2. Mortality Late in Infection

One difficulty with equation (2.26) is that when the mortality rate is very high, the infectious period is substantially reduced. In some cases, we may wish to consider a disease where mortality generally occurs at (or toward) the end of the infectious period. This is often plausible because the onset of disease (and symptoms) may be significantly delayed from the onset of infection. In such cases, the following model would be appropriate:

$$\frac{dS}{dt} = \nu - \beta SI - \mu S, \tag{2.36}$$

$$\frac{dI}{dt} = \beta SI - (\gamma + \mu)I, \tag{2.37}$$

$$\frac{dR}{dt} = (1 - \rho)\gamma I - \mu R. \tag{2.38}$$

For this form of model mortality, we again need to consider the precise form of transmission. For frequency-dependent transmission, the reduction in the recovered population has no effect on the dynamics and hence this model has the same properties as the standard SIR model. However, for density-dependent transmission, we find that $X^* = \frac{\nu(\mu + \gamma(1-\rho))}{(\beta - \gamma\rho)\mu}$ and $Y^* = \frac{\nu(\beta - \gamma - \mu)}{(\beta - \gamma\rho)(\gamma + \mu)}$, although again this is stable and feasible if $R_0 = \frac{\beta\nu}{\mu(\gamma + \mu)} > 1$.

The equilibrium levels of diseases that cause mortality are critically dependent upon whether frequency- or density-dependent transmission is assumed, due to the changes in the total population size that occur. However, we generally find that the endemic equilibrium is feasible and stable as long as $R_0 > 1$.

2.2.3. Fatal Infections

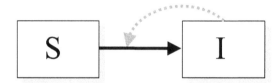

The models of Sections 2.2.1 and 2.2.2 represent cases where infection does not always kill. There are, however, numerous examples of animal and plant pathogens that are always fatal (Feline Infectious Peritonitis (FIP), Spongiform Encephalopathy (BSE), Leishmaniasis, Rabbit Haemorrhagic Disease, and Highly Pathogenic Avian Influenza (H5N1)). In this situation, we can simplify the SIR equations by removing the recovered class, which leads to a family of models known as SI models. Here, infecteds are assumed to remain infectious for an average period of time $(1/\gamma)$, after which they succumb to the infection. Assuming frequency-dependent transmission, the equations describing the SI model are simply:

$$\frac{dX}{dt} = \nu - \beta XY/N - \mu X, \tag{2.39}$$

$$\frac{dY}{dt} = \beta XY/N - (\gamma + \mu)Y. \tag{2.40}$$

It is straightforward to demonstrate that the endemic equilibrium ($X^* = \frac{\nu}{\beta - \gamma}$, $Y^* = \frac{\nu(\beta - \gamma - \mu)}{(\beta - \gamma)(\gamma + \mu)}$) is feasible as long as $R_0 = \frac{\beta}{(\mu + \gamma)} > 1$ and is always locally stable.

Alternatively, if we assume pseudo mass-action transmission, such that the contact rate scales with density, we obtain:

$$\frac{dX}{dt} = \nu - \beta XY - \mu X, \tag{2.41}$$

$$\frac{dY}{dt} = \beta XY - (\gamma + \mu)Y. \tag{2.42}$$

Similarly, for this system the endemic equilibrium ($X^* = \frac{\gamma+\mu}{\beta}$, $Y^* = \frac{\nu}{(\gamma+\mu)} - \frac{\mu}{\beta}$) is feasible as long as $R_0 = \frac{\beta\nu}{(\mu+\gamma)\mu} > 1$ and is again always locally stable.

2.3. WITHOUT IMMUNITY: THE *SIS* MODEL

The *SI* and *SIR* models both capture the dynamics of acute infections that either kill or confer lifelong immunity once recovered. However, numerous infectious diseases confer no long-lasting immunity, such as rotaviruses, sexually transmitted infections, and many bacterial infections. For these diseases, a individuals can be infected multiple times throughout their lives, with no apparent immunity. Here, we concentrate briefly on this class of models, called *SIS* because recovery from infection is followed by an instant return to the susceptible pool.

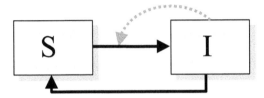

Simply, these *SIS* models are described by a pair of coupled ordinary differential equations:

$$\frac{dS}{dt} = \gamma I - \beta IS, \tag{2.43}$$

$$\frac{dI}{dt} = \beta SI - \gamma I. \tag{2.44}$$

This is online program 2.5

The parameters are as defined in the previous section, but with $S + I = 1$. In this simple example, demography (births and deaths) has been ignored. Despite this lack of susceptible births, the disease can still persist because the recovery of infectious individuals replenishes the susceptible pool. We can, therefore, substitute $S = 1 - I$ into equation (2.44) and simplify to get

$$\frac{dI}{dt} = (\beta - \beta I - \gamma)I = \beta I((1 - 1/R_0) - I), \tag{2.45}$$

where, as usual, $R_0 = \beta/\gamma$. This equation is the equivalent of the logistic equation used to describe density-dependent population growth in ecology (Murray 1989). The equilibrium number of infectives in this population is obtained by setting equation (2.45) to zero and solving for I^*. This gives $I^* = (1 - 1/R_0)$, hence yet again $S^* = 1/R_0$, and the equilibrium will be stable as long as $R_0 > 1$. However, for this class of model, convergence to the equilibrium is monotonic with no oscillatory behavior. The dynamics of *SIS* models and sexually transmitted diseases is described in more detail in Chapter 3.

For an infectious disease that does not give rise to long-term immunity, as long as the infection is able to invade the population, loss of immunity will guarantee its long-term persistence.

2.4. WANING IMMUNITY: THE SIR MODEL

The SIR and SIS models are two behavioral extremes where immunity is either lifelong or simply does not occur. An intermediate assumption is that immunity lasts for a limited period before waning such that the individual is once again susceptible.

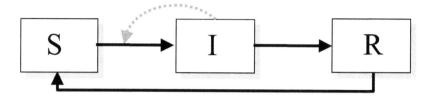

This translates into the following model:

$$\frac{dS}{dt} = \mu + wR - \beta SI - \mu S \tag{2.46}$$

$$\frac{dI}{dt} = \beta SI - \gamma I - \mu I \tag{2.47}$$

$$\frac{dR}{dt} = \gamma I - wR - \mu R, \tag{2.48}$$

where w is the rate at which immunity is lost and recovered individuals move into the susceptible class. Not surprisingly, the dynamics of this model provide a smooth transition between the SIR framework (when $w = 0$) and the SIS model (when $w \to \infty$). As before, $R_0 = \frac{\beta}{\gamma + \mu}$ and we require $R_0 > 1$ to obtain a plausible stable endemic solution. Once again, it is possible to obtain the oscillatory damping toward endemic equilibrium that was seen in the SIR model; the period of these oscillations is given by:

$$T = \frac{4\pi}{\sqrt{4(R_0 - 1)\frac{1}{G_I}\frac{1}{G_R} - \left(\frac{1}{G_R} - \frac{1}{A}\right)^2}},$$

where $A = \frac{w + \mu + \gamma}{(w + \mu)(\beta - \gamma - \mu)}$ is again the average age at first infection, whereas $G_I = 1/(\gamma + \mu)$ is the average period spent in the infectious class, and $G_R = 1/(w + \mu)$ is the average time spent in the recovered class.

Figure 2.9 shows the dramatic effects of changing the level of waning immunity: As the period of immunity $(1/w)$ is reduced (and the dynamics more closely resemble those of the SIS model), we observe a dramatic increase in the prevalence of infectious disease together with a marked drop in the period of the damped oscillations. This waning immunity result was used by Grassly et al. (2005) to explain the 9–10-year cycle observed in syphilis cases in the United States. It was postulated that these cycles were determined by the natural oscillations of syphilis infection if waning immunity was assumed. Grassly et al. (2005) found that oscillations consistent with those observed could be generated by a model with $R_0 = 1.5$, $1/w = 10$ years, $1/\gamma = 2$ months, and $1/\mu = 33$ years—this high birth rate accounts for individuals entering and leaving the pool of sexually active individuals. We note that although the deterministic model predicts that eventually such oscillations

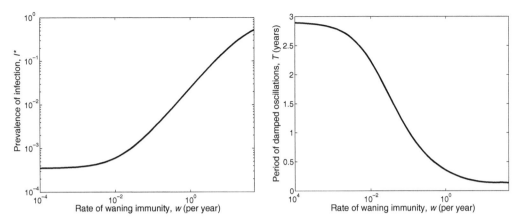

Figure 2.9. The effects of waning immunity on the equilibrium infection prevalence (left graph) and the period of the damped oscillations (right graph). ($R_0 = 10$, $1/\mu = 70$ years, $1/\gamma = 10$ days.)

will decay to zero, the chance nature of transmission can lead to sustained oscillations generally close to the natural period, T (see Chapter 6).

2.5. ADDING A LATENT PERIOD: THE $SEIR$ MODEL

We briefly introduce a refinement to the SIR model to take into account the latent period. The process of transmission often occurs due to an initial inoculation with a very small number of pathogen units (e.g., a few bacterial cells or virions). A period of time then ensues during which the pathogen reproduces rapidly within the host, relatively unchallenged by the immune system. During this stage, pathogen abundance is too low for active transmission to other susceptible hosts, and yet the pathogen is present. Hence, the host cannot be categorized as susceptible, infectious, or recovered; we need to introduce a new category for these individuals who are *infected* but not yet *infectious*. These individuals are referred to as Exposed and are represented by the variable E in $SEIR$ models.

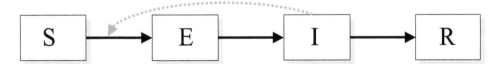

Assuming the average duration of the latent period is given by $1/\sigma$, the $SEIR$ equations are:

$$\frac{dS}{dt} = \mu - (\beta I + \mu)S, \tag{2.49}$$

$$\frac{dE}{dt} = \beta SI - (\mu + \sigma)E, \tag{2.50}$$

$$\frac{dI}{dt} = \sigma E - (\mu + \gamma)I, \tag{2.51}$$

$$\frac{dR}{dt} = \gamma I - \mu R. \tag{2.52}$$

This is online program 2.6

As before, we typically assume $S + E + I + R = 1$ and hence equation (2.52) is redundant. The addition of a latent period is essentially akin to introducing a slight time delay into the system and we may expect that such a feature may act to destabilize and slow the system. As we will demonstrate, the dynamic properties of the $SEIR$ model are qualitatively similar to those of the SIR system.

Standard equilibrium stability analysis proceeds by finding the steady states of the system and determining the criteria for their stability. As with previous disease models, the $SEIR$ model also possesses both an endemic (S^*, E^*, I^*, R^*) and a disease-free $(1, 0, 0, 0)$ equilibrium solution. As usual, the endemic fixed point is of greater interest and is given by

$$S^* = \frac{(\mu + \gamma)(\mu + \sigma)}{\beta \sigma} = \frac{1}{R_0}, \tag{2.53}$$

$$E^* = \frac{\mu(\mu + \gamma)}{\beta \sigma}(R_0 - 1). \tag{2.54}$$

$$I^* = \frac{\mu}{\beta}(R_0 - 1), \tag{2.55}$$

with $R^* = 1 - S^* + E^* + I^*$. The expression for R_0 is now slightly different, due to the death of some individuals in the exposed class who do not contribute to the chain of transmission. However, this difference is often negligible because typically $\sigma/(\mu + \sigma) \sim 1$ as the latent is far smaller that the expected lifespan. As expected, if the latent period is infinitesimally small (i.e., $\sigma \to \infty$), then we recover the same expression for R_0 as for the SIR model ($R_0 = \beta/(\gamma + \mu)$).

For the endemic equilibrium to be feasible and stable (and the disease-free equilibrium to be unstable), equation (2.55) once again requires that the basic reproductive ratio exceed one ($R_0 > 1$). By exploring the Jacobian of the system (equations (2.49)–(2.51)) in the usual way, we obtain a quartic in the eigenvalues Λ. As for the SIR model (Box 2.4), $\Lambda = -\mu$ is an obvious solution, leaving us with a cubic equation:

$$\Lambda^3 + (\mu R_0 + 2\mu + \sigma + \gamma)\Lambda^2 + \mu R_0(2\mu + \sigma + \gamma)\Lambda + \mu(R_0 - 1)(\mu + \sigma)(\mu + \gamma) = 0.$$

Unfortunately, there is no obvious solution to this equation. As pointed out by Anderson and May (1991), however, in many cases σ and γ will be much larger than μ and μR_0. If so, then one solution to this equation is approximately $\Lambda \sim -(\sigma + \gamma)$, which leaves us with a quadratic for the two remaining eigenvalues:

$$\Lambda^2 + \mu R_0 \Lambda + \frac{\gamma \sigma}{\sigma + \gamma}\mu(R_0 - 1) \approx 0.$$

Clearly, this is similar to the analogous equation for the SIR model (equation (2.22)), with a slight correction term due to the death of exposed individuals. By working out the expression for these eigenvalues, the endemic equilibrium is stable given $R_0 > 1$ with perturbations dying out in an oscillatory manner. Again, we can find an expression for the natural period of oscillations in this case, which is given by $T \sim 2\pi\sqrt{AG}$, where G (the "ecological generation length" of the infection) is slightly modified to include the latent period: $G = \frac{1}{\mu + \gamma} + \frac{1}{\mu + \sigma}$.

From this analysis, one may be tempted to conclude that the $SEIR$ model is an unnecessary complication of the SIR model. Given a small death rate ($\mu \ll \gamma, \sigma$), the SIR

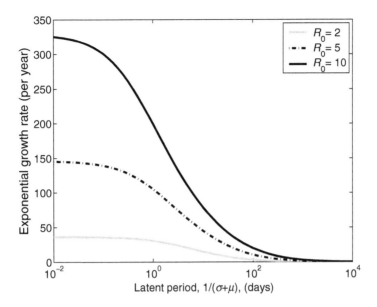

Figure 2.10. The effects of an exposed period on the initial growth rate of an infection in a totally susceptible population. Although a small exposed period leads to dynamics close to those predicted by the SIR model, a long exposed period can dramatically slow the spread of infection, or even prevent the spread if too many individuals die before becoming infectious. ($1/\mu = 70$ years, $1/\gamma = 10$ days.)

and $SEIR$ models behave similarly at equilibrium as long as the basic reproductive ratio ($\beta_{SIR}/\gamma_{SIR} = \beta_{SEIR}/\gamma_{SEIR}$) and average infected period ($1/\gamma_{SIR} = 1/\gamma_{SEIR} + 1/\sigma_{SEIR}$) are identical. However, the two models behave very differently at invasion, with the presence of an exposed class slowing the dynamics. Examining the eigenvalues at the disease-free equilibrium allows us to describe the increase in prevalence during the invasion phase:

$$I_{SEIR}(t) \approx I(0)\exp\left(\frac{1}{2}\left[\sqrt{4(R_0 - 1)\sigma\gamma + (\sigma + \gamma)^2} - (\sigma + \gamma)\right]t\right),$$

$$\left\{\approx I(0)\exp\left([(\sqrt{R_0} - 1)\gamma]t\right) \qquad \text{if } \sigma = \gamma\right\},$$

whereas without an exposed class the dynamics are given by:

$$I_{SIR}(t) \approx I(0)\exp([(R_0 - 1)\gamma]t),$$

where natural mortality has been ignored to simplify the equations. This behavior is exemplified in Figure 2.10, showing how long exposed periods can slow or even prevent the spread of infection. Therefore, if large fluctuations in the prevalence of infection are of interest, or we wish to consider both invasion and equilibrium properties, the exposed class must be realistically modeled.

Although the SIR and $SEIR$ models behave similarly at equilibrium (when the parameters are suitably rescaled), the $SEIR$ model has a slower growth rate after pathogen invasion due to individuals needing to pass through the exposed class before they can contribute to the transmission process.

2.6. INFECTIONS WITH A CARRIER STATE

Although the SIR and $SEIR$ model paradigms are a good approximation to the epidemiological characteristics of many infectious diseases, such as measles or influenza, other infections have a more complex natural history. As an example of how such complexities can be accommodated in the model, will we consider infections such as hepatitis B or herpes, where a proportion of infected individuals may become chronic carriers, transmitting infection at a low rate for many years. The greater biological complexity of these systems can be readily incorporated into our current modeling framework, although accurate parameterization becomes more complex. Here we focus on the inclusion of a single carrier class, using hepatitis B as our prototypic infectious disease.

This general framework of building compartmental models can be readily extended to infections with more complex biology.

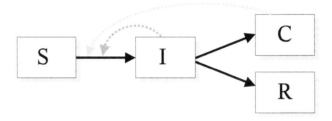

For diseases with carrier states, susceptible individuals can be infected by either carriers or acutely infectious individuals. It is generally assumed that the progress of infection within an individual is independent of their source of infection; that is, those infected by acutely infectious individuals and those infected by carriers are indistinguishable. A recently infected individual is acutely (highly) infectious for a given period and then either recovers completely or moves into the carrier class. Such dynamics lead to the following model:

$$\frac{dS}{dt} = \mu - (\beta I + \varepsilon \beta C)S - \mu S, \tag{2.56}$$

$$\frac{dI}{dt} = (\beta I + \varepsilon \beta C)S - \gamma I - \mu I, \tag{2.57}$$

This is online program 2.7

$$\frac{dC}{dt} = \gamma q I - \Gamma C - \mu C, \tag{2.58}$$

$$\frac{dR}{dt} = \gamma(1-q)I + \Gamma C - \mu R. \tag{2.59}$$

Here, C captures the proportion of carriers in the population, ε is the reduced transmission rate from chronic carriers compared to acute infectious individuals, q is the proportion of acute infections that become carriers while a fraction $(1-q)$ simply recover, and Γ is the rate at which individuals leave the carrier class—all other parameters have their standard meanings.

As with any new epidemiological model, it is important to understand when an epidemic can occur and hence to calculate the basic reproductive ratio, R_0. For infections with

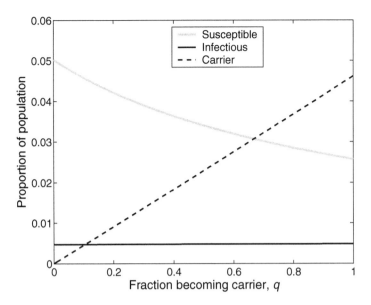

Figure 2.11. The equilibrium values of a model with a carrier class. ($\beta = 73$ per year, $1/\gamma = 100$ days, $1/\Gamma = 1000$ days, $\varepsilon = 0.1$, $1/\mu = 50$ years.) Using these parameters R_0 is relatively large (from 19.9 to 38.8), hence the level of susceptibles is always low and therefore the variation in the level of acute infections is negligible.

a carrier state, R_0 has two components: One comes from acutely infectious individuals, which follow the standard calculation for R_0 that has been given throughout this chapter; the other comes from infections caused while in the carrier state and must take into account the fraction of infecteds becoming carriers:

$$R_0 = \frac{\beta}{\gamma + \mu} + \frac{q\gamma}{(\gamma + \mu)} \frac{\varepsilon\beta}{(\Gamma + \mu)},$$

where the term $\frac{q\gamma}{\gamma+\mu}$ accounts for those individuals that do not die in the infectious class and go on to become carriers. Therefore, as expected, the fact that infected individuals can enter an infectious carrier state rather than simply recovering increases the value of R_0.

The value of R_0 is the sum of separate components from the acutely infected and chronic carriers.

With only a little more work than usual, the equilibrium values of the model can be found:

$$S^* = \frac{\gamma + \mu}{\beta + \frac{q\gamma\varepsilon\beta}{\Gamma+\mu}} = \frac{1}{R_0},$$

$$I^* = \frac{\mu(1 - S^*)}{\gamma + \mu}, \qquad C^* = \frac{\gamma q\mu(1 - S^*)}{(\gamma + \mu)(\Gamma + \mu)},$$

Figure 2.11 shows how these equilibrium values change as q, the fraction of infected individuals that become carriers, varies. Due to their much longer "infectious" period, carriers can easily outnumber acutely infected individuals.

In practice, the epidemiology of diseases such as hepatitis B virus (HBV) is more complex, because the risk of becoming a carrier is age-dependent, with infected children more likely to become carriers. This can potentially lead to some surprising behavior with two stable endemic situations for a given set of parameters due to positive feedbacks between R_0 and the number of carriers (Medley et al. 2001). When R_0 is high, the average age of infection is low, which in turn leads to many carriers and hence a higher R_0; in contrast, a low R_0 means that the average age of infection is high and so few carriers are produced and R_0 remains low. In this manner, and using an age-structured formalism (Chapter 3), it is possible to observe "endemic stability" where both high and low R_0 solutions are stable and the equilibrium prevalence of infection depends on the initial starting conditions.

2.7. DISCRETE-TIME MODELS

This chapter has concentrated exclusively on epidemiological models written as differential equations. This is partly because the vast majority of models in the literature are based on this framework. The inherent assumption has been that the processes of disease transmission occur in real time and that variability in factors such as the infectious period may be dynamically important. Some have argued that if the latent and infectious periods are relatively constant, it is reasonable to construct models phrased in discrete time. Such models were first developed by Reed and Frost in 1928 (Abbey 1952) as probabilistic entities, assuming transmission probabilities are binomial (giving rise to "chain binomial" models; see Bailey 1975 and Daley and Gani 1999 for more details). Here, we introduce these models within a deterministic setting and refer interested readers to Chapter 6 for a discussion of analogous stochastic models.

An important issue that arises in modeling epidemics in discrete time is precisely what a time increment represents. Ideally, units of time should represent the "generation" length of the infection through a host, though in some cases this can lead to some difficulty especially if latent and infectious periods differ markedly. To demonstrate this approach, consider a disease with latent and infectious periods of exactly a week. We are therefore interested in determining the future changes in the fraction of individuals in the population who are susceptible (S_t), exposed (E_t), or infectious (I_t) in week t. The following *difference* equations can be used to represent such a scenario:

$$S_{t+1} = \mu - S_t e^{-\beta I_t}, \tag{2.60}$$

$$E_{t+1} = S_t(1 - e^{-\beta I_t}), \tag{2.61}$$

$$I_{t+1} = E_t, \tag{2.62}$$

where μ now represents the *weekly* per capita births. The exponential term in equation (2.60) represents the per capita probability of *not* contracting the infection given I_t infectives with transmission β (Box 2.1). This formalism inherently assumes that transmission probability per susceptible follows a Poisson distribution, with mean βI_t. Hence, the probability of escaping infection is given by the zero term of this distribution. The transmission parameter (β) is now analogous to the maximum reproductive potential—or R_0—of the infection. Consider the situation where everyone in the population is susceptible and a single infectious individual is introduced. Then, the initial spread of the disease occurs at rate β, very much analogous to R_0 in the continuous-time models. Thus, as before, for the infection to invade, we require $\beta > 1$.

The introduction of discrete time does little to change the analytical approaches used to explore model dynamics. For example, we can easily perform equilibrium analysis on these equations (2.60)–(2.62) by solving for $S^* = S_{t+1} = S_t$, $E^* = E_{t+1} = E_t$, and so forth. Such an exercise yields the following equilibrium solutions:

$$S^* = \frac{\mu}{1 - e^{-\beta\mu}}, \tag{2.63}$$

$$E^* = \mu, \tag{2.64}$$

$$I^* = \mu. \tag{2.65}$$

To establish the criterion for endemicity, we insist that at equilibrium the proportion of susceptibles is less than one ($S^* < 1$). From equation (2.63), this translates into requiring that

$$\beta > \frac{-\log(1 - \mu)}{\mu}.$$

This relationship is nearly always satisfied because (1) $\log|1 - \mu| \sim \mu$ (for $\mu \ll 1$) and (2) for the infection to successfully invade, we require $\beta > 1$.

To establish the stability of this equilibrium, we follow a similar procedure to that outlined in Box 2.4. The Jacobian for this system is simply a matrix whose elements represent the partial derivatives of equations (2.60)–(2.62) with respect to S_t, E_t, and I_t, evaluated at equilibrium (for more details, see Murray 1989):

$$\boldsymbol{J} = \begin{pmatrix} e^{-\beta I^*} & 0 & -\beta S^* e^{-\beta I^*} \\ 1 - e^{-\beta I^*} & 0 & \beta S^* e^{-\beta I^*} \\ 0 & 1 & 0 \end{pmatrix}.$$

To obtain a polynomial for the eigenvalues, we subtract Λ from the diagonal elements and work out the determinant for the Jacobian (i.e. set $\det|\boldsymbol{J} - \Lambda\boldsymbol{I}| = 0$). As for the SIR equations, because we have three variables, we obtain a cubic equation in the eigenvalues. In this instance, because of the discrete time nature of the model, stability of the equilibrium solution requires the dominant eigenvalue to have magnitude less than one (May 1973; Nisbet and Gurney 1982; Murray 1989). For the system described by equations (2.60)–(2.62), this reduces to requiring all roots of the following equation to lie in the unit circle:

$$\Lambda^3 - e^{-\beta I^*}\Lambda^2 - \beta S^* e^{-\beta I^*}\Lambda + \beta S^* e^{-\beta I^*} = 0.$$

Although an analytical solution to this equation is not possible, we can use any number of commands in mathematical software packages to find its roots. In Figure 2.12, we have numerically calculated the dominant eigenvalues for a range of parameter combinations and show that as long as the endemic equilibrium is feasible, it is also stable. Note, however, that the absolute values of Λ are very close to unity, suggesting a weakly stable equilibrium: perturbations eventually decay to zero, but this may happen over a long time.

Recently, some authors (Finkenstädt and Grenfell 2000; Bjørnstad et al. 2002; Xia et al. 2004) have argued that when infection prevalence is relatively small, we may conveniently

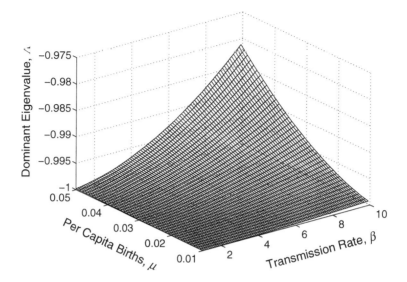

Figure 2.12. The stability of the endemic equilibrium for the SIR model phrased in discrete-time. The magnitude of the dominant eigenvalue is always less than one, but not massively so, suggesting very weakly stable dynamics; the dominant eigenvalue is also real and negative, indicating a 2-week oscillatory cycle in the approach to equilibrium.

rewrite equation (2.60) using a bilinear term to represent the transmission term:

$$S_{t+1} = S_t + \mu - \beta S_t I_t.$$

This is essentially the same as expanding the exponential term in equation (2.60) and ignoring the $\beta^2 I_t^2$ and higher-order terms. (For an ecological example, contrast the host-parasitoid models of Hassell (1978) and Neubert et al. (1995)). If we proceed with an equilibrium analysis, as above, we find that the dominant eigenvalue is *always* equal to unity. In this system, the endemic equilibrium is neutrally stable, with small perturbations to the equilibrium neither decaying nor growing. This is an undesirable and unrealistic property of the model and results largely from the above simplification. To overcome this undesirable structural instability of the model, Finkenstädt and Grenfell (2000) incorporated an exponent into the S_t and I_t terms:

$$S_{t+1} = S_t + \mu - \beta S_t^\alpha I_t^\phi.$$

When α and ϕ are less than one, the endemic equilibrium is stabilized.

The major benefit of formulating a model in discrete time is that it is much easier to parameterize using discrete time data than the associate differential equation counterparts (Bailey 1975; Mollison and Ud-Din 1993; Finkenstädt and Grenfell 2000). Their major drawback, however, is their demonstrated mathematical fragility (Glass et al. 2003).

2.8. PARAMETERIZATION

Although models in isolation can provide us with a deep understanding of the transmission and control of general infectious diseases, if we wish to examine the epidemiology of

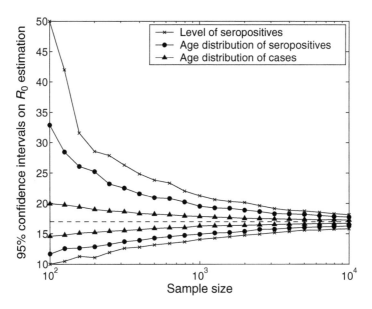

Figure 2.13. The effect of sample size on the estimated value of R_0. The true value of R_0 is 17, and the lines show the 95% confidence intervals associated with various estimation methods. For the serological estimates, the sample size refers to the number of sampled individuals (irrespective of disease status); for the case-based estimates (triangles), the sample size refers to the number of recorded cases. We have assumed no age bias in the reporting of cases.

a specific pathogen, we need to parameterize our models accordingly. Although generic models provide an intuitive explanation, it is only though detailed parameterization that useful public health guidance can be generated. From a modeling perspective, only the parameterization differentiates models of smallpox and measles, or models of chlamydia and HIV.

The standard SIR model has four parameters: the birth rate, the natural death rate, the average infectious period (or, conversely, the rate of recovery), and the transmission rate. In an ideal world, robust statistical inference would provide information on all parameters. In reality, this approach is hampered both by limited data, as well as underdeveloped appropriate inferential methdologies, though a substantial body of ongoing work redresses this shortcoming (see King and Lele, in prep, Ionedes et al. 2006, Vasco and Rohani, in prep). As a result, we often use other relevant information to assist with parameterization. A good understanding of the host's biology, for example, will provide accurate estimates of the birth and death rates (which we have generally assumed to be equal). The infectious period can usually be estimated independently via clinical monitoring of infecteds, either by observation of transmission events or by more detailed techniques measuring the amount of pathogen excreted. This leaves us with a single parameter—the transmission rate, β—to estimate. In practice, researchers generally focus attention on estimating the basic reproductive ratio, R_0, from which the transmission rate can be derived. There are multiple approaches to finding the value of R_0, which if appropriate will depend on the type of data that is available (Figure 2.13).

2.8.1. Estimating R_0 from Reported Cases

An initial response to estimating transmission, β, or the basic reproductive ratio, R_0, would be to record the total number of reported cases of infection within a given community or area. However, if we examine the equilibrium prevalence (equation (2.19)), we find that:

$$I^* = \frac{\mu}{\beta}(R_0 - 1) = \frac{\mu}{\gamma + \mu}\left[1 - \frac{1}{R_0}\right].$$

Therefore, if R_0 is substantially larger than one, then estimation is going to be very difficult because the effect of R_0 on prevalence is relatively small. In addition, we are unlikely to record every case within a population as many infections will go unreported. Even for measles in England and Wales (for which we possess one of the most accurately recorded long-term epidemiological data sets), the reporting rate is only 60% (Clarkson and Fine 1985; Finkenstadt and Grenfell 2000), which swamps the small effects of R_0 that we wish to detect.

An alternative approach is to examine the early behavior of an epidemic. The SIR model predicts that the early growth of an epidemic will be exponential:

$$I(t) \sim I(0)\exp([R_0 - 1](\gamma + \mu)t)$$

Fitting to this type of behavior is possible, and during the early stages of a novel infection (such as the 2001 foot-and-mouth epidemic in the United Kingdom or the 2003 SARS epidemic in China) is the only way of estimating transmission levels. However, in general, three difficulties may exist: (1) this method can work only if there is an epidemic to be observed, and cannot provide information on endemic diseases; (2) in the early stages of an epidemic, due to the low number of cases, the dynamics may be highly stochastic and influenced by large fluctuations (see Chapter 6), which will cause the estimates of R_0 to also fluctuate. By the time stochastic fluctuations become negligible, it is likely that the epidemic behavior is nonlinear and therefore the exponential approximation is no longer valid. (3) The final difficulty with this approach is that unless the pathogen is novel to the population, some individuals are likely to be immune. Although this partial susceptibility can be incorporated into the equations leading to $I(t) \sim I(0) \exp([S(0)R_0 - 1](\gamma + \mu)t)$, it is impossible to separate its effects from the value of R_0. One plausible solution is to fit to the entire epidemic, finding which values of R_0, the initial level of susceptible, $S(0)$, and case identification probability lead to a model epidemic that most closely matches the recorded profile of cases (see Gani and Leach 2001 for an example of this method applied to smallpox data). Additional aspects affecting the use of this approach in estimating R_0 arc discussed in Chapter 3 and Wearing et al. (2005).

At this stage, it is important to distinguish between the variable $Y(t)$ in models and the kinds of longitudinal epidemiological data typically available. The plague case notification data plotted in Figure 2.3, for example, represent the total number of *new* cases diagnosed in a given week—the incidence of infection. This is distinct from $Y(t)$, which is the number of individuals infectious at time t—the prevalence of infection. If the infectious period of a disease happens to be approximately the same as the resolution of the data, then the cases and Y are comparable (Ferrari et al. 2005). Otherwise, we need to introduce a new model variable that can be used in any estimation procedure. In general, it is reasonable to assume that case reports take place once an individual leaves the infectious class. This may be because diagnosis occurs most readily once an individual is symptomatic, at which point either the infection has already been cleared or the individual remains infectious but has reduced transmission opportunities due to quarantining and convalescence. Therefore, $K(T)$, the number of new cases reported at time point T, may be represented by

$$K(T) = \int_{T-1}^{T} \gamma I \, dt.$$

This formulation assumes no reporting error, which may be inappropriate and should be addressed before estimation to ensure bias-free parameter values (see, for example, Finkenstädt and Grenfell 2000; Clarke 2004). The implementation of reporting error is discussed in more detail in Chapter 6.

Finally, one of the most powerful approaches using case reports it to identify the average age at infection in an endemic situation. We have already seen (equation (2.24)) that the average age of infection is given by $A \approx 1/[\mu(R_0 - 1)]$, and this approximation is most reliable when R_0 is large. The average age of infection is generally estimated by simply finding the average age of all reported cases. Figure 2.13, triangles, shows how successful this estimation procedure is at finding the value of R_0 as the number of sampled cases varies. Two potential difficulties exist with this methodology: (1) the age of patients may not be recorded as a matter of course, and therefore it may be impossible to analyze historical data with this technique; and (2), the researcher generally has little control over which individuals report infection, and therefore age-related reporting biases can influence the results.

2.8.2. Estimating R_0 from Seroprevalence Data

Estimating R_0 from case report data is problematic in humans because reporting is often patchy and biased because not all infected individuals seek medical advice. For wildlife diseases, obtaining data on individual cases is even more difficult. An alternative approach is to use standard molecular techniques to detect the presence of an antigen against a particular pathogen. In this way, we can differentiate the population into those who have not experienced infection (susceptibles) and those who have (either recovered or currently infected). The primary advantage of working with such serological data is that the researcher has full control over the sampling of the population. In contrast, the advantage of working with infectious cases (as detailed above) is that a larger amount of data can be collected during normal medical practices.

The simplest way of utilizing serological information is via the relationship that $S = 1/R_0$; this is shown by the crosses in Figure 2.13. Here the concept is relatively simple: R_0 is estimated as the inverse of the proportion of our sample that are seronegative (susceptible). The complication with this approach is that we need to make sure that our sample represents of the entire population, because the level of seroprevalence (proportion of individuals recovered from infection) is expected to increase with age.

An alternative method that can take far better advantage of the information from a serological survey is to once again utilize the age-dependent nature of the likelihood of being susceptible (Anderson and May 1991). For an individual of age a, the standard SIR model predicts that the probability an individual is still susceptible is given by $P(a) \approx \exp(-a\mu(R_0 - 1))$. Therefore, if we know the ages of our serological sample, we can construct the likelihood that the data comes from a disease with a particular R_0 value, and then find the R_0 that maximizes this likelihood. If we have n individuals who are susceptible (seronegative) of ages $a_1, \ldots, a_n$, and m individuals who are seropositive (ages $b_1, \ldots, b_m$), then the likelihood is:

$$L(R_0) = \prod_{i=1}^{n} \exp(-a_i \mu (R_0 - 1)) \prod_{i=1}^{m} \left[1 - \exp(-b_i \mu (R_0 - 1)) \right].$$

This use of additional age-related information together with serology provides a much better prediction of the value of R_0 (Figure 2.13, circles). However, if we are free to choose our serological sample from any members of the population, then we should preferentially select individuals of an age close to (what we initially believe to be) the average age of infection—sampling many older individuals will provide little information because they are all likely to be seropositive. In Figure 2.13, where $R_0 = 17$, then sampling individuals just in the 3- to 7-year-old age range produces estimates of similar accuracy to those from the case-based average age of infection assuming unbiased reporting (triangles).

2.8.3. Estimating Parameters in General

The above likelihood argument provides a general framework in which models can be successfully parameterized from data. Ideally, as much information (and parameterization) should be gained from individual-level observations; often this provides a very accurate parameterization of individual-level characteristics such as average infectious and latent periods, but is generally unreliable for transmission characteristics. For the missing parameters (for example, in Section 2.6, equations (2.56)–(2.59) require a transmission rate for both the infectious and carrier class), a maximum likelihood approach is often the most suitable. For a given set of parameters, we can determine the dynamics predicted by the model and then calculate the likelihood that the observed data came from such dynamics. By finding the set of parameters that maximize this likelihood, we are selecting a model that is in closest agreement with the available data. In the above example, the dynamics were the long-term equilibrium age-distribution of susceptibility, but there is no reason why more dynamic behavior could not be utilized. Finally, this maximum likelihood approach is not confined to parameter selection and can even allow us to distinguish between models, providing us with a means of selecting the most appropriate framework.

The ensuing chapters return to this question of parameterization, where appropriate, and show how the extra levels of heterogeneity can be characterized from the type of observation and available survey data.

2.9. SUMMARY

In this chapter, we have introduced the simplest models of epidemics, whose structures have been determined by the biology of the aetiological agent and its effects on the host. We have met (1) the SI model, which represents infections that are fatal; (2) the SIS model for infections that do not illicit a long-lasting immune response; (3) the $SIRS$ when immunity is not permanent; and (4) the SIR and $SEIR$ models that describe host-pathogen systems with lifelong protection following infection.

This chapter also introduced the epidemiologically important measure of a pathogen's reproductive potential, R_0. It summarizses the number of new individuals infected by a single infectious individual in a wholly susceptible population. Irrespective of model structure, the infection will experience deterministic extinction unless $R_0 > 1$, which is intuitively appealing (and obvious).

By concentrating on these simple models, a number of important epidemiological principles can be demonstrated:

➤ For an infection to invade a population, the initial fraction of the population in the susceptible class has to exceed $\frac{1}{R_0}$.

➤ Initially, the infection grows at a rate proportional to $R_0 - 1$.

➤ In the absence of host demography, strongly immunizing infections will always go extinct eventually. After the infection has died out, some susceptibles remain in the population. Thus, the chain of transmission is broken due to too few infectives, not a lack of susceptibles.

➤ To maintain an immunizing infection, the susceptible pool must be replenished via recruitment.

➤ In the SIR model with host demography, the endemic equilibrium is feasible if $R_0 > 1$ and is always stable. In this system, trajectories converge onto the asymptotic state via damped oscillations, the period of which can be determined as a function of epidemiologically simple measures such as the mean age at infection and the effective infectious period of the disease. This is often referred to as the "natural period" of the system.

➤ For diseases that fail to elicit long-term immunity to subsequent reinfection, the SIS model is appropriate, which is identical to the standard logistic model population growth. The key feature of that model is the presence of an exponentially stable equilibrium point. This means the infection will always be feasible and stable if $R_0 > 1$ and the approach to the equilibrium is not oscillatory.

➤ If we assume that infectious and latent periods are exactly fixed, then it is possible to formulate models in discrete time. The take-home message from these models is qualitatively similar to the standard SIR models, with the endemic equilibrium feasible and stable as long as $R_0 > 1$.

➤ Not surprisingly, discrete-time models are much less stable than their continuous-time counterparts. The endemic equilibria are very weakly stable, with perturbations decaying over long periods.

This chapter has focused on model simplicity, often sweeping known epidemiological and demographic complications under the carpet. In particular, we have focused exclusively on deterministic approaches to understanding disease dynamics. Implicit in this approach is an assumption that chance events (in the transmission process or host population demography) are either unimportant or can be "averaged out," to reveal the true underlying system traits. Chapter 6, relaxes this assumption and explores the consequences of stochasticity in SIR and SIS systems.

Chapter Three

Host Heterogeneities

The standard models introduced in Chapter 2 compartmentalize the population only in terms of infection status and history—classifying individuals as susceptible, infected, or recovered—and modeling the number of individuals in each compartment. As such, there is only one degree of subdivision within the population. In this chapter, we introduce a second degree, further dividing the population into classes with similar behavioral characteristics. These characteristics should be chosen such that all members of a class have comparable risk of both contracting and transmitting infection.

Two clear motivating examples dominate the literature of models dealing with risk: (1) age structure for childhood infections, and (2) risk structure for sexually transmitted infections (STIs). Models for STIs frequently subdivide the population into classes dependent upon the risk associated with the behavior in each class. High-risk individuals, for example, have many sexual partners (or for some STIs could partake in intravenous drug use, frequently sharing needles). As a result, individuals in this class have a higher risk of both contracting and transmitting infection. Given the very clear link between the number of partners and the risk of infection, and the heterogeneity between individuals (Johnson et al. 1994), it seems intuitive to include such variation in models if we wish to understand and predict the patterns of sexually transmitted infections. Although ignoring the behavioral heterogeneity and assuming that everyone has the same number of partners is appealing for its simplicity, we will show that such averaging can produce very misleading results.

For many infections, such as measles, mumps, or chickenpox, a high basic reproductive ratio (R_0) means that the average age at first infection is low (see Chapter 2), and therefore they are most commonly encountered during childhood. The modeling of these so-called "childhood diseases" also requires further partitioning of the population, this time in terms of age. For STIs, the subdivisions are determined by the *number* of contacts. For childhood diseases, in contrast, the subdivisions are due to the *nature* of contacts. Because such diseases are common in childhood but rare in adults, those individuals who mix most with children are at the greatest risk. Due to their aggregation in schools, children predominantly mix with other children, and therefore age acts as the major determinant of risk.

Both of these examples show how differential patterns of mixing influence the likelihood of contracting and transmitting an infection. Some subsets of the population are clearly at greater risk, whereas some are relatively isolated. By incorporating such heterogeneities into models, we gain three distinct advantages: The aggregate behavior of models becomes more accurate, we can determine the prevalence of infection within the different classes, and finally we can use this information to determine more efficient targeted measures for disease control.

Including risk heterogeneities inevitably increases the number of equations. Incorporating high-risk and low-risk classes with STI models doubles the number of equations compared to homogeneous (averaged) models. With age-structured models, which can have tens of age classes, the increase in the number of equations is even more pronounced. However, the equations for each class have a very similar structure, so an increase in their number does not correspond to an increase in the intellectual challenge or computational difficulty, merely an increase in computational time. However, including heterogeneities does increase the number of parameters it is necessary to estimate—this often translates into a need for more biological data. Generally, it is preferable to have case reports (or other information) subdivided into the same classes as used in the model because this simplifies parameterization and the comparison of model results with data.

Finally, in this chapter we consider levels of heterogeneity within the infected class, discriminating between individuals due to the time since infection. Frequently, very accurate data is available to parameterize these forms of models, where the distribution of latent and infectious periods is well known. Although these changes in the shape of the distribution may at first appear trivial, they can often have a profound impact on the infection dynamics.

Ultimately, three applied questions drive the work in this chapter:

1. How does risk structure influence the spread and prevalence of pathogens?
2. Can risk structure be used to more effectively control the spread of infection?
3. How can risk structure be parameterized in realistic scenarios?

To answer these questions, we need to extend the basic models of Chapter 2 to include the various risk groups.

3.1. RISK-STRUCTURE: SEXUALLY TRANSMITTED INFECTIONS

In this section, we introduce the concepts of modeling population heterogeneity with the particular example of sexually transmitted infections and just two groups (high risk and low risk). In practice, there are a vast range of sexual risk groups, from prostitutes to celibate individuals, and each ideally requires its own particular class within the model. However, the two-class model used as the primary example in this chapter is sufficiently complex to demonstrate the necessary tools and techniques, yet simple enough to be intuitively understood.

Sexually transmitted infections are a growing problem in many areas of the world. The AIDS epidemic that began in the 1980s is continuing to increase, and in recent years there has been a rise in the prevalence and incidence of many other less well-known STIs. It is important to understand how our ever-changing sexual practices dictate the prevalence of such infections, and how their increase can be effectively combated. Since HIV and AIDS were first diagnosed in 1983 (Barre-Sinoussi et al. 1983), it has spread worldwide with alarming speed (Mertens and LowBeer 1996). In 2002, this pandemic was estimated to infect around 42 million people, increasing at a rate of 5 million new cases a year, and be responsible for around 3 million deaths per year—the vast majority of which are in Sub-Saharan Africa and the heterosexual community (UNAIDS/WHO 2002). In the United States, over 15 million new cases of sexually transmitted infections are diagnosed every year (Cates et al. 1999). In the United Kingdom, there has been a clear increase in many STIs over the last decade (Figure 3.1), with the greatest growth in syphilis, gonorrhoea,

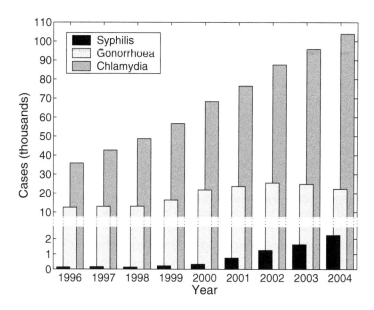

Figure 3.1. The recent trends in three sexually transmitted infections in England, highlighting the dramatic rise in STIs over a nine-year period. The y-axis has been split to show the lower but increasing levels of syphilis. (Data from the Public Health Laboratory Services www.phls.co.uk/ infections/topics_az/hiv_and_sti/epidemiology/dataresource.htm).

and chlamydia. This has prompted a great deal of research activity to be focused on modeling these sexually transmitted infections and asserting effective means of control. (syphilis: Cates et al. 1996; Oxman et al. 1996; Chesson et al. 1999; Morris 2001; Pourbohloul et al. 2003. Gonorrhoea: Kretzschmar et al. 1996; Garnett et al. 1999. Chlamydia: Delamaza and Delamaza 1995; Welte et al. 2000; Kretzschmar et al. 2001; Kretzschmar 2002. General: Boily and Masse 1997; Stigum et al. 1997; Garnett and Bowden 2000.)

The majority of sexually transmitted infections conform to the SIS framework (Chapter 2), which leads to one of the simplest of all disease models. For the purposes of epidemiological modeling, the natural history of sexually transmitted infections is relatively simple: Individuals are born susceptible to infection and remain so until they enter sexually active relationships. Transmission of infection, from an infected to a susceptible individual, occurs during sex—the potential transmission routes are, therefore, more clearly defined and determinable than for airborne infections. Recovery from infection generally occurs only following medical treatment, after which the individual becomes susceptible again. HIV is the notable exception to this rule because individuals never recover from the infection. Although the population-level dynamics of STIs with SIS behavior are relatively simple, the within-host dynamics are far more complex. In many cases, the immune response is suppressed so that the infection persists without clinical symptoms for many months or years, thus increasing the number of partners to which it can be spread. As laid out in Chapter 2, for an unstructured (sexually active) population, where everyone is assumed to be at equal risk, the SIS framework of progression leads to the following equations:

$$\frac{dS}{dt} = -\beta SI + \gamma I$$

$$\frac{dI}{dt} = \beta SI - \gamma I,$$

(3.1)

where the effects of individuals entering and leaving the sexually active population have been ignored. As before, the parameter β determines the contact and transmission rates between susceptible and infectious individuals, and γ is the rate at which treatment is sought. If the population size is constant, the rates of change in the fractions susceptible and infectious are always reflections of each other ($\frac{dS}{dt} = -\frac{dI}{dt}$) and only one needs to be calculated. We now consider how this framework can be extended to include multiple interacting groups.

3.1.1. Modeling Risk Structure

From biological and mechanistic principles, we can derive sets of equations for the various risk groups within the population, and from these equations develop a robust, generic framework to explain the interaction between risk and epidemiological dynamics. We start with a simple two-class model, incorporating high-risk and low-risk individuals.

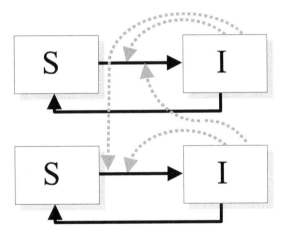

3.1.1.1. High-Risk and Low-Risk Groups

We focus initially on the behavior of the high-risk group, and denote the number of susceptible and infectious individuals within this group by X_H and Y_H, and the total number in the high-risk group by $N_H (= X_H + Y_H)$. Alternatively, it is often simpler to use a frequentist approach, such that S_H and I_H refer to the proportion of the entire population that are susceptible or infectious and also in the high-risk group, in which case n_H is the proportion of the population in the high-risk group: $S_H = X_H/N$, $I_H = Y_H/N$, $n_H = N_H/N$. This is the approach that will be adopted throughout this chapter. With this formulation, a disease-free population has $S_H = n_H < 1$, which is crucial when calculating R_0 or invasion criteria. (A third formulation is to model the proportion of each group that are infected or susceptible, such that $\widehat{S}_H = X_H/N_H, \widehat{I}_H = Y_H/N_H, \widehat{S}_H + \widehat{I}_H = 1$; however, this approach does not correspond intuitively with the earlier unstructured models.)

The dynamics of either group is derived from two basic events, infection and recovery. In this simple formulation we explicitly do not allow the movement of individuals between risk groups; individuals are "born" into a risk group and remain so for life. We initially focus on the dynamics of the high-risk group. Recovery, or the loss of infectious cases, can occur only through treatment and, following the unstructured formulation, we assume this occurs at a constant rate γ. For more generality, we could let the treatment rate be specific to the group (i.e., use γ_H), but at this stage such complexity is unnecessary. New infectious cases within the high-risk group occur when a high-risk susceptible is infected by someone in either the high- or low-risk group. These two distinct transmission types require different transmission parameters: We let β_{HH} denote transmission to high risk from high-risk and β_{HL} represent transmission to high risk from low risk. (Note throughout this book we use the same ordering of subscripts such that transmission is always $\beta_{\text{to from}}$.) Putting these elements together, we arrive at the following differential equation:

$$\frac{dI_H}{dt} = \beta_{HH}S_H I_H + \beta_{HL}S_H I_L - \gamma I_H. \tag{3.2}$$

By a similar argument, we can derive an expression for the low-risk individuals:

$$\frac{dI_L}{dt} = \beta_{LH}S_L I_H + \beta_{LL}S_L I_L - \gamma I_L. \tag{3.3}$$

As demonstrated in Chapter 2, the susceptible equations can be ignored because their dynamics are determined by the number of infectious individuals: $S_H = n_H - I_H$ and similarly for the low-risk group (assuming proportions in each group do not change).

This is online program 3.1

There are now four transmission parameters, and the simplest way to encapsulate this information is in a matrix $\boldsymbol{\beta}$:

$$\boldsymbol{\beta} = \begin{pmatrix} \beta_{HH} & \beta_{HL} \\ \beta_{LH} & \beta_{LL} \end{pmatrix}.$$

These transmission matrices are often termed WAIFW (Who Acquires Infection From Whom) matrices, and provide a convenient means of capturing the mixing between different social groups. This matrix $\boldsymbol{\beta}$ plays a similar role to the scalar parameter β of the unstructured model. We now seek to make that relationship more transparent. The first consideration is the relative magnitudes of the four terms. The individuals in the high-risk group should be at a higher risk of infection, therefore $\beta_{HH} + \beta_{HL}$ should be larger than $\beta_{LH} + \beta_{LL}$. We would also expect *assortative* mixing, where individuals from the high-risk group are more likely to partner with other high-risk individuals, and low-risk individuals are more likely to be in long-term relationships with another member of the low-risk group (see Section 3.1.6 and Box 3.4). This means that the diagonal elements of the matrix dominate, with β_{HH} being the largest term. Finally, we insist that interactions between the groups are symmetric such that the number of interactions between high- and low-risk groups is the same as interactions between low and high; this implies that $\beta_{HL} = \beta_{LH}$, or more generally that the matrix is symmetric. This assumption equates with the fact that individuals in both groups have an equal response to infectious challenge—if one group was, for some reason, more susceptible, then the symmetry property may break down (see Section 3.1.2.2 for an example of such asymmetry).

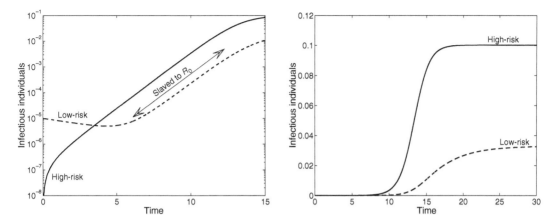

Figure 3.2. The dynamics of an STI introduced into the low-risk class at a small level, showing the prevalence of infection in the high- and low-risk groups relative to the entire population size (I_H and I_L, respectively). Three distinct phases of dynamics can be observed; a transient phase that is determined by the initial distribution of infection, a phase dominated by the value of R_0 when the prevalence of infection increases (or decreases) exponentially, and finally an asymptotic phase when density-dependent factors begin to operate and the equilibrium levels are achieved.

The single transmission parameter in the unstructured models is replaced by a matrix of values for structured models. In general, when all groups have an equal response to infection challenge, this matrix is symmetric.

From the above arguments, a plausible transmission matrix is therefore:

$$\beta = \begin{pmatrix} 10 & 0.1 \\ 0.1 & 1 \end{pmatrix},$$

and we shall use this particular example in the illustrative calculations that follow. We suppose that high-risk individuals make up 20% of the population, with the remaining 80% being low risk. Finally, we make the convenient assumption that $\gamma = 1$.

3.1.1.2. Initial Dynamics

For unstructured models, the single parameter β was vital in determining the basic reproductive ratio, R_0, and hence the rate of increase in infection following invasion. A naive proposition might be to assume that there is a specific R_0 for each class, and this can be calculated from the expected number of secondary cases a primary case (in a particular group) would cause. So, using the above matrix, the R_0 for the high-risk group is 2.08 ($R_0^H = 10 \times 0.2 + 0.1 \times 0.8 = [\beta_{HH}n_H + \beta_{LH}n_L]/\gamma$), whereas the R_0 for the low-risk group is 0.82 ($R_0^L = 0.1 \times 0.2 + 1 \times 0.8$). For the disease-free state $S_H = n_H < 1$ and $S_L = n_L < 1$, and so the size of the risk groups enters the calculation.

Although such formulations provide some insight into the dynamics (generally allowing us to bound the possible rates of increase following invasion), soon both risk groups contain infected individuals and their dynamics become *slaved*, increasing at the same exponential rate (Figure 3.2). The rate of increase in this slaved region is independent of the initial seed of infection and determines whether or not a infection will successfully invade and, as such, it is related to our basic intuition about R_0. R_0 for the entire population lies between the values calculated for members of each group, and is generally greater than the weighted average if the WAIFW matrix is assortative. To calculate the actual value of R_0, an eigenvalue value approach is required to deal with the recursive nature of transmission (Diekmann et al. 1990; Heesterbeek 2002); Box 3.1 outlines this approach in more detail.

To keep the sense of the original verbal definition of R_0 (Chapter 2), for structured models we may wish to consider R_0 as:

the average number of secondary cases arising from an average infected individual in an entirely susceptible population, once initial transients have decayed

We therefore calculate R_0 from the distribution of infection across risk groups in the region of slaved dynamics, where the behavior is independent of the initial conditions. This slaved distribution provides a natural weighting for the number of secondary cases generated by a primary case in each group.

Box 3.1 Eigenvalue Approach

Let us consider infection dynamics in both classes following invasion. The standard mathematical way of doing this is to linearize the equations about the disease-free state. Simply put, during the initial invasion phase, the relative change to the number of susceptibles is small, and therefore we can fix their values at the disease-free equilibrium ($S_H = n_H$ and $S_L = n_L$). This leads to:

$$\frac{dI_H}{dt} \approx (\beta_{HH}n_H - \gamma_H)I_H + (\beta_{HL}\gamma_H)I_L$$

$$\frac{dI_L}{dt} \approx (\beta_{LH}n_L)I_H + (\beta_{LL}n_L - \gamma_L)I_L.$$

Note that for generality, different recovery rates have been allowed. Such a linear system of differential equations is understood by looking at the matrix of coefficients, J:

$$J = \begin{pmatrix} \beta_{HH}n_H - \gamma_H & \beta_{HL}n_H \\ \beta_{LH}n_L & \beta_{LL}n_L - \gamma_L \end{pmatrix} = \begin{pmatrix} 1 & 0.02 \\ 0.08 & -0.2 \end{pmatrix}$$

and its dominant eigenvalue $\lambda_1 \approx 1.0013$ (taking $\gamma_L = \gamma_H = 1$). This eigenvalue then determines the dynamics in the slaved phase:

$$I_H \propto \exp(\lambda_1 t) \qquad \text{and} \qquad I_L \propto \exp(\lambda_1 t).$$

Thus it is clear that when $\lambda_1 > 0$ the infection can successfully invade (c.f. $R_0 > 1$), and when $\lambda_1 < 0$ the infection will always die out (c.f. $R_0 < 1$).

The ratio of I_H to I_L is determined by the eigenvector ($\underline{e}_1$) associated with the maximum eigenvalue. We let I_H^s and I_L^s be the distribution of *infection* in the slaved region as determined by the eigenvector, specifying that $I_H^s + I_L^s = 1$. For the particular matrix under consideration $I_H^s = 0.9376$ and $I_L^s = 0.0624$, such that in the slaved region the high-risk group has about 15 times more infection than the low risk group.

We can now use this slaved distribution to weight our average of R_0 for each of the classes:

$$R_0 = R_0^H I_H^s + R_0^L I_L^s,$$

$$= \left(\frac{\beta_{HH} n_H + \beta_{LH} n_L}{\gamma_H} \right) I_H^s + \left(\frac{\beta_{HL} n_H + \beta_{LL} n_L}{\gamma_L} \right) I_L^s, \quad (3.4)$$

$$\approx 2.08 \times 0.9376 + 0.82 \times 0.0624,$$

$$\approx 2.0013.$$

Equal Recovery Rates

Finally, for the simpler situation where the recovery rates within all classes are equal, the calculation of the basic reproductive ratio can be simplified and a more intuitive relationship to the single-class results can be obtained. If we define a matrix of the number of secondary cases produced in each class:

$$\boldsymbol{R} = \begin{pmatrix} \beta_{HH} n_H/\gamma & \beta_{HL} n_H/\gamma \\ \beta_{LH} n_L/\gamma_H & \beta_{LL} n_L/\gamma \end{pmatrix} = \begin{pmatrix} 2 & 0.02 \\ 0.08 & 0.8 \end{pmatrix},$$

then the dominant eigenvalue of this matrix is simply R_0. In addition, for the case of equal recovery rates, in the slaved region the growth rate is given by:

$$I \propto \exp([R_0 - 1]\gamma t),$$

which is the same relationship as found in the nonstructured models.

The basic reproductive ratio for the entire population is bounded by values for individuals in each group.

The basic reproductive ratio from structured models is generally larger than if the structures were ignored and all individuals had the same average transmission rates.

The basic reproductive ratio is found using an eigenvalue approach.

Figure 3.2 shows the dynamics for our simple example, starting with a few infectious individuals in the low-risk group. Initially, prevalence drops; this is because the majority of infecteds are in the low-risk group and its "reproductive ratio" in isolation is less than one. The initially steep rise of cases in the high-risk group is due to infection spreading from the low-risk class, governed by β_{HL}. The exact behavior in this transient period is determined by the initial conditions. Only by assuming that infection starts in the low-risk group do we obtain this counterintuitive decreasing-then-increasing behavior, the more common assumption that infection starts in the high-risk group leads to an increasing number of cases from the beginning. The converse is also true; it is possible to have a situation in which the total number infected increases before dying away to zero. With the same 20 : 80 ratio of high-to low-risk groups, the WAIFW matrix:

$$\beta = \begin{pmatrix} 1 & 1.5 \\ 1.5 & 0.5 \end{pmatrix}$$

possesses this surprising behavior. If infection starts in the high-risk group, then this on average causes $0.2 \ (= \beta_H H \times n_H)$ cases in the high-risk group and $1.2 \ (= \beta_L H \times n_L)$ in the low-risk group; a net increase to 1.4 cases. However, the eventual dynamics is given by the basic reproductive ratio, and for this matrix $R_0 = 0.9083$—which is less than one—hence in the long term the disease fails to persist.

The dynamics of Figure 3.2 exemplify the three major phases observed in all structured models. During the initial transient phase, the dynamics are often complex and frequently related to the reproductive ratios for each group in isolation. During this early phase, the relative disease prevalence rapidly approaches the distribution predicted by the right eigenvector of the equilibrium—this is analogous to the "stable age distribution" of matrix models used in ecology (Caswell 2000). Once this distribution is achieved, the prevalence of infection in all groups grows at the rate determined by the basic reproductive ratio R_0, until the number of susceptibles is sufficiently depleted and density-dependent effects arise. We then enter the final asymptotic phase where prevalence levels tend to their equilibrium value.

The initial behavior of a structure model depends on the initial conditions, not just the basic reproductive ratio, R_0.

3.1.1.3. Equilibrium Prevalence

Calculation of the prevalence of infection at equilibrium is far from trivial. Mathematically, we need to find where the rates of change are zero; hence, remembering that $S_H = n_H - I_H$, we need to solve:

$$0 = \beta_{HH}(n_H - I_H)I_H + \beta_{HL}(n_H - I_H)I_L - \gamma I_H,$$
$$0 = \beta_{LH}(n_L - I_L)I_H + \beta_{LL}(n_L - I_L)I_L - \gamma I_L.$$

These equations both contain quadratic terms, and therefore finding simple analytical solutions is generally impossible. We therefore have to rely on either numerical solutions of the above equilibrium equations, or more frequently we iterate the model forward to find the equilibrium levels.

For simple unstructured models, a clear relationship exists between the initial growth rate and the equilibrium density, $S(\infty) = 1/R_0$. In our example, we find (looking at Figure 3.2) that $I_H(\infty) \approx 0.1$ and $I_L(\infty) \approx 0.033$; the total proportion of the population that is susceptible is therefore $0.867(= [0.2 - 0.1] + [0.8 - 0.033])$. However, our calculations and the observed rate of increase suggest that $R_0 \approx 2$. It is therefore clear that the simple relationship between equilibrium and invasion dynamics no longer holds in structured populations. In our example, the small high-risk group is responsible for the value of R_0, but due to the large low-risk group, the prevalence of infection is much smaller than unstructured theory would predict. This also implies that density-dependent saturation effects occur far earlier in risk-structured models compared to their homogeneous counterparts.

In many structured models that have high associativity, the equilibrium fraction of susceptibles is much higher, and density-dependent saturation effects occur earlier than in unstructured models (where $S(\infty) = 1/R_0$). Hence, although the equilibrium prevalence of infection is low, the disease may be difficult to eradicate because R_0 is still large.

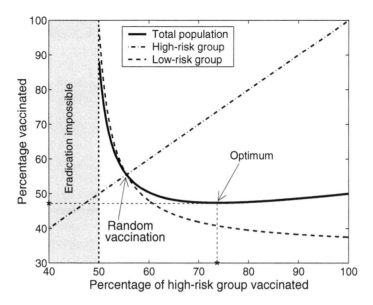

Figure 3.3. The critical level of vaccination needed to eradicate a STI, as a percentage of the entire population, for a range of coverage in the high-risk group. In this example, the transmission matrix is $\beta = \begin{pmatrix} 10 & 1 \\ 1 & 2 \end{pmatrix}$, with $n_H = 0.2$, $n_L = 0.8$, and $\gamma = 1$ as before. For each level of vaccination within the high-risk group, we have searched for the corresponding level of vaccination in the low-risk group that sets R_0 equal to one. Optimal control, which minimizes the amount of vaccine used, is achieved when vaccinating about 75% and 40% of the high- and low-risk groups, respectively.

3.1.1.4. Targeted Control

In unstructured models, a further simple relationship exists between the basic reproductive ratio (or the equilibrium distribution) and the level of control, such as vaccination, required for eradication of the disease. Standard models predict that the critical level of vaccination p_C needed to eradicate infection is given by:

$$p_C = 1 - S(\infty) = 1 - 1/R_0.$$

We now wish to investigate whether this basic formula holds and if it can be improved by targeting the control measures at those individuals who are most at risk. We again want a specific example to illustrate the basic concepts behind targeted control, and this is given in Figure 3.3. We have adjusted the matrix from the previous example to better demonstrate the trade-offs. In this graph, we have taken a series of vaccination levels within the high-risk group (p_H, dot-dash line) and found the level of vaccination in the low-risk group (p_L, dashed line), which forces the effective reproductive ratio R (calculated in a similar manner to R_0) to be one. The total percentage of the population that needs to be vaccinated (solid line) is then a measure of the cost associated with eradication and the prevention of successful re-invasion.

Only for the very simplest of models can the calculation of optimal targeting be performed analytically. In general, for a given level of control (e.g., vaccination, screening, or quarantining), we must search through all possible deployments to find which produces the more desirable results. In Figure 3.3, we explore the minimum level of vaccination that could set R_0 below one and hence eradicate infection. Alternatively, control measures may be optimally targetted to reduce the number of cases for a given control effort, or minimize the duration of an epidemic for given logistical limitations. Whatever the particular scenario, the quantity to be optimized (and any constraints) must be clearly specified by policy makers and planners. More examples of optimal control are discussed in Chapter 8.

For the transmission matrix in Figure 3.3, the basic reproductive ratio is around 2.25. After invasion, each case in the high-risk and low-risk groups causes 2.8 and 1.8 further cases, respectively. If we vaccinate at random (ignoring which group an individual belongs to), then the vaccination threshold is $1 - 1/R_0 \approx 55\%$, so we can always do as well as this standard result. However, as can be seen in Figure 3.3, the optimal strategy is to vaccinate nearly 75% of the high-risk group and only 40% of the low-risk class. A second point is that although excessive targeting of the high-risk group has only moderate adverse effects, failure to meet vaccination targets within this group has severe penalties. In many real scenarios, the difference between high- and low-risk groups is far more extreme, in which case the incentive for optimal targeting is far greater.

In structured models, the critical level of vaccination that eradicates infection is the same as in unstructured models, $1 - 1/R_0$, if vaccination is applied at random.

Targeting vaccination or other control measures generally works far better than random control. It is generally better to over-target the most at-risk groups rather than under-target.

3.1.1.5. Generalizing the Model

The matrix formulation for $\boldsymbol{\beta}$ can be readily adapted to model the interaction of multiple groups (e.g., high-, medium- and low-risk groups). Those infected individuals in group i obey the following differential equation:

$$\frac{dI_i}{dt} = \sum_j \boldsymbol{\beta}_{ij} S_i I_j - \gamma_i I_i, \qquad (3.5)$$

This is online program 3.2

where the matrix form of $\boldsymbol{\beta}$ is again used to parameterize transmission between the groups. The above set of equations can be recast into vector notation (as shown in Box 3.2), which can simplify the formulation and in some cases speed the computational solution. The value of R_0, and therefore whether the infection can successfully invade or not, is again determined by an eigenvalue approach as illustrated in the Box 3.1. When there are a large number of classes, this technique is particularly valuable.

3.1.1.6. Parameterization

The main key to applying such structured models is successful parameterization. For unstructured models there were few parameters to estimate. The infectious period (and latent period, if necessary) can usually be estimated from careful observation of infected

---

Box 3.2 Vector Notation

We have already stated that the contact rates are now most naturally specified as a matrix of values. To correspond to this notational change, we specify the number of infected (or susceptible) individuals in each class as a vector. In this notation the full set of equations becomes:

$$\frac{d\underline{I}}{dt} = \underline{S} \otimes (\boldsymbol{\beta}\underline{I}) - \underline{\gamma} \otimes \underline{I},$$

where $\otimes$ refers to the piecewise multiplication of two vectors.

---

hosts; this leaves a single transmission parameter β. The traditional means of finding this value is through the use of the relationship $S^* = \frac{\gamma}{\beta}$, where the equilibrium level of susceptibles ($S^* = 1 - I^* - R^*$) can be estimated from either serological surveys or long-term case records.

For structured models, the parameterization is far less straightforward. Whereas the low number of parameters needed for unstructured models allows us to sweep through all possible configurations and develop an intuitive understanding, the complexity and variety of structured models means that they are more reliant on good data with which they can be parameterized. The first step is to determine the appropriate risk groups and the proportions within them. The divisional structure of the model will come from good epidemiological evidence and should hopefully correspond with the structure of any sampled data. The infectious/latent period can once again be found from observing infected hosts. However, we now need a matrix of transmission rates and yet we have only a vector of serological or other information. So, for a general structured model with n distinct classes, we require n^2 transmission terms, but we have at most one piece of information for each class (e.g., we might know S_i^*). The usual way to deal with this lack of specificity is to assume a simplified structure for the transmission matrix; this approach is dealt with in more detail in the next section which considers age-structured models.

Because the transmission matrix generally has more terms than the structured data, simplifications are needed to overcome this deficit in information.

For sexually transmitted infections, there is a very natural and parsimonious means of subdividing the population: using the number of sexual partners. We let class i refer to those individuals who have exactly i partners within a given period. If we then assume that individuals form partnerships at random, but in proportion to their expected number of partners, then the form of the matrix $\boldsymbol{\beta}$ is given, with only a scaling parameter to be determined:

$$\boldsymbol{\beta}_{ij} = \beta\frac{ij}{\sum_k kn_k}, \tag{3.6}$$

where, as usual, n_k is the proportion of individuals in class k. This approach has proved extremely popular and highly successful, as it has the huge advantage that only a single parameter (the scalar β) needs to be estimated. We can now examine what effect this form of heterogeneity has on the basic reproductive rate R_0. Because of the random formation

---

Box 3.3 Random Partnership Model

Given the situation where partnerships are formed at random but in proportion to the expected number of partners, the contact matrix is given by:

$$\beta_{ij} = \beta \frac{ij}{\sum_k kn_k}.$$

To calculate R_0 for this epidemic model, it is easier and more informative to go back to first principles rather than use the eigenvalue approach. Irrespective of who is initially infected, due to the random nature of the mixing, the distribution of infection across the classes soon asymptotes:

$$\frac{I_i}{I} \rightarrow \frac{in_i}{\sum_k kn_k}, \qquad \text{where } I = \sum_i I_i$$

and remains in this distribution throughout the slaved period. This means that during this early growth phase the level of infection within a class (I_i/n_i) is proportional to the number of partners. We can thus find R_0 as the number of new cases we expect to be caused by infection with this distribution:

$$R_0 = \text{Number of secondary cases produced per primary case,}$$

$$R_0 = \frac{1}{\gamma} \sum_{i,j} S_i \beta_{ij} I_j / I = \frac{1}{\gamma} \sum_{i,j} n_i \beta \frac{ij}{\sum_k kn_k} \frac{jn_j}{\sum_k kn_k}$$

$$= \frac{\beta}{\gamma} \frac{\sum_i in_i \sum_j j^2 n_j}{\left[\sum_k kn_k\right]^2} = \frac{\beta}{\gamma} \frac{M(M^2+V)}{M^2} = \frac{\beta}{\gamma} \frac{M^2+V}{M},$$

where M and V are the mean and variance of the number of sexual partners.

---

of partnerships, the eigenvalue approach is no longer necessary, and we find:

$$R_0 = \frac{\beta}{\gamma} \frac{M^2+V}{M},$$

where M and V are the mean and variance of the number of sexual partners (greater detail is given in Box 3.3). If the heterogeneities had been ignored, and everyone assumed to have the same average number of partners, then the basic reproductive ratio is reduced to $\frac{\beta M}{\gamma}$. Hence even though we assume that partnerships are formed at random, heterogeneities can play a major role as the infection "focuses" on those high-risk individuals who are both more likely to catch the infection, and also more likely to transmit it.

Assuming individuals form contacts at random and proportional to their expected number of partners provides a very natural means of specifying the matrix β. We find that R_0 is increased due to the variance in the number of partners.

We can make this approach more explicit by returning to our simple model with high-risk and low-risk groups. If we assume that individuals in the high-risk group have an average of 5 new partners per year, whereas those in the low-risk group average 1 only, then the matrix becomes:

$$\boldsymbol{\beta} = \beta \begin{pmatrix} \dfrac{5 \times 5}{0.2 \times 5 + 0.8 \times 1} & \dfrac{5 \times 1}{1.8} \\[2ex] \dfrac{1 \times 5}{1.8} & \dfrac{1 \times 1}{1.8} \end{pmatrix} \approx \beta \begin{pmatrix} 13.9 & 2.78 \\ 2.78 & 0.556 \end{pmatrix},$$

for which the mean and variance are $M = 1.8$, $V = 2.56$, and calculating the basic reproductive ratio by either the eigenvalue or the method outlined above gives $R_0 = 3.22\beta/\gamma$.

Although this random partnership assumption captures many aspects of the transmission of STIs, it loses the property of assortative mixing where individuals are more likely to form partnerships with others in the same or similar classes. Two mechanisms can be used to overcome this problem. First, given a small amount of extra information on disease incidence, we could add a second parameter, α, to the transmission matrix that scales the proportion of within-group mixing. (More precisely, in the example given below, α is the fraction of partnerships that are forced to be made within the same class, with the remainder still being formed at random.) For example,

$$\boldsymbol{\beta}_{ij} = \beta(1-\alpha)\frac{ij}{\sum_k k n_k} \qquad \boldsymbol{\beta}_{ii} = \beta\left[\alpha\frac{i}{n_i} + (1-\alpha)\frac{i^2}{\sum_k k n_k}\right],$$

where although the distribution of transmission from an individual may change, the total amount of transmission remains constant.

In this model, as α increases, the matrix becomes closer to diagonal and the degree of assortative mixing becomes larger. When α is close to one, the eigenvalue of the matrix is dominated by the largest diagonal element; hence

$$R_0 \approx \frac{\beta}{\gamma}\max_i(i),$$

which is always larger than the random partnership formation example. Figure 3.4 illustrates this change in R_0 for the simple high-risk/low-risk example given above. We always expect more assortative mixing to lead to larger values of R_0; intuitively this is because as assortativity increases, spread is focused with the high-risk with less being "wasted" by infecting low-risk individuals. Interestingly, as assortativity increases, the targeting of vaccination may also change. For the first scenario considered in Figure 3.4, it is never worthwhile to vaccinate the low-risk group because they cannot maintain a chain of transmission because each infected individual can infect at most only one susceptible person. In contrast, the second example shows that although greater assortative mixing focuses more of the early growth within the high-risk class, the optimal vaccination strategy is increasingly spread between the two groups. This surprising result can be understood by realizing that when the mixing is random, it is always more efficient to concentrate vaccination on the high-risk individuals; however, when the mixing is completely assortative ($\alpha = 1$), then the risk groups act independently and the vaccination level in each must be sufficient to control transmission. More formally, the degree of assortative mixing for any matrix can be measured (Gupta et al. 1989), the details of which are given in Box 3.4.

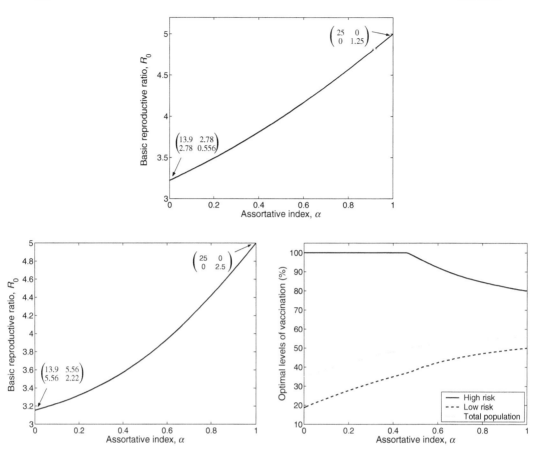

Figure 3.4. The effect of assortativity on the epidemic behavior of a population split into two classes, a high-risk group that comprises 20% of the population and a low-risk group that makes up the remaining 80%. In the top graph, individuals in the high-risk group have an average of five partners per year, whereas the low-risk group averages only one per year. In this situation it is always optimal to focus control exclusively on the high-risk group. For the bottom graphs, members of the high-risk group again average five partners per year, but now the low-risk group averages two partners per year. In this latter case, the optimal control strategy is a complex mix between the two groups. (For simplicity of presentation we have assumed that $\beta = \gamma = 1$.)

Increased assortative mixing increases R_0, but may mean that the optimal vaccination strategy is less targeted toward the high-risk groups.

A second approach to fully parameterize the transmission matrix is to utilize the very detailed data that has been collected on sexual partnership networks. In a few isolated examples this method has allowed researchers to reconstruct the full network of sexual partners within a population (Klovdahl 1985; Potterat et al. 2002). Although this type of network reconstruction may be highly sensitive to the occasional missing connection (which can vastly alter the topology of the network), the transmission matrix that emerges is far more robust, although some elements of the network structure are lost (see Chapter 7).

Box 3.4 Degree of Assortative Mixing

Given a contact matrix β and a vector of the proportion of individuals in each risk-group $\underline{n}$, we first define a new matrix $\widehat{\beta}_{ij} = \beta_{ij}n_i$. We next define the matrix B, which is the relative proportion of transmission from one group to all others:

$$B_{ij} = \frac{\widehat{\beta}_{ij}}{\sum_{i=1}^{M} \widehat{\beta}_{ij}},$$

where M is the number of classes being modeled. The degree of assortative mixing, Q, can be determined by comparing the relative amounts of within-group transmission to what is expected from the random formation of partnerships (Gupta et al. 1989):

$$Q = \frac{\sum_{i=1}^{M} B_{ii} - 1}{M - 1}.$$

As such, a matrix that comes from random formation of partnerships will have degree $Q = 0$, whereas one with complete assortative mixing (all diagonal elements) will have $Q = 1$. However, this weights all risk groups equally, therefore a more appropriate measure might be to utilize the eigenvalues of the matrix.

The values of the minor eigenvalues $(\widehat{\lambda}_2 \ldots \widehat{\lambda}_M)$ relative to the dominant eigenvalue $(\widehat{\lambda}_1)$ of $\widehat{\beta}$ provides a more natural description of the spread of infection between classes. In particular,

$$q = \frac{\widehat{\lambda}_2}{\widehat{\lambda}_1}$$

is an alternative measure of assortative mixing, which effectively weights mixing between similar classes.

Figure 3.5 shows an example of a detailed traced network from Colorado Springs together with the associated transmission matrix that arises. For this matrix, the degree of assortative mixing can be calculated as $Q = -0.0776$ or $q = 0.2452$ (see Box 3.4), hence showing that the matrix is either almost random, or only moderately assortative. This type of approach has been advocated by Garnett and Anderson (1993b), where individuals are classified into three basic groups (core, adjacent, and peripheral), and where the partnerships within and between the group are determined by detailed interviews (Rothenberg 1983).

Detailed information on the exact network of sexual partners can be used to parameterize the transmission matrix.

3.1.2. Two Applications of Risk Structure

We now consider two applications from the literature that complement the methodology detailed above. The first comes from the vast body of work done on the early spread of the HIV epidemic in developed countries, and using detailed epidemiological and sociological data illustrates many of the points already made. The second example of an STI comes from the work on chlamydia in koalas, because this provides a radical departure from the standard models and demonstrates some more unusual attributes associated with wildlife diseases.

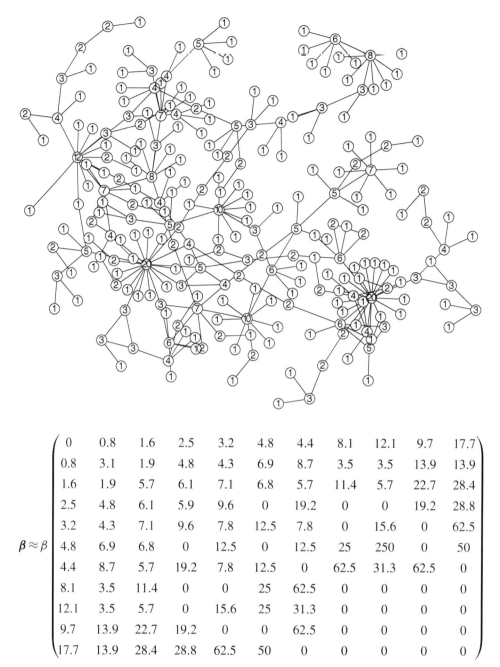

$$\boldsymbol{\beta} \approx \beta \begin{pmatrix} 0 & 0.8 & 1.6 & 2.5 & 3.2 & 4.8 & 4.4 & 8.1 & 12.1 & 9.7 & 17.7 \\ 0.8 & 3.1 & 1.9 & 4.8 & 4.3 & 6.9 & 8.7 & 3.5 & 3.5 & 13.9 & 13.9 \\ 1.6 & 1.9 & 5.7 & 6.1 & 7.1 & 6.8 & 5.7 & 11.4 & 5.7 & 22.7 & 28.4 \\ 2.5 & 4.8 & 6.1 & 5.9 & 9.6 & 0 & 19.2 & 0 & 0 & 19.2 & 28.8 \\ 3.2 & 4.3 & 7.1 & 9.6 & 7.8 & 12.5 & 7.8 & 0 & 15.6 & 0 & 62.5 \\ 4.8 & 6.9 & 6.8 & 0 & 12.5 & 0 & 12.5 & 25 & 250 & 0 & 50 \\ 4.4 & 8.7 & 5.7 & 19.2 & 7.8 & 12.5 & 0 & 62.5 & 31.3 & 62.5 & 0 \\ 8.1 & 3.5 & 11.4 & 0 & 0 & 25 & 62.5 & 0 & 0 & 0 & 0 \\ 12.1 & 3.5 & 5.7 & 0 & 15.6 & 25 & 31.3 & 0 & 0 & 0 & 0 \\ 9.7 & 13.9 & 22.7 & 19.2 & 0 & 0 & 62.5 & 0 & 0 & 0 & 0 \\ 17.7 & 13.9 & 28.4 & 28.8 & 62.5 & 50 & 0 & 0 & 0 & 0 & 0 \end{pmatrix}$$

Figure 3.5. An example of a sexual contact network, taken from the study of HIV transmission in Colorado Springs (Potterat et al. 2002). The matrix below the network, is the associated mixing matrix for individuals with 1, 2, 3, 4, 5, 6, 7, 8, 10, 12, and 20 partners, and is calculated as $\boldsymbol{\beta}_{ij} \propto$ (number of $i-j$ partnerships)$/(n_i n_j)$.

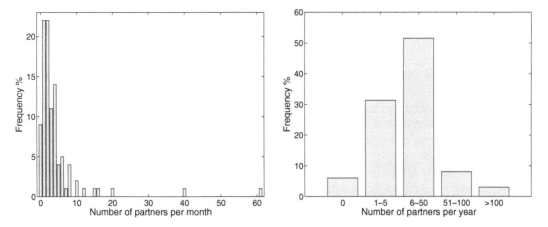

Figure 3.6. Frequency distribution of the number of sexual partners of male homosexuals from two studies in and around London in the mid-1980s. (Data from Anderson and May 1991, taken from original surveys by Carne and Weller (unpublished), and McManus and McEvoy (1987).)

3.1.2.1. *Early Dynamics of HIV*

To illustrate the use of the methodology already described, we focus on the early work that attempted to predict the spread of HIV in male homosexual communities (Anderson et al. 1986; Kolata 1987; May and Anderson 1987; Jacquez et al. 1988; May and Anderson 1989; Bongaarts 1989). This narrow focus has the distinct advantage that the models used at the time had a small number of risk classes and so the parameterization and dynamics can be shown explicitly.

Recent models, which focus on the more immediate problem of HIV spread in sub-Saharan Africa and elsewhere, are often much more complex either with a large number of risk classes or explicitly model the transmission network (Anderson et al. 1992; Arca et al. 1992; Garnett and Anderson 1993a; Downs et al. 1997; Koopman et al. 1997; Artzrouni et al. 2002; Koopman 2004). Therefore, although such models essentially follow the same principles outlined above, the degree of complexity and number of parameters required makes them infeasible as illustrative examples. A complete description of the spread of an STI would involve multiple overlapping classes, such as gender, age, level of sexual activity, sexual preference, drug use, condom use, visits to sex workers, and so forth, as well as transmission parameters that depend on time since infection (Longini et al. 1989). As such, for any complete model, there could be thousands of classes and at least as many parameters. Here, we will concentrate on the work of Anderson et al. (1986) and May and Anderson (1987), who used data on the number of partners per year together with the assumption of random partnership formation to develop simple mechanistic models. Many model assumptions are clearly vast simplifications of the underlying system, but these models help to demonstrate how an element of risk may be included.

The underlying mixing matrix comes from detailed social studies of the distribution of the number of sexual partners over a given period. The data we were recollected from homosexual males attending STI clinics in and around London, and shows a significantly skewed distribution, with a significant proportion of those questioned having a very large number of partners (Figure 3.6). Clearly these data will be biased by the sampling methods and are probably focused toward the high-risk groups that are more likely to be infected with an STI and therefore more likely to attend a clinic. However, similar patterns of partners have been recorded in other studies (McKusick et al. 1985). Since then, more detailed and large-scale studies have been performed, most notably by Johnson et al. (1994), who attempted to document the range and distribution of sexual practises in the United Kingdom—such data could be used to study the spread of STIs in the population as a whole, rather than a restricted class of high-risk individuals. The presence of a small core group of high-risk individuals makes the use of risk-structured models necessary for understanding the spread and persistence of sexually transmitted infections in general and HIV in particular.

We are now in a position to develop a model of this epidemic. For simplicity and brevity we adopt the data from the right-hand graph of Figure 3.6, and assume that there are five classes with 0, 3, 10, 60, and 100 partners per year in each class. This gives a matrix and population distribution

$$\boldsymbol{\beta} = \beta \begin{pmatrix} 0 & 0 & 0 & 0 & 0 \\ 0 & 0.65 & 2.15 & 12.9 & 21.5 \\ 0 & 2.15 & 7.17 & 43.1 & 71.8 \\ 0 & 12.9 & 43.1 & 258 & 431 \\ 0 & 21.5 & 71.8 & 431 & 718 \end{pmatrix} \qquad \underline{n} = \begin{pmatrix} 0.06 \\ 0.31 \\ 0.52 \\ 0.08 \\ 0.03 \end{pmatrix}.$$

The full set of equations is therefore:

$$\frac{dS_i}{dt} = v_i - \sum_j \beta_{ij} S_i I_j - \mu S_i,$$

$$\frac{dI_i}{dt} = \sum_j \beta_{ij} S_i I_j - \mu I_i - \gamma I_i, \qquad (3.7)$$

$$\frac{dA_i}{dt} = \gamma d I_i - \mu A_i - m A_i.$$

For greater realism, the class A of infected individuals that have developed AIDS is modeled separately, with d being the proportion of infected individuals who develop AIDS. It is assumed that individuals with AIDS curb their sexual behavior and therefore no longer contribute to the spread of infection. "Births" or recruitment, (v_i), have been added to the susceptible equation, and natural deaths, μ, and AIDS-induced mortality, m, have also been included. This precise form of the equations is specific to HIV, a disease that does not obey the assumptions of the standard SIS framework typical of sexually transmitted infections. Anderson et al. (1986) quote a value for R_0 of around 5; this comes from examination of the doubling time of the epidemic during the early stages and allows us to fix the multiplicative parameter which is part of the mixing matrix. Using the theory developed earlier (Box 3.3), we can calculate the value of R_0 from the given matrix:

$$R_0 = 5 \approx 46.14 \frac{\beta}{\gamma},$$

where 46.14 is the largest eigenvalue of the associate matrix. Other parameters for the model are $\mu = 0.0312$, $v_i = \mu n_i$, $\gamma = 0.2$, $m = 1$, and $d = 0.3$, with all rates in years (Anderson et al. 1986). These high birth and natural death rates merely reflect the rate at which individuals enter and leave the sexually active class. From these parameters, we find that $\beta \approx 0.0217$.

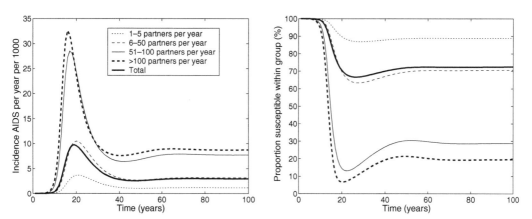

Figure 3.7. The dynamics of HIV infection from the structured model, equation (3.7), showing the behavior within the four classes (the zero class is ignored) and the average behavior. The simulations are begun with 1 infected individual in a population of 100,000. The left-hand graph shows the incidence of AIDS (those entering the A_i classes) per year—this can be substantially different from the prevalence of the infection but corresponds more closely to the information that is recorded. The right-hand graph shows the percentage of each class that are susceptible, S_i / n_i, which provides some indication of the force of infection experienced.

Figure 3.7 shows the dynamics of this system; the results for the class with no partners have not been plotted because there is no risk of these individuals becoming infected. From the left-hand graph, the slaved exponential growth phase lasts for about 10–15 years. The peak in incidence occurs first in the highest-risk group, and because the low-risk groups are primarily infected from more high-risk groups, the other peaks lag by about 3 years. As expected, the high-risk groups show a far greater incidence of HIV. After the peak, the incidence drops, showing damped oscillations toward an equilibrium solution. This type of dynamics is uncharacteristic of sexually transmitted infections and is because the mortality (removal) of infected individuals and recruitment of new susceptibles effectively makes the dynamics SIR (see Chapter 2). This oscillatory nature makes it difficult to assess the impact of control strategies from case report data; the natural decline predicted in years 20 to 40 could easily mask any control efforts. Switching attention to the proportion of susceptibles within each class, we find that although only 20% of the high-risk group are susceptible at equilibrium, for the population as a whole the susceptibility is more than 70%. In a homogeneous model, this would be in direct conflict with the R_0 value of 5, whereas in this structured population R_0 is driven by the small high-risk group.

The parameters used in this model are taken from a select subset of the population (male homosexuals), whose behavior puts them at greater risk than the norm. Therefore, all results pertain to this small subset and not the population as a whole. However, two interesting and robust conclusions can be ascertained from such models: (1) the expected peak numbers of AIDS cases will be much higher (double for the parameters used here) than the equilibrium level, and (2) the time-scale of the epidemic dynamics is long, taking 40–50 years before equilibrium is reached. Over such long time scales social changes may have a significant impact on the dynamics.

Although inherently simple and easy to parameterize from available data, this model demonstrates the power of modeling as a tool for understanding disease dynamics and predicting long-term trends. Obviously, more accurate predictions require more complex models with more classes; however, although such models are relatively easy to formulate, their parameterization is more awkward and obtaining the necessary social and sexual-behavior data is a difficult and time-consuming task (Johnson et al. 1994). However, several key features could be included to make this a more realistic model of HIV transmission. Infection risk for HIV is known to be highly dependent of time since infection, which can be modeled by subdividing the infectious class (see Section 3 of this chapter). Transmission through shared needles by intravenous drug users is another important route and one that could be modeled by a similar risk-structured approach. Finally, greater realism of sexual behavior could be included, capturing both the assortative nature of contacts, as well as both the heterosexual and homosexual communities.

3.1.2.2. Chlamydia Infections in Koalas

Localized koala (*Phascolarctos cinereus*) populations have undergone significant declines in recent years (Phillips 2000). One possible cause of this decline is infectious disease, which has been demonstrated to regulate other natural populations (Hudson et al. 1998). Chlamydia infections in koalas are known to lead to sterility of females and increased mortality. It therefore seems likely that chlamydia will have an impact on koala population dynamics. Here, we follow the work of Augustine (1998), although we consider a much reduced model to highlight the factors we have already discussed. The original paper used a discrete-time model to mimic the seasonal reproduction of koalas (see Chapter 5), but also included age-structure (see Section 3.2) and stochasticity (see Chapter 6) to achieve greater realism.

This example considers the interaction between epidemiological and ecological interactions. For human populations, there are little or no density-dependent effects; in contrast, most wildlife populations experience considerable density-dependence, limiting their numbers (e.g., Clutton-Brock et al. 1997; Gaillard et al. 2000). The interplay between the epidemiological and ecological factors can substantially complicate the infection dynamics and negate some of our understanding based upon human disease models (Hudson et al. 2001). This issue is further discussed in Chapter 8.

For this sexually transmitted infection, we do not structure the population by the number of sexual partners, but instead males and females are treated as the two risk groups. This gender-based structuring is necessary due to the very different behavior of males and females, both in terms of transmission and in terms of response to infection. A similar argument could be made for models of chlamydia transmission in humans. Finally, the assumption of heterosexuality within the koala population means that the transmission matrix β takes a relatively simple form.

We first assume that the underlying population dynamics of koalas can be described by a continuous-time logistic growth model, with density-dependent mortality, and that the carrying capacity is normalized to one. We now structure the population into males and females, such that in the absence of infection the number of susceptible males and females obeys the following differential equations:

$$\frac{dX_F}{dt} = rX_F - rX_FN, \qquad \frac{dX_M}{dt} = rX_F - rX_MN, \qquad N = X_F + X_M.$$

Thus, in an uninfected population, the koalas have a carry-capacity of 1 ($N = 1$) and an equal ratio of males and females. Infection with chlamydia is assumed to be lifelong, and transmission is modeled as frequency dependent (mass-action), which agrees with observations of koala having a fixed number of mates per year irrespective of population size or density. We initially consider only horizontal transmission (between male and female koala) during intercourse, which leads to the following set of equations:

$$\frac{dX_F}{dt} = r(X_F + \alpha Y_F) - rX_F N - \beta_{FM} X_F Y_M / N,$$

$$\frac{dX_M}{dt} = r(X_F + \alpha Y_F) - rX_M N - \beta_{MF} X_M Y_F / N,$$

$$\frac{dY_F}{dt} = \beta_{FM} X_F Y_M / N - rY_F N - mY_F,$$

$$\frac{dY_M}{dt} = \beta_{MF} X_M Y_F / N - rY_M N - mY_M,$$

$$N = X_F + X_M + Y_F + Y_M.$$

where α is the reduction in fertility due to disease, and m is the additional disease-induced mortality rate. There is negligible vertical transmission, so a mother does not pass infection to her offspring and all newborns are susceptible. It is interesting to consider the structure of the transmission matrix $\boldsymbol{\beta}$ in more detail. Based on the work of Augustine (1998), a plausible matrix is:

$$\boldsymbol{\beta} = \begin{pmatrix} 0 & 1.0 \\ 1.2 & 0 \end{pmatrix}.$$

First, we note that the diagonal terms β_{MM} and β_{FF} are zero, which breaks the common assumption of assortative mixing (in fact $Q = -1$, because there are only two classes)—this is because we are interested in transmission between males and females instead of "risk groups," per se. The second is that the matrix is nonsymmetric because females are more likely to catch chlamydia from an infected male than vice versa, although this is outweighed by the fact that breeding males are likely to have more mates. When the koala population is at its disease-free equilibrium, $X_F = X_M = \frac{1}{2}$; this simple, but unconventional, transmission matrix leads to a basic reproductive ratio of:

$$R_0 = \frac{\sqrt{1.0\frac{1}{2} \times 1.2\frac{1}{2}}}{r + m} = \frac{0.5477}{r + m}.$$

Thus, although high disease mortality (large m) may be problematic for the koala population, it can be devastating for chlamydia because it can easily reduce R_0 below one.

Figure 3.8 illustrates the effect the disease can have on the breeding population of koalas, by plotting $X_F + \alpha Y_F$ as a function of both α and m. We observe that if infected females have sufficiently reduced fecundity (as in humans, chlamydia infection is likely to cause sterility), then the entire population can be driven extinct. In contrast, the behavior with respect to disease mortality, m, is more complex and intermediate levels of mortality have the most detrimental effects on the total population size. This is because with very high mortality the infection has a low R_0 and cannot spread. At the other extreme, although low mortality allows the pathogen to spread further it does little to reduce the population size. These results echo the findings of Augustine (1998) and illustrate why detailed mathematical models are often necessary and insightful.

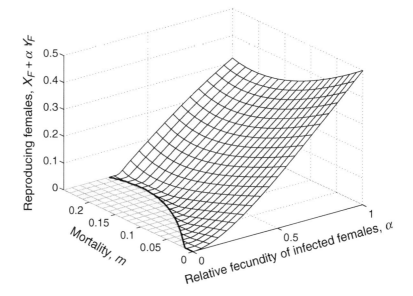

Figure 3.8. The level of reproducing females, $X_F + \alpha Y_F$, as a function of two infection attributes: disease-induced mortality rate, m, and reduced fertility, α. When infected females have low fertility it is possible for the infection to drive the population to extinction. We take $r = 0.2$; all parameters are within the ranges given by Augustine (1998).

3.1.3. Other Types of Risk Structure

The formulation of risk-structured models and the WAIFW matrices provide a useful distinction between two types of important individuals in airborne infections: *super-shedders* and *super-spreaders* (Austin and Anderson 1999b; Riley et al. 2003). Although rare, both of these types of individuals can be responsible for a disproportionately large number of secondary cases and, therefore, can severely hamper control efforts. Let us consider a population in which only a small proportion are either super-shedders or super-spreaders and consider how the transmission matrix β reflects their intrinsic differences.

Here, we define super-shedders as individuals who once infected excrete large amounts of the infectious agent. This super-shedding is often due to genetic attributes of the host or a compromised immune system. Although able to produce many more secondary cases, super-shedders are not necessarily at any greater risk of coming into contact with the infections disease. Therefore, labeling super-shedders with a subscript S and the rest of the population with a subscript R, a suitable transmission matrix would be:

$$\beta = \begin{pmatrix} \beta_{SS} & \beta_{SR} \\ \beta_{RS} & \beta_{RR} \end{pmatrix} = \begin{pmatrix} f\beta & \beta \\ f\beta & \beta \end{pmatrix},$$

where $f > 1$ reflects the greater transmission *from* super-shedders in comparison with the rest of the population. Because super-shedders are assumed to have different epidemiological responses to infection compared to the rest of the population, the WAIFW matrix is no longer symmetric.

In contrast, we define super-spreaders as individuals with a very high number of contacts, often due to their occupation. Hence these individuals could generate many secondary cases but are also at a much higher risk of being infected. Therefore, using subscript S to now represent super-spreaders, a suitable transmission matrix would be:

$$\boldsymbol{\beta} = \begin{pmatrix} \beta_{SS} & \beta_{SR} \\ \beta_{RS} & \beta_{RR} \end{pmatrix} = \begin{pmatrix} f^2\beta & f\beta \\ f\beta & \beta \end{pmatrix},$$

where $f > 1$ reflects greater transmission from and to super-spreaders. It should be clear from this formulation that super-spreaders generate a much greater risk than super-shedders, because their behavior means that they are much more likely to be infected and subsequently much more likely to transmit.

A recent approach to capture heterogeneities in transmission has been proposed by Lloyd-Smith et al. (2005). They examined contact tracing data for eight directly transmitted diseases, including measles, smallpox, pneumonic plague, and SARS, and found the distribution of individual infectiousness around R_0 to be often rather skewed. To study the epidemiological implications of this, Lloyd-Smith et al. (2005) used results from branching process theory (see Chapter 6) by assuming that the number of secondary cases resulting from each infected is described by an "offspring distribution" given by $Z \sim$ negative binomial (R_0, k), where the parameter k captures the skew in the transmission distribution (note: $k \to \infty$ gives the Poisson distribution and $k = 1$ yields the geometric). The values of k estimated from data ranged from 0.01 to approximately 0.1, highlighting the large variance in individual infectiousness. These results suggest the 20/80 "rule" (whereby 20% of individuals are responsible for 80% of transmission)—which was previously thought to apply to STIs and vector-borne diseases—may also apply to directly transmitted infectious diseases.

The epidemiological implications of these observations are interesting and important. For example, compared to models that assume little individual variation in transmission (k very large), this documented heterogeneity implies an increased disease extinction risk, and a reduced likelihood of epidemics, though outbreaks are more severe when they do occur. Additionally, the public health lessons highlight the importance of individual-specific control measures, rather than population-wide approaches. Let us assume control effort c is imposed ($c = 0$ represents no control and $c = 1$ reflects the full blockage of transmission). Under population-wide control, the infectiousness of every individual in the population is reduced by c—control is homogeneous. Alternatively, with individual-specific control, a proportion c of infecteds (chosen at random) are traced and isolated completely such that they cause zero infections (also see Chapter 8)—control is heterogeneous. Individual-specific control raises the degree of heterogeneity in the outbreak as measured by the variance-to-mean ratio of Z, whereas population-wide control reduces heterogeneity. Both approaches yield the effective reproductive number $R = (1 - c)R_0$, so the threshold control effort for guaranteed disease extinction is $c \geq 1 - 1/R_0$, as in conventional models (compare Chapter 8). For intermediate values of c, however, the individual-specific approach always works better because higher variation favors disease extinction.

3.2. AGE-STRUCTURE: CHILDHOOD INFECTIONS

The dynamics of childhood infections, such as measles, mumps, chickenpox, rubella, or whooping cough, have been extensively studied by epidemiologists and mathematical modelers (Hamer 1906; Bartlett 1957; Black 1966; Yorke and London 1973; Cliff et al.

1981; Fine and Clarkson 1982; Schenzle 1984; Schwartz 1983; McLean and Anderson 1988a,b; Olsen and Schaffer 1990; Rand and Wilson 1991; Grenfell 1992; Grenfell et al. 1994; Bolker and Grenfell 1995; Ferguson et al. 1996a,b; Keeling and Grenfell 1997a; Finkenstädt and Grenfell 1998; Earn et al. 2000; Grenfell et al. 2001; Bjørnstadt et al. 2002; Rohani et al. 2002). From a theoretical perspective, the study of such diseases is facilitated by the relative simplicity of the natural history of the infection, the large number of cases, the high quality of the recorded data, and the interesting dynamics that emerge. From an epidemiological and public health standpoint, these diseases are important to understand because they have high R_0 and can therefore produce large epidemic outbreaks and, although usually fairly benign, complications and mortality can occur (Rohani et al. 2003).

Although the standard theory for unstructured models provides many insights into the dynamics of such infections—especially when seasonal forcing is included (see Chapter 5) —these diseases predominantly affect a subset of the population (children) and hence any models developed should reflect this. The formalism of such age-structured models is superficially the same as the risk-structured models described above, but with the addition of individuals aging. Thus, whereas with risk-structured models hosts are generally assumed to stay in the same risk class for their entire lives, with age-structured models hosts move sequentially through the age classes.

In many countries, these childhood diseases are notifiable with local and national statistics being compiled. The primary reason for collecting these data is public health motivated, enabling health agencies to determine the effects of control measures and providing an insight into the future epidemic behavior. However, these statistics are also a rich source of data for model parameterization and time-series analysis. In England and Wales, the number of cases reported weekly in over 1,400 locations has been collected since 1944, providing the largest such epidemiological data set in existence (Cliff et al. 1981; Finkenstädt and Grenfell 1998; Grenfell et al. 2001). To date, most interest has been focused on measles, which has been recorded with an efficiency of around 60% (Finkenstädt and Grenfell 1998), however data on other childhood diseases is now being examined (Rohani et al. 1999, 2000). In addition, the use of age-structured models is particularly useful when considering pandemic infections, because a significant amount of transmission could occur within schools, and closing schools may be a relatively easy way of restricting disease spread (Ferguson et al. 2006).

3.2.1. Basic Methodology

The standard approach is again to subdivide the population into a number of discrete compartments, classified by the age of the host. Although for sexually transmitted infections there was some arbitrary choice to the qualities used to divide the population, here the situation is much more clear cut and only the age range within each compartment has to be determined. In principle, age is a continuous variable (which would suggest a partial-differential-equation approach, see Box 3.5), but because children are generally grouped into school classes of a given age cohort, the compartmental approach is often more realistic.

Although a continuous parameter, age-structured models usually group individuals into a limited number of classes—often representing school years.

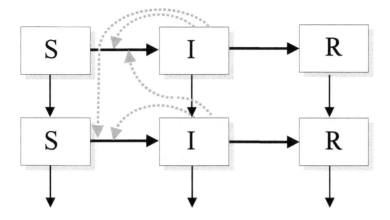

Box 3.5 PDE Model

This type of model considers the variables S, I, and R to be functions of both age and time. As such, $S(a, t)$ refers to the probability-density of susceptibles of age a at time t. The age-dependent SIR equations can now be written as a PDE:

$$\frac{\partial S(a)}{\partial t} = \nu \delta(a) - S(a) \int_0^\infty \beta(a, a') I(a', t) da' - \mu S(a) - \frac{\partial S}{\partial a},$$

$$\frac{\partial I(a)}{\partial t} = S(a) \int_0^\infty \beta(a, a') I(a', t) da' - \mu I(a) - \gamma I(a) - \frac{\partial I}{\partial a},$$

where the delta-function forces all births to be age zero, and the last partial derivative term in each equation accounts for the population getting older. Although this formalism has its mathematical elegance, it is computationally difficult to implement and would be solved numerically by discretization into multiple age classes. Finally, the contact rate β, which is now a function of two continuous ages, would be very difficult to parameterize from real data.

We again start by considering a model that subdivides the population into two classes, in this case children and adults identified by subscripts C and A, respectively. Although most childhood diseases have a clearly defined latent-period leading to $SEIR$-type dynamics, for clarity we begin with an age-structured version of the SIR equations:

$$\frac{dS_C}{dt} = \nu - S_C (\beta_{CC} I_C + \beta_{CA} I_A) - \mu_C S_C - l_C S_C,$$

$$\frac{dI_C}{dt} = S_C (\beta_{CC} I_C + \beta_{CA} I_A) - \gamma I_C - \mu_C I_C - l_C I_C,$$

$$\frac{dS_A}{dt} = l_C S_C - S_A (\beta_{AC} I_C + \beta_{AA} I_A) - \mu_A S_A,$$

$$\frac{dI_A}{dt} = l_C I_C + S_A (\beta_{AC} I_C + \beta_{AA} I_A) - \gamma I_A - \mu_A I_A.$$

This is online program 3.3

The parameters μ_C and μ_A are the age-specific death rates, l_C is the rate that individuals mature (leave the childhood class and move to the adult class), and ν is the birth rate (assuming no maternally derived protection). In this model, the interaction between the two classes comes from transmission between them (as captured by the off-diagonal elements of the matrix β) and due to the slow trickle of individuals moving from the childhood to the adult class. Again, we are dealing with parameters that represent proportions of the entire population, such that $S_C = X_C/N$ is the proportion of the population that is susceptible children. We now seek to compare the general dynamics observed for this type of model with those of the risk-structured approach, contrasting with the behavior of nonstructured models.

Age-structured models differ from the earlier risk-structured models due to the regular progression of individuals into increasingly older age classes.

3.2.1.1. Initial Dynamics

The first observation is that when the transmission dynamics are rapid in comparison with both host demography and the aging process, the initial behavior and in particular the basic reproductive ratio are largely unchanged from the risk-structured results given earlier, where there was no movement between risk classes. Although this assumption is generally true for common childhood infections of humans, if the infection dynamics are particularly slow relative to the life expectancy (for example BSE in cattle, see Section 3.2.2), we need to use the eigenvalues approach for the full system of equations, which includes the effects of movement between age classes while infected. This can be done in a similar manner to described in Box 3.1, finding the eigenvector distribution of infecteds in the slaved growth phase and using this to weight the individual number of secondary cases expected for each age class.

3.2.1.2. Equilibrium Prevalence

When contrasting risk-structured models of STIs and age-structured models of childhood infections, two distinct elements lead to differences in the equilibrium distribution of infection: (1) the nature of the infectious disease and the inherent differences between SIS- and SIR-type models, and (2) the sequential progression through the age classes, which means that at equilibrium, the proportion of susceptible individuals in the older age classes must be lower,

$$\frac{S_i^*}{n_i} \geq \frac{S_j^*}{n_j} \qquad \forall i < j.$$

For diseases conferring life-long immunity, this must be true because hosts in older age classes must have been subjected to at the very least the same risks of infection as hosts in younger classes. An extreme version of this is the case where the force of infection experienced is independent of age, in which case the fraction susceptible decreases exponentially with age. We note, however, that if immunity to infection can wane, then this result no longer holds, because older individuals may have had sufficient time to recover.

For SIR-type infections at equilibrium, the proportion recovered (measures by seroprevalence) must increase with age.

From this understanding, the equilibrium serological profile (which measures the proportion of the population that have been exposed to the infection) with respect to age can be used to estimate both the incidence of infection and the force of infection acting within that age class. Let us consider the ith age class, and assume that this class lasts for an average of T_i years. Individuals who enter the ith age class are susceptible with probability S_{i-1}/n_{i-1}, whereas when individuals leave the ith age class the probability of being susceptible has dropped to S_i/n_i. Thus, ignoring mortality, during the T_i years a fraction $S_{i-1}/n_{i-1} - S_i/n_i$ must have been infected; this sets the force of infection in this age class to be:

$$\lambda_i = \frac{1}{T_i}\left(\frac{S_{i-1}}{n_{i-1}} - \frac{S_i}{n_i}\right)\frac{n_i}{S_i} = \frac{1}{T_i}\left(\frac{S_{i-1}n_i}{S_i n_{i-1}} - 1\right)$$

and the equilibrium disease prevalence is:

$$I_i^* = \frac{1}{\gamma T_i}\left(\frac{S_{i-1}n_i - S_i n_{i-1}}{n_{i-1}}\right). \tag{3.8}$$

As such, the gathering of age-structured seroprevalence information is a vital step in understanding the dynamics and relative transmission strengths of childhood infections (Figure 3.10).

3.2.1.3. Control by Vaccination

For risk-structured models, it was clear that targeting controls toward the high-risk groups was the most efficient means of combating infection. However, this was because individuals in the high-risk group were assumed to remain there indefinitely. For age-structured models, most individuals make it through all the age classes and therefore experience, and contribute to, the full range of dynamic behaviors. In such situations, and assuming the vaccine provides lifelong protection, it is always best to vaccinate as early as possible so that immunity covers the greatest proportion of the host's lifespan.

For age-structured models, when vaccination offers lifelong protection, it is always best to target the youngest age groups.

Figure 3.9 shows the equilibrium results from an age-structured model with just two classes, children and adults. To allow comparisons to previous work, we take the contact matrix to be a multiple of the risk-structured matrix seen previously:

$$\beta = \begin{pmatrix} 100 & 10 \\ 10 & 20 \end{pmatrix},$$

with children having the higher transmission rate, and an infectious period of length 0.1 years. (This set of parameters gives dynamics very similar to those in Figure 3.2, but operating ten times faster). Finally, we also assume that the childhood class is from ages 0 to 15 years, and that life expectancy is 75 years. Setting $\mu_C = 0$, this implies that $n_C = 0.2$, $n_A = 0.8$, $l_C = 0.0667$, and $\mu_A = 0.0167$ per year. Clearly, epidemiological dynamics are on a much faster time scale than the demography—and so the birth and death rates have an insignificant impact on $R_0 \approx 2$.

Figure 3.9 also illustrates a few basic points that were made earlier. First, a smaller proportion of adults are susceptible compared to children, although obviously the two

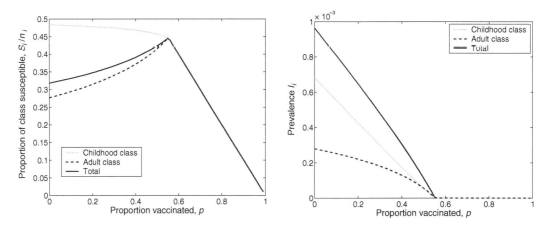

Figure 3.9. Equilibrium results from an age-structured model with vaccination at birth, such that the birth rate of susceptible children is reduced to $v(1-p)$. The left-hand graph shows the proportion of each age class that are susceptible (S_i^*/n_i). The right-hand graph shows the absolute prevalence of infection in each class as well as the total.

levels are equal once the disease has been eradicated ($p > 0.55$). However, notice that vaccination actually increases the proportion of the adult class who are susceptible to infection (S_A/n_A) and hence leads to an increase in the proportion of the entire population that is susceptible ($S_C + S_A$). These results can be explained as follows: With increasing vaccination at birth, the decrease in prevalence is greater than the decrease in newborn individuals who are susceptible—although there are fewer susceptible births and even fewer infectious cases and hence the total fraction of susceptibles increases. Considering disease prevalence (right-hand graph), although the adult class contains four times as many individuals as the childhood class, it contains fewer infecteds. Vaccination can also be seen to have a stronger effect on prevalence among children. This is comparable with earlier results (see Chapter 2)—that a lower reproductive ratio (in this case $(1-p)R_0$) corresponds to a higher average age of infection, which means that proportionally fewer cases occur in children. Finally, we find that vaccinating 55.35% of children at birth is sufficient to control the epidemic, which is just slightly less than the standard $1-1/R_0$. Pragmatically, it is reassuring that despite substantial heterogeneity and assortative mixing in the transmission matrix, the simple eradication threshold remains a surprisingly accurate approximation when vaccinating newborns. An intuitive explanation for this close agreement is that, as the newborns age, eventually the entire population is vaccinated at a proportion p, and hence this is asymptotically equivalent to random vaccination. We note, however, that for vaccination at birth, the dynamics just after the start of a vaccination campaign may be very different from the standard random-vaccination models (see Chapter 8).

3.2.1.4. Parameterization

Although risk-structured models for STIs often have a natural parameterization of the contact matrix, age-structure models are more difficult to parameterize (Grenfell and Anderson 1985). We return to the problem stated earlier of having an $n \times n$ matrix of values to find, and only a vector of n observed values to use. The standard way to cope with this sort of problem is to reduce the number of terms needed in the matrix, usually focusing on the age classes that are most responsible for infection transmission.

In our simple two-class example, we may have data on the proportion of susceptibles in each class. Thus, given two pieces of information, we need to specify the matrix with just two parameters. The most natural way to do this is to isolate the childhood interactions, such that the contact matrix becomes

$$\beta = \begin{pmatrix} \beta_1 & \beta_2 \\ \beta_2 & \beta_2 \end{pmatrix}.$$

Clearly we have lost some of the model structure, but it is now possible to estimate this reduced matrix. As shown in equation (3.8), it is possible to estimate the equilibrium prevalence of infection from the fraction of susceptibles:

$$I_C^* = \frac{1}{\gamma\, y_C}\left(n_C - S_C^*\right) \qquad I_A^* = \frac{d_A}{\gamma}\left(\frac{S_C^* n_A - S_A^* n_C}{n_C}\right),$$

where y_C and $1/d_A$ are the average times spent in the childhood and adult classes. For this solution to be at equilibrium, we require the transmission matrix parameters to satisfy

$$\frac{dI_C}{dt} = 0 = S_C^*\left(\beta_1 I_C^* + \beta_2 I_A^*\right) - \gamma I_C^* - m_C I_C^*,$$

$$\frac{dI_A}{dt} = 0 = S_A^*\left(\beta_2 I_C^* + \beta_2 I_A^*\right) + m_C I_C^* - \gamma I_A^* - \mu_A I_A^*.$$

After some rearranging, this gives:

$$\beta_2 = \frac{\gamma I_A^* + \mu_A I_A^* - m_C I_C^*}{S_A^*(I_C^* + I_A^*)}, \qquad \beta_1 = \frac{\gamma + m_C}{S_C^*} - \frac{I_A^* \beta_2}{I_C^*}.$$

We can test this result using the transmission matrix defined earlier. The results in Figure 3.9 show that in the absence of vaccination, $S_C^* \approx 0.097$ ($S_C^*/n_C \approx 0.485$) and $S_A^* \approx 0.22$ ($S_A^*/n_A \approx 0.275$). This translates into infection levels of $I_C^* \approx 6.8 \times 10^{-4}$ and $I_A^* \approx 2.8 \times 10^{-4}$, which agrees well with the full numerical calculations, and produces a reduced matrix with $\beta_1 \approx 98.8$ and $\beta_2 \approx 12.9$, which again is in line with the actual values.

The choice of matrix is a question of good epidemiological judgment. Any matrix that contains two free parameters can be made to fit the serological data, but may poorly reflect the general understanding of how transmission operates. In the above example, we chose to focus our attention on transmission between children as distinct from all other transmission events, however we could have equally made the assumption that all transmission involving children was equal and that only transmission between adults was different. Finally, there are a range of bizarre and unlikely matrices:

$$\beta_{unlikely} = \begin{pmatrix} \beta_1 & \beta_2 \\ \beta_2 & \beta_1 \end{pmatrix}, \begin{pmatrix} \beta_1 & 0 \\ 0 & \beta_2 \end{pmatrix}, \begin{pmatrix} 0 & \beta_1 \\ \beta_2 & 0 \end{pmatrix}, \dots$$

all of which could be made to fit the available data, but would not agree with our intuition about the transmission of childhood diseases.

This method can, of course, be extended to deal with higher dimensional models with more age classes, although the algebra gets increasingly ugly (Grenfell and Anderson 1985). However, one particular case is worth describing in some detail. The most common formulation for age-structured models of childhood diseases is to divide the population into four age-classes: pre-school (0–5 years), primary school (5–11 years), secondary school (11 16 years), and adults (16 or over). Thus, the division of the population reflects the standard epidemiological knowledge that it is the mixing of school children (mainly at primary schools) that is responsible for driving the progression of most childhood diseases (Fine and Clarkson 1982; Schenzle 1984; Finkenstädt and Grenfell 2000). There is some degree of freedom in the way that the reduced transmission matrix β is chosen; clearly, primarily school contacts must play a dominant role, but there is no clear general format. Two possible formulations are

$$\beta = \begin{pmatrix} b_2 & b_2 & b_3 & b_4 \\ b_2 & b_1 & b_3 & b_4 \\ b_2 & b_3 & b_3 & b_4 \\ b_4 & b_4 & b_4 & b_4 \end{pmatrix} \qquad \beta = \begin{pmatrix} b_2 & b_4 & b_4 & b_4 \\ b_4 & b_1 & b_4 & b_4 \\ b_4 & b_4 & b_3 & b_4 \\ b_4 & b_4 & b_4 & b_3 \end{pmatrix},$$

where $b_1 < b_2 < b_3 < b_4$. In the first formulation, the interaction strength is determined by the weakest member, and it is assumed that primary-school children followed by preschool and secondary school children are most responsible for transmitting infection. In the second formulation, the degree of assortative mixing is high, and transmission to outside the age class is considered to be less important. When seasonal forcing is included to mimic the effects of opening and closing schools, generally only the b_1 parameter is forced (Chapter 5).

Age-structured data on the proportion of seropositives can be used to determine n terms of the $n \times n$ transmission matrix. The way these n terms are used to define the entire matrix should match the underlying biology.

3.2.2. Applications of Age Structure

We again consider two examples from the recent literature. The first is an amalgamation of the huge body of research into the dynamics of measles. This is one of the best understood of all childhood diseases, and as we will show, the implications of age structure are vital if we are to model the observed dynamics. The second example is the control of BSE (*Bovine Spongiform Encephalopathy*), where age-structured models are again necessary if we are to understand the dynamics of this infection and curtail its spread.

3.2.2.1. Dynamics of Measles

As noted in Chapter 2, the distinguishing feature of measles and other childhood diseases is their large basic reproductive ratio, R_0. For a non-age-structured population of constant size, this translates into an average age at first infection of (life expectancy)/$(R_0 - 1)$, hence a large basic reproductive ratio coupled with life-long immunity is synonymous with childhood infections. As with all infections, the value of R_0, estimated from age-structured seroprevalence data, varies between locations. For measles in modern era England and Wales, several studies have estimated R_0 to be around 17 (Anderson and May 1982; Grenfell 1982; Schenzle 1984; Grenfell and Anderson 1985), and this value is used throughout this section. Age structure is important in the modeling of childhood diseases because of the greater mixing between susceptible and infected children that occurs at schools. In general, although preschool children are the most likely age group to be susceptible, they generally mix only with a small number of other children so the potential to catch and transmit infection is low. However, as soon as children enter school, the number of potential contacts increases to at least class size (20–30 children), and hence the risk of transmission also rises.

The seminal work of Schenzle (1984) modeled measles dynamics using an $SEIR$ framework (Chapter 2) and 21 age classes, where the members of each of the four school groups (preschool school $0 \rightarrow 6$, primary school $6 \rightarrow 10$, secondary school $10 \rightarrow 20$, and adults) share a common mixing pattern. Thus, the mixing matrix is comprised of 16 blocks of identical transmission terms. This early work into realistic age-structured (RAS) models has been greatly extended and expanded by Grenfell and coworkers (Bolker and Grenfell 1993; Bolker 1993; Grenfell et al. 1994; Keeling and Grenfell 1997a; Finkenstädt and Grenfell 1998) using the extensive spatio-temporal record of measles cases available for England and Wales. The values given in Bolker and Grenfell (1993), for a RAS model similar to Schenzle's, are:

$$\beta = \gamma \begin{pmatrix} 1.875 \ldots 1.875 & 2.175 \ldots 2.175 & 0.975 \ldots 0.975 & 0.6 \\ \vdots & \vdots & \vdots & \vdots \\ 1.875 \ldots 1.875 & 2.175 \ldots 2.175 & 0.975 \ldots 0.975 & 0.6 \\ 2.175 \ldots 2.175 & 10.74 \pm 8.56 \ldots 10.74 \pm 8.56 & 0.975 \ldots 0.975 & 0.6 \\ \vdots & \vdots & \vdots & \vdots \\ 2.175 \ldots 2.175 & 10.74 \pm 8.56 \ldots 10.74 \pm 8.56 & 0.975 \ldots 0.975 & 0.6 \\ 0.975 \ldots 0.975 & 0.975 \ldots 0.975 & 0.975 \ldots 0.975 & 0.6 \\ \vdots & \vdots & \vdots & \vdots \\ 0.975 \ldots 0.975 & 0.975 \ldots 0.975 & 0.975 \ldots 0.975 & 0.6 \\ 3 \ldots 0.6 & 0.6 \ldots 0.6 & 0.6 \ldots 0.6 & 0.6 \end{pmatrix},$$

where individuals are assumed to be exposed (but noninfectious) for an average of eight days and infectious for an average of five days—leading to $SEIR$-type dynamics. Rather than maturation being a continual process, to better mimic the school structure, all individuals except adults are assumed to move up an age class at the start of each school year. Finally, the ± 8.56 terms in the primary-school interactions refer to the changes in mixing that occurs between school terms and holidays (Chapter 5). In Schenzle's original formulation, Sundays are treated as school holidays, although this high-frequency variation has little effect on the dynamics.

Figure 3.10 shows model results together with actual data for measles. The results shown, using Schenzle's formulation, ignore temporal forcing and instead show the equilibrium level of seropositives (effectively the proportion recovered within each age class R_i / n_i) for the term time and holiday matrices separately. The boxes show the maximum and minimum levels over each year, and hence correspond to the seropositive levels just before and just after everyone moves up into the next age class. The high degree of mixing between primary school children during school terms means that

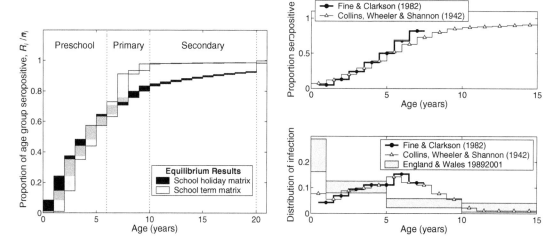

Figure 3.10. Left-hand graph: results from an age-structured model of the same form as Schenzle (1984), showing the level of seropositives across age classes. The boxes show the maximum and minimum levels across the yearly cycle of aging. The results ignore the complexities of temporal forcing and are given for the equilibrium values using the term-time and holiday matrices separately. The right-hand graphs show the measured level of seropositives and the derived distribution of cases from two studies of measles in Europe before vaccination: Collins et al. (1942) and Fine and Clarkson (1982). The lower graph also shows the recent age-structured distribution of cases in England and Wales from the Public Health Laboratory Service, giving some insight into the role of vaccination.

the risk of infection is greatest in these age classes, and therefore the proportion of seropositives increases dramatically. The difference between the school holiday and school term distributions reflects the impact of the extra mixing that occurs within primary schools.

The two right-hand graphs of Figure 3.10 present data from England and Wales (Fine and Clarkson 1982; PHLS 1989–2001) and Germany (Collins et al. 1942). The top graph gives the fraction of seropositives in each age class (R_i/n_i) before the onset of mass vaccination against measles; both data sets show a larger increase in seropositives during the primary school years, with the increase in England and Wales being more pronounced. This information can be translated into prevalence levels ($I_i \propto R_{i+1}/n_{i+1} - R_i/n_i$), and the lower graph shows the distribution of infection ($I_i / \sum_k I_k$) across all age classes for the two prevaccination samples. Prevalence clearly peaks in children around age 5–6. Also plotted on the same axes are the maximum and minimum of the modern case-report data for England and Wales (taken from www.phls.co.uk); mass vaccination has clearly driven a substantial shift in the age structure of those individuals who become infected. In recent years, infection has been concentrated in children under one year of age, and presumably occurs in the window between the loss of maternally derived immunity and the age at which the child is vaccinated.

The age-structured nature of the population can be simplified even further, by just dealing with four distinct groups. This was the approach used by Keeling and Grenfell (1997a), with the four classes: preschool (0–5), primary school (6–9), secondary school (10–19) and adult (20+). In this formulation, each age class contains several yearly cohorts; therefore, only a fraction of each class moves to the class above each school year. The

$SEIR$-type model is therefore:

$$\frac{dS_i}{dt} = \nu_i n_4 - \sum_j \boldsymbol{\beta}_{ij} I_j S_i - \mu_i S_i,$$

$$\frac{dE_i}{dt} = \sum_j \boldsymbol{\beta}_{ij} I_j S_i - \sigma E_i - \mu_i E_i,$$

$$\frac{dI_i}{dt} = \sigma E_i - \gamma I_i - \mu_i I_i,$$

$$\frac{dR_i}{dt} = \gamma I_i - \mu_i R_i,$$

(3.9)

This is online program 3.4

and at the start of the school year, moving up an age group is controlled by:

$$Q_1 = Q_1 - Q_1/6$$

$$Q_2 = Q_2 + Q_1/6 - Q_2/4$$

$$Q_3 = Q_3 + Q_2/4 - Q_3/10 \qquad \text{where } Q \in \{S, E, I, R\}.$$

$$Q_4 = Q_4 + Q_3/10$$

Here $\nu_1 = \mu_4 = (365 \times 55)^{-1} = 4.98 \times 10^{-5}$ per day and all other ν and μ terms are zero, such that all individuals survive until adulthood (aged 20), the average life expectancy is 75 years, and birth and death are equal. We also set $1/\sigma = 8$ days and $1/\gamma = 5$ days to capture the known latent and infectious periods.

Although the use of just four age classes has conceptual and computational appeal, this simplicity comes at a price. In Schenzle's model with 21 age classes, individuals remain with a particular age cohort throughout their school lives, and each individual spends an exact number of years in each school group (i.e., exactly four years in the preschool group). In contrast, when just four age classes are used, the time spent in each age class and therefore in each school grouping is exponentially distributed. This in turn leads to effectively extra mixing between the age cohorts, and a smoother age-related serology pattern. (This difference between the 4 and 21 age-class models can be compared to the distinction between exponential and constant distributions discussed in Section 3.3 of this chapter). Clearly the 21 age-class model is a more realistic representation of the actual movement of children through the school system, however this requires the use of an extra 17 equations. It should also be noted that although the Schenzle model contains 21 age classes, only four distinct transmission values are still used in the transmission matrix $\boldsymbol{\beta}$. The ideal model would use the full 21 age classes and a transmission matrix with (at least) 21 independent transmission values—however, the historical data on the incidence of measles infection are not sufficiently detailed to support such a parameterization with so many degrees of freedom. We are therefore left with a choice between simplicity or greater realism.

For the reduced four age-class models (equation (3.9)), the matrix of parameters was estimated by fitting the deterministically predicted biennial epidemic curve from the temporally forced model to the available prevaccination (1948–1968) data from England

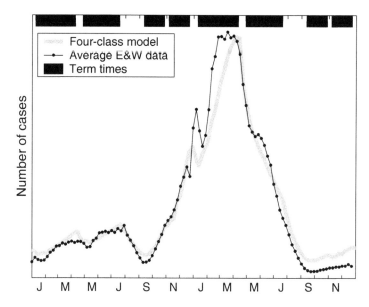

Figure 3.11. Results from a reduced age-structured model of a similar form to Keeling and Grenfell (1997a), where seasonal forcing is included to mimic the opening and closing of schools (Chapter 5). The dots show the average data from England and Wales over the duration of biennial epidemics, the thick gray line gives the deterministic attractor, and the black rectangles show the timing of school terms when the mixing matrix is increased.

and Wales. The best-fit matrix was found to be

$$\beta = \gamma \begin{pmatrix} 2.089 & 2.089 & 2.086 & 2.037 \\ 2.089 & 9.336 \pm 4.571 & 2.086 & 2.037 \\ 2.086 & 2.086 & 2.086 & 2.037 \\ 2.037 & 2.037 & 2.037 & 2.037 \end{pmatrix},$$

such that only the primary-school class shows any significant difference from the norm. It was assumed that only the primary-school class experiences temporal forcing due to the opening and closing of schools (see Chapter 5). Aggregate results ($I = \sum_i I_i$) from this model are compared to the average biennial cycle in Figure 3.11. There is reasonable agreement between the deterministic model and the available data. We note, however, that the observed average epidemic peak is far broader than predicted by the deterministic simulations. In part, this is because different communities and different biennial segments of the data experience peak infection at slightly different times, and the averaging process smears the sharp peaks. We would need to include both stochasticity (Chapter 6) and spatial effects (Chapter 7) to fully correct this discrepancy.

This approach of fitting to the aggregate data is only successful for measles due to the high degree of synchrony between all the communities in England and Wales. For other childhood diseases (e.g., whooping cough), the spatial dynamics in the prevaccination era are far less synchronized (Rohani et al. 1999) and hence the average dynamics are an amalgamation of a variety of epidemic curves. In such cases, we would need to fit simultaneously to the entire time-series of all communities taking into account the stochastic nature of transmission (Keeling and Grenfell 2002). A more practical approach that is generally taken is to match to some aggregate quality of the epidemic data, such as the power spectrum, so that model and data have a similar frequency of epidemics (Bolker 1993). However, it is probable that the transmission matrix for measles could act as a template for other childhood infections because similar routes of transmission are involved in their spread (Chapter 5).

Although more recent advances have shown that in some situations the age structure may be effectively replaced by more complex seasonality (Earn et al. 2000; Finkenstädt and Grenfell 2000; Bjørnstadt et al. 2000), these models lack the mechanistic framework of the more traditional RAS models. The complex seasonality in these models mimics changes in the average transmission rate that are due to changes in the distribution of infection across age classes over the biennial cycle. RAS models will therefore always be an essential tool in predicting the number of cases of childhood diseases, and the effects of vaccine uptake. The extra age-structured information is often particularly useful in targeting vaccination campaigns toward those age classes that are most at risk and modeling the short-term dynamics of control measures.

3.2.2.2. Spread and Control of BSE

In 1986, a disease known colloquially as "mad-cow-disease" was observed spreading through the cattle farms of the United Kingdom. Later identified as *bovine spongiform encephalopathy*, or BSE for short, this disease is caused by the transmission of a prion and is thought to have entered the cattle population via supplementary food stuffs containing meat and bonemeal from other cows. This feedback loop enabled the pathogenic agent to spread through the food chain. As well as a problem to the cattle industry, BSE also has important, but as yet unquantified (Ghani et al. 2000; Ghani et al. 2003a,b; Hilton et al. 2004) implications for human health, as consumption of infected meat can result in new variant *Creutzfeldt-Jacob Disease* (CJD) (Caughey and Chesebro 1997; Almond 1998; Narang 2001).

Mathematical models, explicitly including the age structure of the cattle population, were developed and played a major role in determining policy (Donnelly et al. 1997; Ferguson et al. 1997b). Here, we consider a simplified version of these models that illustrates the main mechanisms and results. The original equations took the form of complex integro-differential equations (see Box 3.5) and accounted for both the age of cattle, the time since infection, and farming practices in the United Kingdom. For greater clarity, the model presented here ignores the time since infection, and aggregates cattle into discrete age groups.

There are three main routes of infection for each cow: maternal (vertical) transmission, horizontal (cow-to-cow) transmission, and consumption of infected feed—cattle were routinely fed on high-protein food supplements containing meat and bonemeal from other cows. We again use an age-structured framework to study this infection, modeling the three transmission routes separately. The dynamics of infection is best captured as SEI, with infected cattle being destroyed as soon as clinical symptoms emerge. Due to the changing population size, we formulate the equations in terms of the number of cattle in each state, and model transmission as mass-action (frequency dependent) due to the density-independent nature of contacts within the farming industry. Separating out newborn and

TABLE 3.1.

Parameter	Value
ν_a	1 if $a > 2$ years, zero otherwise.
μ_a	0 if $a < 2$ years, 0.25 otherwise.
m	3
σ	0.2
l_a	4, using 3-month cohorts
β^M	0.082
$s_a^H = s_a^F$	0.3×0.55^a
τ_a^H	12
τ_a^F	160

older cattle, the equations are:

$$\frac{dX_0}{dt} = \sum_b \nu_b(X_b + W_b + (1 - \beta^M)Y_b) - \sum_b(\beta_{0b}^H Y_b + \beta_{0b}^F \mu_b(W_b + Y_b))X_0/N$$

$$- \mu_0 X_0 - l_0 X_0,$$

$$\frac{dW_0}{dt} = \sum_b \nu_b \beta^M Y_b + \sum_b(\beta_{0b}^H Y_b + \beta_{0b}^F \mu_b(W_b + Y_b))X_0/N - \mu_0 W_0 - l_0 W_0 - \sigma W_0,$$

$$\frac{dY_0}{dt} = \sigma W_0 - \mu_0 Y_0 - l_0 Y_0 - m Y_0,$$

$$\frac{dX_a}{dt} = l_{a-1}X_{a-1} - \sum_b(\beta_{ab}^H Y_b + \beta_{ab}^F \mu_b(W_b + Y_b))X_a/N - \mu_a X_a - l_a X_a,$$

$$\frac{dW_a}{dt} = l_{a-1}W_{a-1} + \sum_b(\beta_{ab}^H Y_b + \beta_{ab}^F \mu_b(W_b + Y_b))X_a/N - \mu_a W_a - l_a W_a - \sigma W_a,$$

$$\frac{dY_a}{dt} = l_{a-1}Y_{a-1} + \sigma W_a - \mu_a Y_a - l_a Y_a - m Y_a.$$

where, at age a, ν_a is the rate at which a cow gives birth, μ_a is the rate of slaughter, and l_a is the rate that an individual cow leaves an age class and matures into the next. The latent period of BSE is $1/\sigma$; m is the death rate due to infection; and β^M, β^H, and β^F refer to maternal transmission, horizontal transmission, and transmission via food supplements. The total population size is $N(= \sum_b X_b + W_b + Y_b)$. It is assumed that each transmission matrix (for horizontal and food supplements) β_{ab} is the product of an age-dependent susceptibility vector s_a and an age-dependent transmission vector τ_b. Parameter values are given in Table 3.1, where rates are all measured in months. Note that this is a highly infectious disease, with R_0 for the food supplement route alone estimated to be around 10 (Ferguson et al. 1999b).

We emphasize that the three transmission routes occur at very different stages. Maternal transmission can pass only from a mature cow (older that 2 years) to a newborn calf. Horizontal transmission can occur throughout the lifetime of a cow, but susceptibility declines with age. Finally, transmission via food supplements only occurs once a cow dies; we assume only cattle that die of "natural" causes (i.e., those slaughtered) are processed into food supplements, those that obviously die of BSE are not fed back into the food

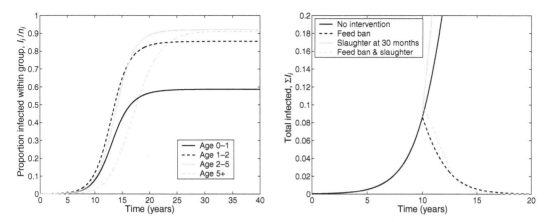

Figure 3.12. Results from a simplified age-structured model of BSE in cattle populations, showing the proportion infected (latent +infectious). The left-hand graph shows the distribution of infection across age-classes measured in years; the infection takes off sooner in young cattle, but reaches higher prevalence in older ones. The right-hand graph shows the effects of controls on feed and age-structured culling.

chain. This age-based heterogeneity in transmission and susceptibility makes the use of age-structured models vital in understanding this infection.

Figure 3.12 shows examples of the infection dynamics predicted by this comparatively simple model. These dynamics display the characteristics that are expected from any such structured model. There is an early short phase (not discernible on the graphs) that depends on the initial distribution of infecteds; during this phase prevalence in young cows increases most rapidly due to their greater susceptibility. In the second phase, the prevalence of infection within the age classes becomes "slaved", increasing at the same exponential rate. Finally, density-dependent forces begin to act and the infection levels settle to equilibrium values. Interestingly, although young cows dominate the early stages of the epidemic, at equilibrium older cattle are most likely to be infected because they will have had a longer exposure. Due to the very high transmission by infected feed and the long duration of the exposed period, the equilibrium prevalence of infecteds is very high. This clearly poses both a considerable hazard to human health, as well as devastating consequences for the farming industry. Control measures therefore needed to be implemented to wipe out the epidemic as quickly as possible, while not creating too large an economic burden. The costs associated with BSE control in Europe during 2001 were estimated at 7 billion euros. Therefore, determining a cost-effective and efficient policy is vital.

Three different control measures are investigated for this model (right-hand graph); these are the slaughter of all animals over 30 months of age ($\mu_{10} = \infty$), the ban of cattle meat and bonemeal in supplementary food ($\tau^F = 0$), and the adopted policy which is both of these measures. Surprisingly, although older animals are the most likely to be infected, a policy that involves just culling these animals actually leads to an increased epidemic. This is for two major reasons: (1) young animals are most susceptible to the infection, so the cull of older animals raises the average susceptibility of the population; and (2) by slaughtering animals much earlier there has been a significant shortening of the generation time for this infection, so infected animals are quickly converted into food supplements which speeds up the epidemic rise.

The major transmission route for BSE is via infected feed, therefore, it is intuitive that a ban on the inclusion of meat and bonemeal, effectively setting $\tau_F = 0$, should rapidly bring the epidemic under control. However, it is more surprising that a combination of early slaughter and a feed ban results in a faster eradication of the disease. This is again attributable to the fact that early slaughter of cattle reduces the infection generation time and thus the controls operate at a faster rate. This faster route to eradication is offset by a less rapid initial decrease, which is again due to the rise in average susceptibility caused by the slaughter of older less-susceptible animals.

Several other features were included in the predictive models used at the time (Donnelly et al. 1997; Ferguson et al. 1997b). The seasonal and demographic trends for cattle herds in the United Kingdom were incorporated. Also, most notably, infectious status was modeled in terms of time-since-infection (see Section 3.3 of this chapter), rather than the standard compartmental (SIR or $SEIR$) approach. This adds an extra dimension to the calculations, because each cow is now indexed by two variables (age and time-since-infection) rather than simply by age. Hence the full model is both age-and stage-structured. This complication allows a great deal more flexibility and realism compared to the simple model illustrated here. The distribution of incubation and latent periods can be modeled explicitly, and the transmission rate can be modified as the infection progresses within each animal. These more realistic features mean that the speed of infection reduction and timing of eradication can be predicted with far greater accuracy.

It is interesting and important to question what benefits can be obtained from such modeling approaches. The ban on meat and bonemeal in feed is an intuitive measure given the strengths of the various transmission routes. So at this level, models of any kind are clearly unnecessary. However, the real power of models comes when investigating issues such as the culling of older cattle, where feedback from nonlinear processes can produce counterintuitive results. Models are also useful in determining whether a control measure is likely to be sufficient; although it was clear that a feed ban was necessary, whether it would control the epidemic, in either the short or long term, could be addressed only with predictive models. Similarly, more or less drastic culls on cattle of different ages could be simulated to assess the benefits they produced. Implicit in all the calculations is a desire to minimize the risks to human health and eradicate infection to re-open export markets, while not placing unnecessary burdens on the farming industry. This type of modeling approach ideally needs to tie into economic models to produce a detailed cost-benefit analysis of various control options.

Finally, we raise the question of how complex the models need to be, what are the advantages of moving from the simple age-structured model developed here to the full (age-and stage-structured) model developed by Ferguson and coworkers (Ferguson et al. 1997b; Ferguson et al. 1999b). Although both models produce qualitatively similar results, the inclusion of time-since-infection allows far more accurate predictions of the future course of the epidemic—this degree of accuracy is required if economic factors are to be assessed. We can be convinced of the accuracy of our simple model only because it agrees with the more complex models that are believed to be a better representation of reality. In conclusion, given the economic and human health importance of the epidemic, it is vital that the predictive models used are believed to be the most accurate description available, while still maintaining sufficient transparency that their results can be interpreted. This trade-off between accuracy and transparency is one that permeates the entire field of mathematical modeling.

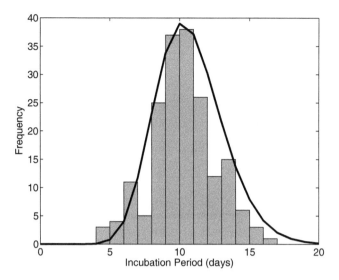

Figure 3.13. Frequency distribution of incubation periods for measles from Hope Simpson's (1952) classic household study. The data (gray bars) demonstrate that on average, the incubation period following measles infection is around 10 days, with some individual variation around this figure. The solid line represents the maximum likelihood fit to a gamma distribution, with the shape parameter (n) given by 20.

3.3. DEPENDENCE ON TIME SINCE INFECTION

In this section, we introduce another aspect of "risk structure" into the standard SIR model. Specifically, we are concerned with an individual's probability of remaining infectious as a function of time since infection, as well as the possibility that the risk of transmission may vary with the time since infection. To motivate this discussion, consider the standard SIR framework (Chapter 2) which assumes that the recovery rate is constant, independent of the time since infection. Thus, if we consider only the recovery and death processes of individuals infected at time $T = 0$, we have the classic equation:

$$\frac{dI}{dT} = -\gamma I - \mu I.$$

Upon integrating this equation, we see that the assumption of a constant recovery rate leads to the infectious period being exponentially distributed:

$$\mathbb{P}(\text{Infectious after time T}) = \exp(-(\gamma + \mu)T). \tag{3.10}$$

Biologically, this means that some individuals are infectious for only a very short period of time and contribute little to transmission, whereas others may be infectious for much longer. This assumption contradicts many empirical observations for a number of infectious diseases, especially those affecting the respiratory system. Classic work by Sartwell (1950) and Hope Simpson (1952) demonstrated that incubation period distributions for infections such as measles, chicken pox, polio, and the common cold typically show a strong central tendency. As demonstrated for measles in Figure 3.13, the incubation period has a pronounced mode, with some individual variation; the infectious period shows a similar type of distribution. As we show in Section 3.3.1, the precise assumptions made concerning the distribution of latent and infectious periods have profound epidemiological implications, especially in the early stages of an epidemic when by chance an individual having a long infectious period will promote invasion (Keeling and Grenfell 1999).

3.3.1. *SEIR* and **Multi-Compartment Models**

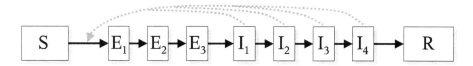

To begin this analysis, we start by comparing and contrasting the dynamics of *SIR* and *SEIR* models (Chapter 2). We can view the *SEIR* model as a multi-compartment version of the standard *SIR* model, in which the infected class has been subdivided into an exposed and an infectious class. As described in Chapter 2, the *SEIR* model is given by:

$$\frac{dS}{dt} = \nu - \beta SI - \mu S,$$

$$\frac{dE}{dt} = \beta SI - \sigma E - \mu E, \tag{3.11}$$

$$\frac{dI}{dt} = \sigma E - \gamma I - \mu I.$$

For this system of equations, the probability of being infectious at time T after becoming infected is given by:

$$\mathbb{P}(\text{infectious after time } T) = \sigma \exp(-\mu T)\frac{\exp(-\gamma T) - \exp(-\sigma T)}{\sigma - \gamma}. \tag{3.12}$$

This accounts for the exponential distribution of the exposed period, the exponential distribution of the infectious period, and the risk of natural mortality. For the *SIR* and *SEIR* models, Figure 3.14 shows the two infectious probability distributions (equations (3.10) and (3.12)) and the epidemic curves generated assuming equal R_0 and equal infected (infectious + exposed) periods in both cases. This highlights an important issue; these two models have comparable equilibria (equilibrium traits in terms of the prevalence of infection and the level of susceptibles), however the *SEIR* has a much slower growth rate. This slower growth is attributable to the delay that infected individuals wait before they can start transmitting caused by the exposed period. Chapter 2 provides a full description of this phenomenon.

We can extend this concept of compartmentalization still further, by subdividing the infected class:

$$\frac{dS}{dt} = \nu - \beta S \sum_{i=m+1}^{n} I_i - \mu S,$$

$$\frac{dI_1}{dt} = \beta S \sum_{i=m+1}^{n} I_i - \gamma n I_1 - \mu I_1, \tag{3.13}$$

$$\frac{dI_i}{dt} = \gamma n I_{i-1} - \gamma n I_i - \mu I_i \qquad \forall i = 2, \dots, n,$$

This is online program 3.5

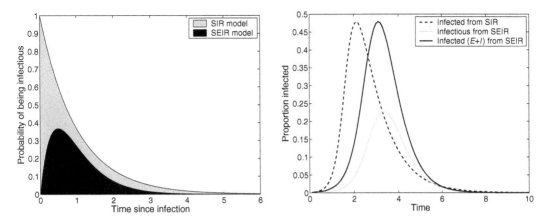

Figure 3.14. Comparison of SIR and $SEIR$ models. The left-hand graph shows the probability of being infectious at time T after being first infected, as given in equations (3.10) and (3.12). The right-hand graph shows the epidemic curve associated with these two distributions. Here we have assumed that the average time from infection to recovery is equal in both models $\frac{1}{\gamma_{SIR}} = \frac{1}{\gamma_{SEIR}} + \frac{1}{\sigma_{SEIR}}$, and $R_0 = 5$ for both models. (SIR model: $B = \mu = 5.5 \times 10^{-5}$, $\beta = 5$, $\gamma = 1$. $SEIR$ model: $B = \mu = 5.5 \times 10^{-5}$, $\beta \approx 10$, $\gamma = 2$, $\sigma = 2$.)

where it is assumed that only individuals in $I_{m+1}, \ldots, I_n$ are infectious, with the rest being in an exposed state. These equations are explicitly formulated to ensure that the average time between infection and recovery remains constant, so as the number of subclasses increases so does the rate at which individuals move between them. When $n = 1$, we return to the standard SIR model and when $n = 2$, $m = 1$, we obtain the $SEIR$ model with equal exposed and infectious periods. For general n (Figure 3.15), the exposed period, infectious period, and infected period (exposed plus infectious periods) are all gamma distributed (Lloyd 2001):

$$\mathbb{P}(\text{infected after time } T) = \int_T^\infty \frac{(\gamma n)^n}{(n-1)!} \tau^{n-1} \exp(-\gamma n \tau) d\tau.$$

This probability distribution means that the variance in the length of the infected period decreases to zero as n increases, such that in the limit when $n \to \infty$ all individuals spend exactly the same amount of time in the infected class.

Subdividing the infected period allows some control over the distribution of this period, scaling from exponential (when there is just one class) to constant (when there are many classes)

Associated with this change in the distribution of the infectious period at the individual level is a change in the epidemic profile at the population level (Figure 3.15). Despite having identical values of R_0 and the same average infectious period, changing the number of subdivisions (n) has a profound impact on the dynamics. As shown by Wearing et al. (2005), when n is large, the epidemic takes off more quickly and ends more abruptly. This difference is due to the different generation times in the two approaches. When $n = 1$, those individuals with long infectious periods produce the majority of the secondary cases and a high proportion of these are produced when the individual has been infectious

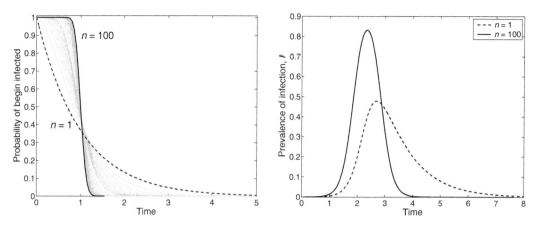

Figure 3.15. The change in the probability of infection and the epidemic profile as the number of subdivisions within the infected class increases from $n = 1$ to $n = 100$. Throughout we assume that all individuals in the infected class are infectious. Left-hand graph: for $n = 1$ the distribution of infected periods is exponential, whereas as n increases the infectious period becomes closer to a constant length ($\gamma = 1$). The right-hand graph shows the consequences of this change for the SIR-type epidemic without births or deaths. For the same basic reproductive ratio, $R_0 = 5$, and the same average infectious period, $\gamma = 1$, more subdivisions of the infectious period lead to a steeper rate of increase and an epidemic of shorter duration. Here we have ignored the exposed class by assuming that $m = 0$.

for a long time. Thus, when n is small, the rate of increase is much slower than when n is large (Kermack and McKendrick 1927; Metz 1978; Keeling and Grenfell 1997b). Conversely, during the latter stages of the epidemic when $n = 1$, the decrease in prevalence cannot be faster than the rate at which individuals recover, so $I(t) \sim \exp(-\gamma t)$. In contrast, when n is large, individuals infected at the peak of the epidemic always recover quickly, rapidly decreasing the prevalence of infection. Such changes to the dynamics may require a different parameterization (in terms of R_0 and average infectious period) for different n values to fit the same epidemic profile (Keeling and Grenfell 2002; Wearing et al. 2005), in particular a longer average infectious period may be required as n increases.

A greater subdividing of the infected population leads to more rapid growth rate and a shorter epidemic, necessitating different parameters for different models.

To demonstrate this point more fully, the growth rate of an outbreak in a totally susceptible population can be explored by determining the dominant eigenvalue of the (unstable) disease-free equilibrium for equations (3.13). This is the same basic methodology that was used to determine the growth rate in Box 3.1. As shown by Anderson and Palmer (1980), one can find an exact expression for the characteristic equation (see Chapter 2), where λ specifies the growth rate:

$$\lambda(\lambda + \gamma n)^n \left[\lambda(\lambda + \sigma m)^m - R_0 \gamma (\sigma m)^m \left(1 - \left(\frac{\lambda}{\gamma n} + 1 \right)^{-n} \right) \right] = 0.$$

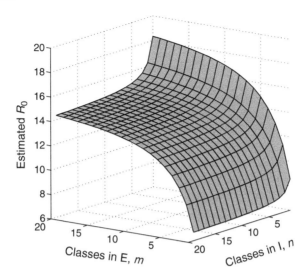

Figure 3.16. Estimates of R_0 from data on the initial growth rate, of an epidemic. The figure shows the effects of changing the distributions of the exposed and infectious periods on the value of R_0 estimated, with λ assumed to be 100 per year, $1/\sigma = 1/\gamma = 1$ week. The precise shape of each surface is independent of the exact value of λ.

During the early phase of an epidemic, the exponential growth rate, λ, satisfies the above equation.

Solution of this equation for a given set of epidemiological parameters (R_0, γ, σ, n, and m) is complex. We therefore look at the reverse problem—given we have data on an exponentially increasing epidemic, how does the subdivision of the exposed and infectious classes modify our estimation of R_0? Such an exercise reveals that for any empirically determined λ, the precise value of R_0 estimated depends on the fundamental assumptions made concerning the distributions of incubation and infectious periods. Specifically, the following equation determines the relationship between R_0 and the epidemic growth rate, λ:

$$R_0 = \frac{\lambda \left(\dfrac{\lambda}{\sigma m} + 1 \right)^m}{\gamma \left(1 - \left(\dfrac{\lambda}{\gamma n} + 1 \right)^{-n} \right)}, \tag{3.14}$$

where m and n represent the number of subclasses in the exposed and infectious categories, respectively. The mean exposed and infectious periods are represented by $1/\sigma$ and $1/\gamma$, respectively, and are assumed to be known or estimated from independent data. This equation establishes how the value of R_0 estimated from data is influenced by a priori assumptions concerning the distributions of latent and infectious periods. This relationship is demonstrated in Figure 3.16 (Wearing et al. 2005). It reveals a very startling result: For a fixed epidemic growth rate, the value of R_0 estimated may vary from 6 to almost 20 as the number of subclasses changes.

In general, as the infectious period becomes more tightly distributed (increasing n), lower values of R_0 are estimated for any given growth rate, λ. On the other hand, as the variance in the exposed period is reduced (increasing m), *higher* values of R_0 are estimated. Indeed, we may use the relationship given by equation (3.14) to arrive at the following general principle: If we ignore the exposed period, then models with an exponentially distributed infectious period will *always* overestimate the basic reproductive ratio. When the latent period is included, however, this finding is *reversed* whenever the growth rate is large (Wearing et al. 2005). By closely examining equation (3.14), we note that the basic reproductive ratio estimated from a model without an exposed class ($\sigma \to \infty$) is always smaller than the estimate from the corresponding model when an exposed period is included ($1/\sigma > 0$). Therefore, when faced with a rapidly spreading infection, either entirely ignoring the exposed period or assuming exponential distributions will lead to an underestimate of R_0 and therefore will underestimate the level of global-control measures (such as mass vaccination) that will be needed to control the epidemic.

The distribution of infectious periods can have a profound impact on the epidemic behavior when the models are stochastic and individual-based (Keeling and Grenfell 1997a; Chapter 6). When $n = 1$, and the infectious period is exponentially distributed, transmission relies on the few individuals with longer than average periods—this increases the variability in transmission. In contrast, when n is large all individuals contribute equally to transmission, thus lowering the amount of variation. Such effects are most pronounced when the number of infectious individuals is low, such as at the start of an epidemic.

When births and deaths are included into this formalism, the equilibrium levels are independent of n (Hethcote and Tudor 1980):

$$S^* = \frac{1}{R_0} \qquad I^* = \frac{B}{\gamma + \mu} - \frac{\mu}{\beta},$$

and therefore the standard vaccination thresholds apply. However, the stability of this fixed point is reduced as n increases (Grossman 1980; Lloyd 2001); conceptually, this is because the fixed duration of infection adds a natural period of oscillation to the dynamics.

3.3.2. Models with Memory

The above models have achieved different infectious period distributions by modifying the number of compartmental subdivisions used to partition the infectious class; however, a similar but more flexible distribution of periods can be achieved by explicitly accounting for the time since infection. Hence, these models need to keep track of the history of infection—they are models with memory. Frequently we may wish to specify the probability distribution of both the exposed (latent) and infectious periods that have been derived from data, and are labeled $P_E(t)$ and $P_I(t)$ respectively. The probability that an individual infected at time 0 is still in the exposed class at time t is given by:

$$\text{Prob(exposed at } t) = \mathbb{P}_E(t) = \int_t^\infty P_E(s) ds,$$

that is, the probability that the exposed period is longer than t. The probability it is infectious is calculated as:

$$\text{Prob(infectious at } t) = \mathbb{P}_I(t) = \int_0^t P_E(s) \int_{t-s}^\infty P_I(\tau) d\tau ds,$$

which is the probability that the individual is in the exposed class for less that t, but is still in the infectious class.

We now let $C(t)$ be the rate at which individuals are infected at time t, so that $\int_a^b C(t) dt$ is the fraction of new cases (relative to the population size) between times a and b. This variable C, together with the distributions P_E and P_I, allows us to calculate the number of exposed and infectious individuals at any time. Ignoring the exposed class for the moment, the basic SIR equations now become:

$$\frac{dS}{dt} = \nu - C(t) - \mu S,$$

$$C(t) = \beta S \int_0^\infty C(t-s) \mathbb{P}_I(s) ds,$$

(3.15)

where the integral gives the proportion of individuals who are currently infectious, and so plays a similar role to I in the standard transmission term. Although one differential equation and one integral equation (3.15) may not look much more complex than the standard two differential equations needed to solve the SIR model (Chapter 2), the need to keep track of the historical values of C means that this technique is far more computationally challenging to implement.

A similar technique can be adapted for when we have data on the transmission rate as a function of the time since infection. For deterministic models, this produces a similar formulation of equations to those given above:

$$\frac{dS}{dt} = \nu - C(t) - \mu S,$$

$$C(t) = S \int_0^\infty \beta(s) \exp(-\mu s) C(t-s) ds,$$

(3.16)

where the integral is over the time since infection, s, with $\beta(s)$ defining the transmission rate during the infectious period. For this formulation we do not need to explicitly model an exposed and infectious period; we simply allow the parameter $\beta(s)$ to capture to the known infection profile. This model also includes a term $\exp(-\mu s)$ that accounts for the death of individuals since infection.

This type of distributed infection model has been fit to several recent epidemics including SARS (Donnelly et al. 2003; Riley et al. 2003), foot-and-mouth (Ferguson et al. 2001a,b) and BSE (Ferguson et al. 1997b). These detailed analyses all confirm that, in general, the infectious period is rarely exponentially distributed, which has profound implications for the timing of control strategies. In particular, consider a control measure that isolates an individual some time after he or she has become infectious; although late control would still remove the tail of an exponential distribution, it may not do so for constant periods. Chapter 8 discusses this point in detail.

3.3.3. Application: SARS

Taken from the work of Donnelly et al. (2003) and Riley et al. (2003), we consider the dynamics and parameter estimates for the SARS epidemic in Hong Kong. The host-level lifecycle of the virus is as follows. After the initial infection, it takes about 6 days before individuals start to show clinical symptoms and become infectious. In the early part of the epidemic, in then took a further 4 days before the individual was hospitalized—although this delay dropped once more strict control measures were in place. Finally, individuals remained in hospitals for about 23 days if they recovered, or 36 days if the disease proved fatal. For modeling purposes, the population can therefore be partitioned into five distinct classes with respect to their infection or disease status: Susceptible, S; Exposed, E; Infectious, I; Hospitalized, H, and Recovered, R. The time spent in the exposed, infectious, and hospitalized classes can all be captured with a gamma distribution (Donnelly et al. 2003), and therefore are comparable with further subdividing each of these classes.

As shown above, if we wish to correctly estimate R_0, and therefore the level of control needed, it is important to correctly capture the distribution of the exposed and infectious periods. For the SARS epidemic, there was fortunately sufficiently detailed information on individual patients for the distributions to be estimated with some degree of confidence. When dealing with such public health issues as the control of SARS, it is generally important that models are as accurate as possible—reflecting the known pathogen and host behavior. Here we contrast the dynamics of standard models with distributed period models, assuming either equal R_0 or parameterized to fit to the same initial exponential growth. Although the qualitative dynamics are comparable, the quantitative differences could be very important to public health planning.

Given that the periods are all approximately gamma distributed, two equivalent formulations of the model are possible—one using memory and the other using multiple classes. Both models are given below to contrast the two mechanisms.

If $C(t)$ is the rate of infection at time t, then the proportions of the population in each of the classes are given by:

$$E(t) = \int_0^\infty C(t-s) \int_s^\infty P_E(\tau) d\tau ds,$$

$$I(t) = \int_0^\infty C(t-s) \int_0^s P_E(\tau) \int_{s-\tau}^\infty P_I(T) dT d\tau ds, \qquad (3.17)$$

$$H(t) = \int_0^\infty C(t-s) \int_0^s P_E(\tau) \int_0^{s-\tau} P_I(T) \int_{s-\tau-T}^\infty P_H(W) dW dT d\tau ds.$$

So the proportion of the population that are hospitalized is all those who have been infected, have left the exposed class, have left the infectious class, but have not yet left hospitals. Throughout we will ignore births and deaths due to the rapid time-scale of the epidemic. The dynamics of the SARS infection are then governed by similar equations to before:

$$\frac{dS}{dt} = -C(t),$$

$$C(t) = S(\beta_I I + \beta_H H), \qquad (3.18)$$

where the transmission rate within hospitals (β_H) is allowed to be different from that in the community (β_I).

Alternatively, we may wish to formulate the model by subdividing the exposed, infectious, and hospitalized classes as illustrated in Section 3.1:

$$\frac{dS}{dt} = -S(\beta_I I + \beta_H H),$$

$$\frac{dE_1}{dt} = S(\beta_I I + \beta_H H) - \sigma m E_1,$$

$$\frac{dE_i}{dt} = \sigma m E_{i-1} - \sigma m E_i \qquad i = 2, \ldots, m,$$

$$\frac{dI_1}{dt} = \sigma m E_m - \gamma_I n I_1, \tag{3.19}$$

$$\frac{dI_i}{dt} = \gamma_I n I_{i-1} - \gamma_I n I_i \qquad i = 2, \ldots, n,$$

$$\frac{dH_1}{dt} = \gamma_I n E_n - \gamma_H q H_1,$$

$$\frac{dH_i}{dt} = \gamma_H q H_{i-1} - \gamma_H q H_i, \qquad i = 2, \ldots, q$$

where m, n, and q are the number of subdivisions for the exposed, infectious, and hospitalized classes, respectively.

Figure 3.17 compares the predicted dynamics of the SARS epidemic (without control) under two different modeling assumptions. The dashed lines give the results when the periods are exponentially distributed, which is equivalent to there being just one exposed, one infectious, and one hospitalized compartment. The solid lines correspond to using the distribution of periods taken from clinical observations; the periods were fit with a gamma distribution so the model is equivalent to having three exposed, three infectious, and ten hospitalized compartments. Following Riley et al. (2003), we assume that individuals in the hospitalized class transmit infection at a fifth of the rate of those in the infectious class ($\beta_H = \frac{1}{5}\beta_I$).

As seen in the generic model (Figure 3.15), the assumption of tighter distributions for the periods within the SARS model leads to a faster increase of the epidemic and a more pronounced epidemic peak for the same value of R_0. The differences between the results assuming either exponential and gamma-distributed periods are substantial, with significant public health implications associated with the discrepancies—even when both models are parameterized to match the same initial growth rate. The more accurate gamma-distributed periods consistently predict a larger peak number of hospitalized cases, which could place more strain on health services. In addition, the gamma-distributed periods are associated with a far more rapid decline in the epidemic, and a more rapid eradication of infection.

This work on SARS illustrates a crucial issue in modeling. Although the standard compartmental models, with their associated exponential distributions, are a useful tool for understanding the dynamics of infection, when very precise predictions are required

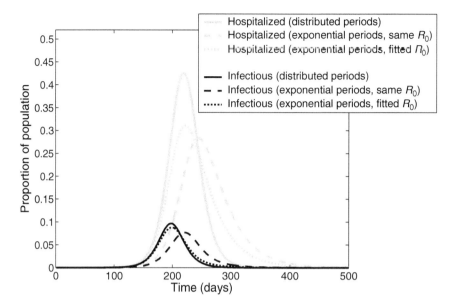

Figure 3.17. The predicted dynamics of SARS infection under two modeling assumptions: The dashed lines and dot-dashed lines correspond to a model with no memory and just one compartment for each of exposed, infectious, and hospitalized; the solid lines correspond to a model with a gamma-distributed period for the time spent in each class parameterized from the data. For the dashed line the model has equal R_0 to the distributed periods, for the dot-dashed line the model has the same initial growth rate as the distributed period model. (Exposed period distribution $1/\sigma = 6$ days, $m = 3$. Infectious period distribution $1/\gamma_I = 4$ days, $n = 3$. Hospitalized period distribution $1/\gamma_H = 30$ days, $q = 10$. Solid and dashed lines $R_0 = 2.8 \Rightarrow \beta_I = 0.28$ per day, $\beta_H = 0.056$ per day. Dot-dashed line $R_0 = 3.03$ with β_I and β_H similarly scaled.) Initialized with $S(0) = 1$, $E(0) = 10^{-6}$, $I(0) = H(0) = R(0) = 0$.

it becomes necessary to include much more of the available information; details often matter.

3.4. FUTURE DIRECTIONS

Although we have concentrated on the risk structure in sexually transmitted infections, almost all populations and associated infections have some form of risk structure. Thus even though infections such an influenza appear to spread randomly through the population, some individuals (super spreaders) have much more contact with the general public and therefore are more likely to catch and transmit infection. In the coming years, it is likely that more public-health motivated predictive modeling will incorporate this level of heterogeneity.

A second cause of risk structure may be due to genetics or the general health of the host, so that some individuals are much more susceptible to infection or more much infectious than others. This form of modeling has already been attempted, by considering a separate hospitalized, highly susceptible core group in addition to the general population (Austin and Anderson 1999b; Lloyd-Smith 2003). Such heterogeneities will be difficult to observe in practice, but again can play a determining role in the dynamics of infection and the ability to control or eradicate the disease. Therefore, although determining this form of risk structure may be difficult, incorporating some generic form of heterogeneity may provide a better match between models and data.

An important future development may be the direct measurement of the general contact matrix (β) for airborne infections as has been attempted for sexually transmission infections (Potterat et al. 2002). One method of ascertaining the network of potential transmission routes is through a diary-based approach (Edmunds et al. 1997), where contacts are recorded over a short period by volunteers. However, with the increased used of mobile phones and GPS technology, it will soon be possible to directly track the movement of a large proportion of the population and determine their position to within a few meters; this provides the opportunity to remotely gather vast quantities of information on the potential mixing patterns of the population. Dealing with this volume of data and using it to parameterize suitable models clearly has many potential public health applications, from estimating the likely spread of influenza to the real-time control of a deliberate or accidental release of an infectious agent.

Finally, as we have seen with the example of the SARS infection, the detailed natural history of infection at an individual level can be very important in determining the population-level dynamics. In particular, without capturing the correct distribution of exposed and infectious periods, it may be impossible to match to both the initial and long-term epidemic behavior. The inclusion of more patient-level information derived from detailed medical observations is ever more important as models are increasingly being expected to provide accurate quantitative predictions.

3.5. SUMMARY

➤ When structuring the population into different classes, the single transmission parameter in the unstructured models is replaced by a matrix of values.

➤ Because the transmission matrix generally has more terms than the structured data, simplifications are needed to overcome this deficit.

➤ The basic reproductive ratio, R_0, for structured populations is found using an eigenvalue approach. In general, this is greater than if the structure were ignored, and bounded by the basic reproductive ratios for each class.

➤ Assortative mixing is common, such that high-risk individuals mix more frequently with other high-risk individuals. Increased assortative mixing tends to increase R_0.

➤ The initial growth of a structure model depends on the initial conditions, not necessarily the basic reproductive ratio.

➤ It is no longer true that $S(\infty) = 1/R_0$; in many structured models that have high associativity, the equilibrium level of susceptibles is much higher and density-dependent saturation effects occur earlier. Hence, although the equilibrium prevalence is low, the infection may be difficult to eradicate because R_0 is still large.

➤ Targeting vaccination or other control measures is more efficient than random control. It is generally better to over-target rather than under-target.

➤ Although age is a continuous parameter, age-structured models usually group individuals into a limited number of classes—often representing school years.

➤ Age-structured models differ from the risk-structured models due to the regular progression of individuals into increasingly older age classes.

➤ At equilibrium, seroprevalence of SIR-type infections must increase with age. This allows the force of infection within an age class to be calculated.

➤ Subdividing the infected period allows some control over the distribution of this period, scaling from exponential (when there is just one class) to constant (when there are many classes).

➤ A greater subdividing of the infected population leads to more rapid growth rate and a shorter epidemic, necessitating different parameters for different models.

Chapter Four

Multi-Pathogen/Multi-Host Models

In this chapter, we follow some recent attempts to develop a theory of "community epidemiology," whereby the traditional framework examining one pathogen spreading through a single host population is extended in two distinct ways: (1) to consider multiple infectious diseases (or strains) spreading through one host species, or (2) a single infectious disease that can be transmitted between different species.

The scientific community is only just beginning to understand the range of complex outcomes when multiple strains compete for a limited supply of susceptibles or when a number of host species share a pathogen. These conceptual extensions of the simple one-pathogen one-host paradigm are becoming increasingly recognized as fundamental to our understanding of key questions in modern epidemiology and public health. Does a reduction in species diversity amplify or reduce the intensity of outbreaks? What role does biodiversity play in the maintenance of pathogens? What are the conditions that determine the spillover of "emerging" pathogens to new host species? How do these community epidemiology issues impact the invasion of species in new environs?

A complete description of models that examine strain dynamics requires a detailed knowledge of the host immune systems, and how different strains interact via the immune response that they elicit. Interest in this area stems from the fact that competition between strains is fundamental to disease evolution. Thus, if we are to understand and ultimately predict the emergence of novel pathogens or, for example, the strain of influenza that will circulate next season, we must better grasp the role of competition between cross-reacting strains of infection.

When one pathogen is transmitted between multiple host species, the models and dynamics are very much akin to those developed to deal with subdivided or structured populations (Chapter 3). The main distinction is that with different species the disease can have dramatically different infectious periods, mortality, and other characteristics, as well as the underlying demographic differences between the species. In general, multi-host models can be partitioned into two different classes: either a secondary obligate host is required for transmission (such as vector-borne diseases), or transmission can occur both within and between the various host species. Vector-borne diseases, such as malaria, dengue fever, and leishmania, are among the most challenging from a public-health perspective, and so models are frequently required to optimize their control. Finally, zoonoses form a special class of multi-host diseases, where animals form a reservoir of pathogens that can be spread to humans. Examples include such high-profile diseases as avian influenza, bubonic plague, hantavirus, Lyme disease, Q-fever, rabies, SARS, toxoplasmosis, trypanosomiasis, West Nile virus, and a range of macro-parasitic infections. In general, the effects of these diseases in humans tends to be quite severe; this is because humans act as a dead end for such infectious diseases and, therefore, there has been little evolutionary pressure for benign infection compared to within the reservoir host species. Although in principle the models for zoonoses have the same structure as other multi-host

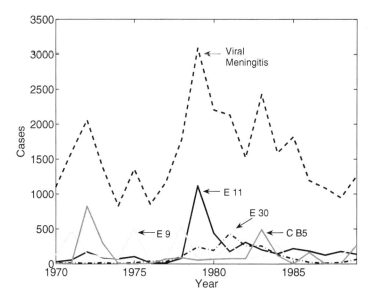

Figure 4.1. Results from the Nonpolio Entrovirus Surveillance scheme in the United States, from 1970–1989. Shown is the total recorded cases of viral aseptic meningitis (dashed line), as well as the constituent cases for echovirus 9, 11, and 30, and coxsackie B5. The interaction between these constituent viral types is complex, but must be understood if we are to successfully understand and predict the incidence of infection (Strikas et al. 1986). (Data from CDC–the Center for Disease Control, USA.)

models, the implications of infection and therefore the need for control measures will be far greater.

4.1. MULTIPLE PATHOGENS

We first focus on the interaction of two, or more, pathogens within a population. Many infectious diseases that we consider as a single disease are in reality comprised of multiple strains, which interact through the cross-immunity that is invoked within a host. An accurate understanding of such diseases, which include malaria, dengue fever, and influenza, requires the consideration of strain structure (Figure 4.1) (Ferguson et al. 1999a; Gog and Grenfell 2002; Ferguson et al. 2003a). Models incorporating multiple pathogens allow us to investigate questions of disease evolution, from theoretical questions such as understanding current disease behavior in terms of an optimal strategy for transmission, to more applied issues such as predicting the influenza strains for the coming year or understanding the effects of strain-specific control. Finally, multi-strain models offer insights into the increasing prevalence of drug-resistant bacteria, and how to limit their spread (Baquero and Blàzquez 1997; Bohannan and Lenski 2000).

The presence of two or more infectious diseases within a population increases the possible number of compartments into which the population can be subdivided. In the most general form, this formulation should uniquely identify the entire infection history of individuals within each compartment and their immunological status with respect to the various pathogens/strains under consideration. Such completely general models are more difficult to study due to the large number of degrees of freedom and therefore the difficulty in both parameterization and analysis. Instead, models usually focus on particular assumptions about conferred immunity and the interaction between pathogens. As such, two completely independent infectious diseases provides the simplest model, although it is of little biological interest. We now consider a range of models based on more complex assumptions, where either transmission of, or susceptibility to, one strain is modified due to resistance to another strain. In some cases, it will be assumed that co-infection (simultaneous infection with two pathogens) is so rare that it can be ignored. This simplification is often true for many airborne infections, though for sexually transmitted diseases, infection with one pathogen may increase the susceptibility to others and hence co-infection is promoted (Renton et al. 1998; Chesson and Pinkerton 2000).

4.1.1. Complete Cross-Immunity

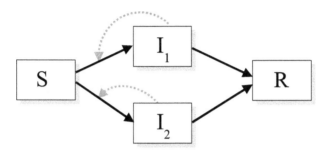

For illustrative purposes, we start with a model of two co-circulating strains within a simplified SIR framework, assuming complete cross-immunity. This means infection by either strain confers lifelong immunity to both. The four distinct compartmental classes are: susceptible to both strains, infectious with strain 1, infectious with strain 2, and recovered and therefore immune to both. Mathematically this leads to the following differential equations which are a simple modification to the standard SIR equations:

$$\frac{dS}{dt} = \nu - \beta_1 SI_1 - \beta_2 SI_2 - \mu S,$$

$$\frac{dI_1}{dt} = \beta_1 SI_1 - \gamma_1 I_1 - m_1 I_1 - \mu I_1,$$

$$\frac{dI_2}{dt} = \beta_2 SI_2 - \gamma_2 I_2 - m_2 I_2 - \mu I_2,$$

$$\frac{dR}{dt} = \gamma_1 I_1 + \gamma_2 I_2 - \mu R.$$

$$(4.1)$$

Here, for generality, it has been assumed that the two strains have different transmission (β), recovery (γ), and mortality (m) rates. The strain-specific basic reproductive ratio is given by $R_0^i = \beta_i / (\gamma_i + \mu + m_i) \ (i = 1, 2)$. It is natural to attempt an understanding of this system by deriving and examining its equilibria. This approach soon leads to the paradoxical situation where the coexistence of both strains requires the fraction susceptible to be at once $1/R_0^1$ and $1/R_0^2$! Closer examination by simulation of equations (4.1) shows that only one strain can persist in such a scenario. Both are competing for the same limited

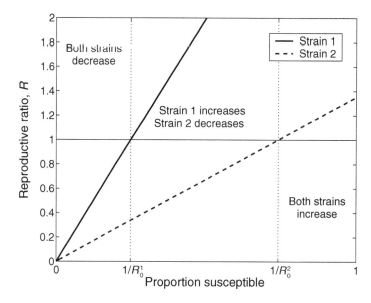

Figure 4.2. The reproductive ratio, $R = S \times R_0$, of two competing strains that offer complete cross-immunity. When the level of susceptibles is low the prevalence of both strains decreases, whereas when a high proportion are susceptible both strains can increase, although this will eventually lead to a decrease in the proportion susceptible, changing the dynamics. In the intermediate region, only strain 1 can increase with the weaker strain 2 being driven to extinction. ($\beta_1 = 4$, $\beta_2 = 1.35$, $\gamma_1 + \mu = \gamma_2 + \mu = 1$, $m_1 = m_2 = 0$.)

resource (susceptibles), and following the ecological tenet, whichever strain utilizes that resource more efficiently will dominate. In particular, the strain with the largest basic reproductive ratio will drive the other to extinction.

When competing strains provide complete protection for each other, the strain with the largest R_0 will force the other strain to extinction.

This competition scenario can be readily understood by considering the growth rate of each strain, captured by the reproductive ratio R^i, as the level of susceptibles varies (Figure 4.2). For convenience, we assume that strain 1 has a higher basic reproductive ratio than strain 2, that is, $R_0^1 > R_0^2$. Standard theory tells us that the fraction of infecteds will increase whenever $S > 1/R_0$ (so that the effective reproductive ratio, $R_0 \times S$, is greater than one), and will decrease whenever $S < 1/R_0$. This behavior leads to three separate regions and two points where one of the strains is at equilibrium ($R_0 S = 1$). If strain 2 is at equilibrium (and therefore $S = 1/R_0^2$), then because the growth rate of strain 1 is positive, this cannot be a stable equilibrium solution for both strains. However, if strain 1 is at equilibrium (and $S = 1/R_0^1$), then because the growth rate of strain 2 is negative, this strain cannot invade and is always forced to extinction. The stable solution is therefore $S = 1/R_0^1$, $I_1 = (\nu R_0^1 - \mu)/\beta_1$, $I_2 = 0$ (assuming births exactly balance deaths).

Although R_0 determines the eventual competitive outcome, pathogens with a more rapid life cycle may be favored in the short term. If two strains invade a highly susceptible population ($S \sim 1$), then the strain with the largest growth rate ($\beta - \gamma - m - d$) will be initially favored. Figure 4.3 shows an example of this type of behavior, starting with two

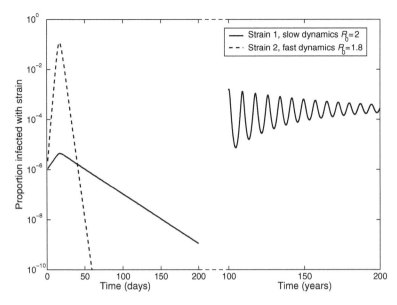

Figure 4.3. The level of two competing strains introduced into a totally susceptible population. During the early epidemic phase (days 0–50), the faster strain 2 dominates, whereas at far later times strain 1 with the largest R_0 is the eventual winner. ($\nu = \mu = 5 \times 10^{-5}$; $\beta_1 = 0.2$, $\gamma_1 = 0.1$, $R_0^1 \approx 2$; $\beta_2 = 1.8$, $\gamma_2 = 1$, $R_0^2 \approx 1.8$; $m_1 = m_2 = 0$. All rates are in days.)

strains invading a totally susceptible population. Although strain 1 has a larger R_0 and therefore eventually dominates (at a time scale of many years), because the life cycle of strain 2 is so much faster it wins out during the initial epidemic lasting about 50 days.

Although the relative R_0 values determine the long-term competitive success, a rapid life cycle may allow short-term dominance.

4.1.1.1. Evolutionary Implications

Now consider what the above results mean in terms of the direction that natural selection would be expected to act. Any mutation that generates a new strain with a larger basic reproductive ratio will be favored, and over time such mutations should accumulate such that R_0 increases. Relating this behavior to more mechanistic parameters implies that both the transmission rate, β, and the infectious period, $1/\gamma$, should increase and where applicable the disease-induced mortality or virulence, m, should decrease (May and Anderson 1983; Bremermann and Thieme 1989). Hence, when no other constrains are present, all infections should become highly transmissible, lifelong infections that are benign or even beneficial to the host (Mann et al. 2003).

Evolution will favor mutants with higher R_0, in theory leading to higher transmission rates, and long-lasting infections associated with a low probability of mortality.

Interestingly, most infectious diseases of public-health concern do not fit this predicted pattern. The reason for this discrepancy has been the subject of much research, generally focused on the existence of trade-offs between the transmission rate of the pathogen and the duration of the infectious period (Bremermann and Thieme 1989; Levin et al. 1999;

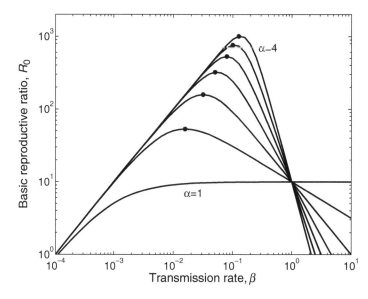

Figure 4.4. Using the power-law trade-off $\gamma + m = k\beta^{\alpha}$, intermediate values of β produce the maximum basic reproductive ratio, R_0. When $\alpha \le 1$, we predict runaway evolution to ever-larger transmission rates. For $\alpha > 1$, the optimal transmission rate and associated R_0 value (indicated with a dot) increases as α becomes larger. ($k = 0.1$, $\mu = 10^{-4}$, rates given per day.)

Boots and Sasaki 1999). The transmission-virulence trade-off is based upon the notion that any infection producing lots of pathogen particles—while readily transmitted—is likely to be harmful to the host, resulting in rapid host death (Anderson and May 1991; Boots and Sasaki 1999). This trade-off could also apply to the infectious period, such that highly transmissible pathogens are most often of short duration. A crude reflection of this can be seen in the low-transmissibility and long-infectious period of sexually transmitted diseases compared to the generally more transmissible airborne diseases that have much shorter infectious periods. Given a relationship between the duration of infection, $1/(\gamma + m + \mu)$, and the transmission rate, β, the disease will evolve to the set of parameters that maximize R_0:

$$R_0 = \frac{\beta}{\mu + \gamma + m}.$$

The trade-off is frequently assumed to be power-law in shape such that:

$$\gamma + m = k\beta^{\alpha} \quad \Rightarrow \quad R_0 = \frac{\beta}{\mu + k\beta^{\alpha}}.$$

Hence, when $\alpha > 1$, R_0 has a well-defined maximum and the transmission rate should evolve toward the intermediate value of $\beta = \sqrt[\alpha]{\mu/(k(\alpha - 1))}$. By applying a trade-off between transmission and infection longevity, we can eliminate runaway evolution and predict intermediate infection parameters and dynamics of the type observed (Figure 4.4).

Trade-offs between transmission rates and duration of infection mean that R_0 is maximized for intermediate values and runaway evolution is prevented.

Although these forms of analysis provide an intuitive understanding of the selection pressures that affect disease evolution, several issues still remain unsolved. Most immediate is the precise form of the trade-off (Levin 1996). Although the assumed power-law relationship is quite flexible, the true trade-off is likely to be more complex. To date, there is little experimental or observational evidence to support the existence of a trade-off and clearly insufficient data to determine a functional form (Ebert and Bull 2003). For this formulation to be of practical benefit, the trade-off curve associated with every pathogen would have to be determined. Finally, the models implicitly assume that infection parameters occupy a continuous spectrum, whereas in reality their behavior is controlled by a discrete set of genes. However, the mapping from genotype to phenotypic behavior is still a major challenge and we are many years from a detailed empirical understanding.

As well as providing an evolutionary perspective on disease behavior, this theory can also be used to investigate the plausible evolutionary changes in response to social and medical changes. The most applied use is the investigation of antibiotic resistance—when it arises and how its spread can be controlled (Austin and Anderson 1999a; Lipsitch et al. 2000). Consider two bacterial strains, a wild type (I_w) and a resistant strain (I_r), that compete for susceptibles. In general, resistance is assumed to be costly, meaning that the wild type is naturally fitter (has a higher R_0^w). It is this assumption that allows the wild type to dominate in the natural environment. The resistant strain, however, is not eliminated by antibiotics and therefore maintains its R_0^r even when medical treatment is provided. By modeling the effects of treatment on the reproductive ratio of the wild type strain, treatment regimes can be found that minimize the evolution of resistant strains. For example, we might suppose that treatment with antibiotics acts to reduce the infectious period of the wild type strain, allowing individuals to recover more quickly. This leads to the following set of equations describing the dynamics:

$$\frac{dS}{dt} = \nu - S(\beta_w I_w + \beta_r I_r) - \mu S,$$

$$\frac{dI_w}{dt} = \beta_w S I_w - (\gamma + T) I_w - \mu I_w,$$

$$\frac{dI_r}{dt} = \beta_r S I_r - \gamma I_r - \mu I_R,$$

where $\beta_r < \beta_w$, and treatment with antibiotics, T, acts to increase the recovery rate. The results are shown in Figure 4.5.

As predicted from previous models, it is clear that due to the complete cross-immunity only one strain can persist; when the treatment level is low the wild type dominates, but above a critical treatment level, T_C, the resistant mutant takes over. This critical level is observed when both strains have equal reproductive ratios:

$$\frac{\beta_w}{\gamma + T_c} = \frac{\beta_r}{\gamma} \quad \Rightarrow \quad T_C = \gamma \left(1 - \frac{\beta_w}{\beta_r}\right).$$

Minimizing the prevalence of infection in the population occurs for treatment levels just below T_C; therefore, there is a careful balance between reducing infection and not producing conditions favorable to the resistant strains. The greater the difference between wild-type and resistant strains (in terms of R_0), the greater the reduction in prevalence before resistant strains are favored. Two biological factors complicate this picture: (1) compensatory mutations can arise that counteract the reduction in fitness suffered by the resistant strain, and (2) although mutation from wild type to resistant is relatively common, back mutations are far rarer. An additional, though often ignored, subtlety in such systems is the horizontal transfer of antimicrobial resistance genes mediated by plasmids (Sørensen et al. 2004).

Application of antibiotic treatments requires a careful balance between combating infection and not providing suitable conditions for resistant mutants to out-compete the wild type.

4.1.2. No Cross-Immunity

At the opposite extreme are infectious diseases that result in no cross-immunity. The modeling of such infections only differs from the case of two independent infections in that co-infection is assumed not to occur. Often, this assumption has a strong epidemiological basis because those individuals infected or convalescing from one infection are not mixing with enough individuals to catch subsequent infections. In this instance, the equations become:

$$\frac{dN_{SS}}{dt} = \nu - \beta_1 N_{SS} I_1 - \beta_2 N_{SS} I_2 - \mu N_{SS},$$

$$\frac{dN_{IS}}{dt} = \beta_1 N_{SS} I_1 - \gamma_1 N_{IS} - \mu N_{IS},$$

$$\frac{dN_{RS}}{dt} = \gamma_1 N_{IS} - \beta_2 N_{RS} I_2 - \mu N_{RS},$$

$$\frac{dN_{SI}}{dt} = \beta_2 N_{SS} I_2 - \gamma_2 N_{SI} - \mu N_{SI},$$

$$\frac{dN_{RI}}{dt} = \beta_2 N_{RS} I_2 - \gamma_2 N_{RI} - \mu N_{RI}, \tag{4.2}$$

$$\frac{dN_{SR}}{dt} = \gamma_1 N_{IS} - \beta_1 N_{SR} I_1 - \mu N_{RS},$$

$$\frac{dN_{IR}}{dt} = \beta_1 N_{SR} I_1 - \gamma_1 N_{IR} - \mu N_{IR},$$

$$\frac{dN_{RR}}{dt} = \gamma_1 N_{IR} + \gamma_2 N_{RI} - \mu N_{RR},$$

$$I_1 = N_{IS} + N_{IR}, \qquad I_2 = N_{SI} + N_{RI},$$

where N_{AB} refers to the proportion of the population that are in state A with respect to disease 1 and state B with respect to disease 2, so N_{SI} refers to individuals that are susceptible to disease 1 and infections with disease 2. In this situation, co-existence of the two strains/diseases is generally possible whenever the two basic reproductive ratios are greater than one.

Even when there is no cross-immunity, the absence of multiply infected individuals is epidemiologically plausible, reflecting the reduced number of contacts when ill.

4.1.2.1. Application: The Interaction of Measles and Whooping Cough

One of the clearest applications of this type of model has focused on the dynamics of measles and whooping cough. Measles is caused by a virus and whooping cough is caused by a bacterium, hence we would not expect any specific immune-mediated interaction.

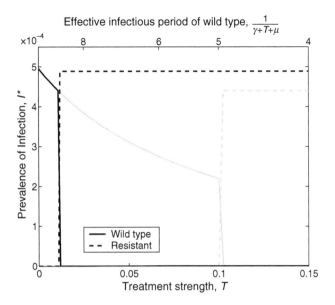

Figure 4.5. The prevalence of wild type, I_w (solid lines), and antibiotic resistant, I_r (dashed lines), infection in the population as differing amounts of treatment, T, are applied. The upper x-axis converts the treatment level into a measure of the infectious period. Due to complete cross-immunity, only one strain can persist. ($\beta_w = 1$, $\gamma = 0.1$, which means that in the absence of treatment, $R_0^w = 10$.) Two resistant scenarios are considered: When the resistant strain is only slightly less fit, $\beta_r = 0.9$, $R_0 = 9$ (black), and when the resistant strain is far less fit, $\beta_w = 0.5$, $R_0 = 5$ (gray). ($\nu = \mu = 5.5 \times 10^{-5}$, all rates are per day.)

However, Rohani et al. (1998) argued that the quarantining that occurs during the later stages of the infectious period and subsequent convalescence means that a fraction of potentially susceptible individuals might be temporarily unavailable to catch any other cocirculating infections. Thus, simple behavioral considerations predict a possible interaction between the epidemics of unrelated infectious diseases. In addition, when pathogens are associated with a considerable probability of mortality following infection, some of those infected become permanently unavailable to contract other diseases (similar to the effects of cross-immunity). This effectively increases the competition for hosts, resulting in "interference" between the epidemics of unrelated infectious diseases.

Rohani et al. (1998) modified equations (4.2) by incorporating a convalescent class and demonstrated that the isolation of individuals during their convalescence period is sufficient to alter the dynamics of competing infectious diseases. They compared the epidemics of measles and whooping cough predicted by two independent single-disease $SEIR$ models to those observed in the modified two-disease model described above. In the two-disease model, the rigidly annual epidemics of whooping cough predicted by the standard $SEIR$ model give way to a range of biennial dynamics, mimicking the epidemics of measles. The presence of whooping cough in the two-disease model results in a reduced effective amplitude of seasonality for measles. As explained by Huang and Rohani (2005), this is because measles has the higher transmission rate, thereby its epidemic patterns become imprinted on whooping cough dynamics as a result of their interaction via the susceptible pool. The most obvious signature of disease interference is, however, out-of-phase epidemics of the two infections when dynamics are multiennial.

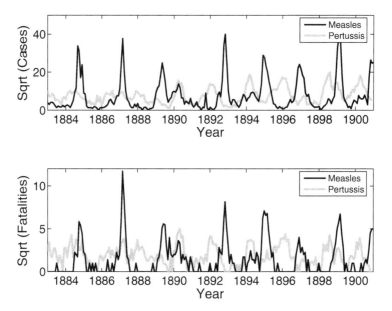

Figure 4.6. Long-term patterns in measles (black lines) and whooping cough (gray lines) epidemics in Aberdeen from 1883 to 1901 (data from Laing and Hay 1902). The top panel contains monthly case notifications; the bottom panel shows monthly case fatalities. The time series for both infections exhibit a strong biennial component, with a striking phase difference between measles and whooping cough outbreaks. (Note that the data is plotted on a nonlinear scale to better illustrate the cyclic nature.)

The most convincing empirical support for this phenomenon has been found in European case fatality data for measles and whooping cough in the early decades of the twentieth century (Rohani et al. 2003). As shown in Figure 4.6, the biennial epidemics of measles and whooping cough are clearly out of phase, particularly from 1891–1902, consistent with the predictions of the two-disease model.

Notice, however, that children are typically exposed to many more infectious diseases than just measles and whooping cough. These other infections include mumps, polio, rubella, or chickenpox, among others. Given the generality of the interaction envisaged by Rohani et al. (1998), would we expect *all* of these infectious diseases to dynamically interact? By extension, ideally, would any modeling work need to include *all* of these pathogens? We believe the answer to this question is likely to be "no." Using simple homogeneous models without age structure, it has been demonstrated that dynamical interference effects are most pronounced between infections with a similar basic reproductive ratio (Rohani et al. 2003, 2007; Huang and Rohani 2006). A complete understanding of this issue will, however, need age-dependence in contact rates to be taken into account. This is because the interference concept relies on "competition" for resources (hosts) between pathogens. For the dynamical effects of this competitive interaction to be noticeable, the pathogens should be infecting largely the same cohort of hosts. Using a two-disease age-structured model, Huang and Rohani (2006) showed that the extent of interference effects is closely determined by the relative distributions of age at infection. Hence, given their similar R_0s, measles and whooping cough are likely to have been strongly "competing" for children in the same age cohorts, whereas their interaction with the other childhood infections is likely to have been less intense. This logic suggests that infectious diseases such as rubella and polio, or chickenpox and mumps, may also be good potential candidates for the study of interference effects.

4.1.2.2. Application: Multiple Malaria Strains

A second example of such models comes from the epidemiology of *Plasmodium falciparum*, the malaria-causing parasite. Section 4.2.2 of this chapter provides a detailed description of a modeling approach for malaria that includes the transmission via mosquito vectors—however, here we simplify the dynamics by approximating malaria by a directly transmitted infection. The basic reproductive ratio, R_0, of malaria is generally estimated from the increase with age, a, of the proportion of the population seropositive, (i.e., the proportion that is no longer susceptible, $1 - S(a)$). Standard theory (Chapters 2 and 3) predicts that when all ages suffer equal exposure to the pathogen, the level of susceptibles decays exponentially with age:

$$S(a) = \exp\left(-\frac{R_0 - 1}{L} a\right),$$ (4.3)

where $L = 1/\mu$ is the average life expectancy. This leads to an average age of first infection $A = L/(R_0 - 1)$. Because the average age of infection is around one year old, such calculations lead to estimates of R_0 in the range 50 to 100, making the eradication of malaria extremely difficult. However, these calculations assume that malaria is a single strain infection and that following infection there is complete immunity. However, detailed serological evidence suggests that a diverse range of antigenically distinct strains may cause the disease that is labeled malaria (Day and Marsh 1991).

Following the work of Gupta et al. (1994), we consider several strains of *Plasmodium falciparum*, each of which causes symptoms that are diagnosed as malaria, and suppose that the strains offer no cross-immunity, to each other. A relaxation of this assumption, so that strains confer partial immunity, does not radically change the general conclusions. The proportion of the population that is still susceptible to strain i at age a is given by:

$$S^i(a) = \exp\left(-\frac{R_0^i - 1}{L} a\right).$$

However, because there is no cross-immunity, strains act independently. Therefore, the proportion of individuals who are totally susceptible and have no malaria antibodies against *any* strain can be calculated as the independent (and therefore multiplicative) probability of being susceptible to each strain:

$$S^{\text{total}}(a) = \prod_i S^i(a) = \exp\left(-\sum_i (R_0^i - 1)\frac{a}{L}\right).$$ (4.4)

Thus, comparing equations (4.3) and (4.4), we see that the value of R_0 that is derived under the assumption of a single malaria strain is related to the sum of the separate R_0 values when multiple strains co-circulate.

$$R_0^{\text{estimated}} = 1 + \sum_i (R_0^i - 1)$$

Hence, the true value of R_0 for each strain is likely to be greatly reduced compared to standard estimates; Gupta et al. (1994) calculated that R_0 may be as low as 6 or 7. So that, instead of having one infectious disease that is very hard to eradicate, this analysis suggests that we have multiple strains, each of which may be substantially easier to eliminate. A control measure that could reduce the overall transmission rate to 10%–15% of its current value would be predicted to eradicate all strains simultaneously. Clearly such results have a profound impact on our understanding of the dynamics and potential control of malaria.

When there is limited cross-immunity, the individual values of R_0 for each strain are lower than estimated from the seropositive level that ignore strain structure.

4.1.3. Enhanced Susceptibility

We now focus on the situation where co-infection with two or more strains is more likely than pure chance would dictate. The classic example here is sexually transmitted infections, where the presence of one infection can increase the *susceptibility* of the host to other infections (Coggins and Segal 1998). This enhanced susceptibility can lead to some surprising results, as discussed below. In addition to these physiological factors, the risk structuring of the population also increases the level of co-infection as the high-risk core group is exposed to a higher force of infection for many sexually transmitted diseases (Chapter 3).

Sexually transmitted infections usually conform to the SIS paradigm, where after treatment infectious individuals are once again susceptible. There are four differential equations correspond to the two possible states (S and I) and the two infections:

$$\frac{dN_{SS}}{dt} = -\beta_1 N_{SS} I_1 - \beta_2 N_{SS} I_2 + \gamma_1 N_{IS} + \gamma_2 N_{SI} + \gamma_3 N_{II},$$

$$\frac{dN_{IS}}{dt} = \beta_1 N_{SS} I_1 - \gamma_1 N_{IS} - \widehat{\beta_2} N_{IS} I_2,$$

$$\frac{dN_{SI}}{dt} = \beta_2 N_{SS} I_2 - \gamma_2 N_{SI} - \widehat{\beta_1} N_{SI} I_1, \qquad (4.5)$$

$$\frac{dN_{II}}{dt} = \widehat{\beta_1} N_{SI} I_1 + \widehat{\beta_2} N_{IS} I_2 - \gamma_3 N_{II},$$

$$I_1 = N_{IS} + N_{II}, \qquad I_2 = N_{SI} + N_{II}.$$

These equations assume that those individuals with both infections would be treated for both simultaneously (at rate γ_3); we also make the simplifying assumption that infections are passed on independently, such that those with both infections do not necessarily pass both on to each individual they infect. (The converse assumption does not affect the qualitative results discussed below.) We are now particularly interested in the case where being infected with one disease increases the susceptibility to the other, which translates to $\widehat{\beta_1} > \beta_1$ and $\widehat{\beta_2} > \beta_2$.

One interesting feature of such enhanced susceptibility is its effect on the invasion and persistence of the two infections. For either infection to invade a naive, totally susceptible population we require, as usual, that R_0 is greater than one; in particular, for each infection to be able to invade we need:

$$\beta_1 > \gamma_1 \qquad \text{and} \qquad \beta_2 > \gamma_2.$$

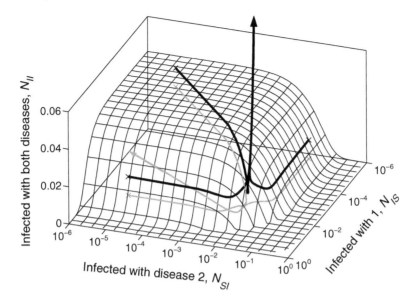

Figure 4.7. Example of six trajectories from the enhanced susceptibility model, equation (4.5), clearly demonstrating the Allee effect. The surface separating persistence from extinction is also shown as a mesh. Gray orbits start just below the surface and lead to extinction, whereas black orbits start just above the surface and tend to a fixed point with a high prevalence of both infections. $(\gamma_1 = \gamma_2 = \gamma_3 = 1, \beta_1 = 0.9, \beta_2 = 0.85, \widehat{\beta}_1 = 8, \widehat{\beta}_2 = 7.)$

At invasion, because prevalence is low and hence co-infection very rare, the terms $\widehat{\beta}_1$, $\widehat{\beta}_2$, and γ_3 do not enter into the invasion criterion. However, once the infections are established and co-infection common, these terms can play a pivotal role in maintenance of both infections. In particular, if β_1 and β_2 are small but $\widehat{\beta}_1$ and $\widehat{\beta}_2$ are large, we can experience what is known in ecology as an Allee effect (Courchamp et al. 1999), whereby the infections cannot invade but may persist once they become established. This occurs when the basic reproductive ratio for the infections is below one so that neither can invade a totally susceptible population; however, given a substantial prevalence of infection 1, the average susceptibility of the population to infection 2 is increased and so infection 2 can persist—and vice versa. Figure 4.7 shows an example of this; trajectories (in black) that start above a critical prevalence tend to an endemic equilibrium, whereas orbits (in gray) that start below tend to zero. The mesh separates the regions of persistence and extinction.

Having one sexually transmitted infection can often increase the susceptibility to others, promoting co-infection. In such circumstances the Allee effect may operate, and reducing the prevalence of one infection may lead to a reduction of the other.

The somewhat extreme example of Figure 4.7, where susceptibility is greatly enhanced by infection, is chosen primarily to illustrate the effects of such nonlinear behavior. In practice the increase in susceptibility is generally far less, although the basic results may still hold. The presence of one sexually transmitted infection may make it difficult to eradicate others compared to a combined control strategy; small changes in sexual practices may cause significant changes in prevalence due to the nonlinear nature of the system and Allee effects may be observed. In conclusion, it may be erroneous to study sexually transmitted infections in isolation and a multi-disease approach may frequently be necessary.

4.1.4. Partial Cross-Immunity

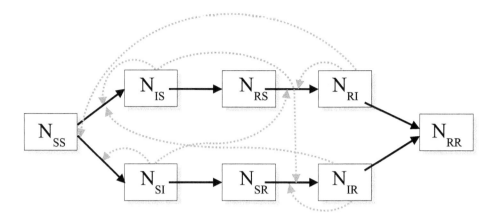

Partial cross-immunity refers to the situation where by having experienced and recovered from one infection, or strain, provides some form of limited protection against other infections, or related strains. This form of immunity is the most commonly studied and most widely applicable formulation of the two-strain model. However, a range of differing assumptions can be made about the nature of cross-immunity. Protection can operate either through reduced susceptibility, reduced transmissibility, or a mixture of the two. There is the additional complication that protection might be conferred only to a faction of those individuals, but for the moment we shall assume a homogeneous response, with all recovered individuals experiencing the same reduction. Following a similar notation to before:

This is online program 4.1

$$\frac{dN_{SS}}{dt} = \nu - \beta_1 N_{SS} I_1 - \beta_2 N_{SS} I_2 - \mu N_{SS},$$

$$\frac{dN_{IS}}{dt} = \beta_1 N_{SS} I_1 - \gamma_1 N_{IS} - \mu N_{IS},$$

$$\frac{dN_{RS}}{dt} = \gamma_1 N_{IS} - \alpha_2 \beta_2 N_{RS} I_2 - \mu N_{RS},$$

$$\frac{dN_{SI}}{dt} = \beta_2 N_{SS} I_2 - \gamma_2 N_{SI} - \mu N_{SI},$$

$$\frac{dN_{RI}}{dt} = \alpha_2 \beta_2 N_{RS} I_2 - \gamma_2 N_{RI} - \mu N_{RI}, \qquad (4.6)$$

$$\frac{dN_{SR}}{dt} = \gamma_1 N_{IS} - \alpha_1 \beta_1 N_{SR} I_1 - \mu N_{RS},$$

$$\frac{dN_{IR}}{dt} = \alpha_1 \beta_1 N_{SR} I_1 - \gamma_1 N_{IR} - \mu N_{IR},$$

$$\frac{dN_{RR}}{dt} = \gamma_1 N_{IR} + \gamma_2 N_{RI} - \mu N_{RR},$$

$$I_1 = N_{IS} + a_1 N_{IR}, \qquad I_2 = N_{SI} + a_2 N_{RI},$$

where α_i is the proportional reduction in the susceptibility to strain i and a_i is the proportional reduction in the transmission of strain i, for individuals who have recovered from the other strain. Clearly, with so many parameters a full description of all possible scenarios is a lengthy undertaking (White et al. 1998); instead we will highlight a few interesting and epidemiologically important points.

Suppose that one strain of the disease has already entered the population and reached equilibrium. The equilibrium fractions of susceptible, infectious, and recovered individuals are those of a single disease in isolation (Chapter 2):

$$N_{SS}^* = \frac{\gamma_1 + \mu}{\beta_1}, \qquad N_{IS}^* = \frac{\mu}{\gamma_1 + \mu} - \frac{\mu}{\beta_1}, \qquad N_{RS}^* = \frac{\gamma_1}{\gamma_1 + \mu} - \frac{\gamma_1}{\beta_1}.$$

Whether or not the second strain can invade can be judged by looking at the growth rate of a small seed of infecteds, seeing whether strain 2 can increase ($\frac{dI_2}{dt} > 0$) when strain 1 is at equilibrium:

$$\frac{dI_2}{dt} = \frac{dN_{SI}}{dt} + a_2\frac{dN_{RI}}{dt} = \beta_2 N_{SS}^* I_2 + a_2\alpha_2\beta_2 N_{RS}^* I_2 - \gamma_2 I_2 - \mu I_2,$$

$$= \beta_2\frac{\gamma_1 + \mu}{\beta_1}I_2 + a_2\alpha_2\beta_2\left(\frac{\gamma_1}{\gamma_1 + \mu} - \frac{\gamma_1}{\beta_1}\right)I_2 - \gamma_2 I_2 - \mu I_2,$$

$$= \beta_2\left[\frac{\gamma_1 + \mu}{\beta_1} - \frac{\gamma_2 + \mu}{\beta_2}\right]I_2 + a_2\alpha_2\left(\frac{\gamma_1}{\gamma_1 + \mu} - \frac{\gamma_1}{\beta_1}\right)I_2.$$

So strain 2 can invade if it has a higher reproductive ratio than strain 1, which implies that $\beta_2(\gamma_1 + \mu) > \beta_1(\gamma_2 + \mu)$; or if the effects of cross-immunity are not particularly strong:

$$a_2\alpha_2 > \frac{\beta_1(\gamma_1 + \mu)\left[\beta_1(\gamma_2 + \mu) - \beta_2(\gamma_1 + \mu)\right]}{\gamma_1(\beta_1 - \gamma_1 - \mu)\beta_2} \approx \frac{R_0^1(R_0^1 - R_0^2)(\gamma_1 + \mu)}{R_0^2(R_0^1 - 1)}. \qquad (4.7)$$

By symmetry, we can obtain a similar set of conditions by allowing strain 2 to reach equilibrium and then seeing whether strain 1 can invade. We assume that the strain with the larger R_0 is labeled strain 1; then if condition (4.7) holds, both strains can invade when the other is at equilibrium and hence coexistence is possible. In fact, more detailed analysis shows that under these assumptions any initial condition (where both strains are present) leads to the two strains coexisting at some equilibrium.

Coexistence of strains is possible when their respective R_0 values are close and cross-immunity is weak.

The inclusion of partial immunity bridges the gap between complete cross-immunity where only one strain persists and no cross-immunity where both strains always coexist. Because the coexistence condition (4.7) contains only the product $a_2\alpha_2$, it is irrelevant for a two-strain model, whether partial cross-immunity acts on transmission (a), susceptibility (α), or a mixture of the two. However, when the competition between more than two strains is considered, differences between reduced transmission and reduced susceptibility do occur. To understand these differences, consider a three-strain model, in a triangular arrangement, where strain 1 confers complete immunity against strain 2, strain 2 confers complete immunity against strain 3, and strain 3 confers immunity against strain 1, with no other interactions. In the reduced susceptibility model, an individual recovered from strain 1 cannot catch strain 2 when challenged, but can later catch strain 3. However, in the reduced transmissibility model, once an individual has recovered from strain 1 he or she can catch (but not transmit) strain 2, which provides immunity against strain 3. This simple conceptual example shows how a detailed understanding of the host's immunological responses is necessary to deal with realistic multi-strain models because the implications of the basic assumptions are profound.

When many cross-reactive strains are circulating within the population, different assumptions about the nature of cross-reactivity can have profound effects on the modeling outcome. Detailed immunological studies are required to clarify the typical behavior.

4.1.4.1. Evolutionary Implications

The partial cross-immunity model can again be considered within an evolutionary setting, by examining competition between strains. In general, most interest has focused on the evolutionary dynamics of influenza, where new strains continually arise and face little herd immunity within the population (Fitch et al. 1997; Earn et al. 2002); models for this type of dynamics therefore must include the interaction of multiple strains (Andreasen et al. 1996; Andreasen et al. 1997; Gomes et al. 2001; Gog and Swinton 2002; Gog and Grenfell 2002; Ferguson et al. 2003a). However, as shown above, this competition does not have to lead to the exclusion of one strain; coexistence is possible. What makes the inclusion of partial cross-immunity so relevant to evolutionary modeling is that when multiple strains are considered, it is plausible to assume that closely related strains have a high level of cross-immunity whereas distantly related strains display little or no cross-immunity. This basic rule has much supporting evidence (de Jong et al. 2000; Earn et al. 2002). Significant jumps in strain structure, due to recombination, may produce strains with high virulence and little population-level immunity; these jumps are epidemiologically important because they may lead to pandemics, as seen 1918 (Gibbs et al. 2001).

The assumption of high cross-immunity for closely related strains has a strong evolutionary implication. The vast majority of new strains, which are genetically and phenotypically close to their parent strain, will face the same level of immunity in the population as their parent and therefore are unlikely to rise to dominance. It is only the rare distant mutation that, facing little conferred immunity, can increase dramatically. Thus, evolution does not lead to the gradual change, but instead progresses by a series of jumps whenever sufficiently distant mutants occur.

The great difficulty with multi-strain models that attempt to track the evolutionary progress of diseases is the rapid proliferation in the number of equations with the number of strains (Gomes et al. 2001). To completely capture the dynamics of n interacting strains within the SIR framework requires 3^n differential equations, or $(n+2)2^{n-1}$ equations if co-infection is ignored. Clearly this sort of exponential growth in model complexity places severe restrictions on the number of strains that can be simulated, even with modern computational technologies. However, substantial simplifications to the model can be made given two reasonable assumptions (Andreasen et al. 1996; Gog and Swinton 2002; Gog and Grenfell 2002). The first is that immunity acts to reduce transmission, not susceptibility; this means that all individuals are equally at risk to any strain irrespective of their epidemic history. Physiologically, this assumption implies that even when an individual has immunity to a particular strain, if challenged by that strain they still become infected and mount a full immune response, but clear the pathogen before transmission can occur. The second assumption is that partial immunity acts heterogeneously, by making a proportion of recovered individuals totally immune, rather than the homogeneous assumption when all recovered individuals have an equally reduced transmission rate. These two assumptions mean that we need to consider only the proportion of individuals "susceptible to" and infectious with each strain. Hence for a model with n strains there are only $2n$ equations:

$$\frac{dS_i}{dt} = \nu - \sum_j \beta_j c_{ij} S_i I_j - \mu S_i,$$

$$\frac{dI_i}{dt} = \beta_i S_i I_i - \gamma_i I_i - \mu I_i,$$

(4.8)

where $0 \leq c_{ij} \leq 1$ provides information on the gain of complete immunity to strain i due to infection (or challenge) with strain j. We would normally insist that $c_{ii} = 1$ so that the single strain dynamics are SIR. S_i refers to the fraction of the population who are not immune and therefore able to transmit strain i. Hence, the susceptibility classes are not mutually exclusive; an individual susceptible to (i.e., able to catch and subsequently transmit) strains 1 and 2 will belong to both S_1 and S_2. Therefore, the population components no longer sum to one.

This change of emphasis, from discrete compartments to overlapping classes, can at first be difficult to grasp. The reasoning behind equation (4.8) can be explained as follows. The I_i equation is relatively straightforward: Only those individuals without immunity to the strain (S_i) can be infectious following transmission. Those individuals who are immune and become infected, but are unable to transmit, do not play a role in the spread of infection and can be ignored. The number of individuals who are susceptible to strain i are not just reduced due to infection with strain i, but other strains can also affect the immunity (captured by c_{ij}). By assuming that *all* individuals can catch *all* strains, we do not need to worry about the immune status of an individual with respect to strain j, just the effect of strain j on the immunity with respect to strain i.

This simple, but powerful, model can now be used to investigate antigenic drift which is evolution driven by the immunity of the population with no discernible change in infection characteristics (so $\beta_i = \beta$ and $\gamma_i = \gamma$). Following Gog and Grenfell (2002) (see also Andreasen et al. 1996), we position strains on a one-dimensional line and assume that immunity is conferred most strongly to nearby strains:

$$c_{ij} = \exp(-A[i-j]^2).$$

It is also presumed that random mutation, at a rate ε, can lead to the spontaneous creation of adjacent strains. This modifies the basic equation:

$$\frac{dI_i}{dt} = \beta S_i I_i - \gamma I_i - \mu I_i - \varepsilon I_i + \frac{1}{2}\varepsilon I_{i+1} + \frac{1}{2}\varepsilon I_{i-1}.$$

(4.9)

Figure 4.8 shows an example of the type of drift dynamics predicted by this model. With 100 strains, a traditional approach that tracks all infection histories would require over 6×10^{31} equations—this is computationally infeasible. Two clear features emerge from this model that mimic the observed behavior of influenza (Fitch et al. 1997; Earn

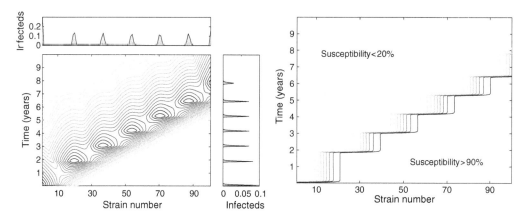

Figure 4.8. Dynamics of a strain-structured model of influenza (equations (4.8) and (4.9)), where partial immunity is conferred to nearby strains and is assumed to act heterogeneously on the transmission rate. The left-hand composite of three graphs show the prevalence of infection within the population. The main graph gives the strain-specific level over time $I_i(t)$ (contours plotted on logarithmic scale, with darkest contours referring to highest prevalences of infecteds); the top graph shows the total prevalence of infection with each strain $\int I_i(t)dt$ over the 10-year simulation; the right graph shows the total infected against time $\sum_i I_i(t)$. The right-hand graph gives the level of "susceptibility" to each strain $S_i(t)$; here the contours are linearly spaced. ($\nu = \mu = 5 \times 10^{-5}$ per day, $1/\gamma = 7$ days, $R_0 = 3$, $A = 0.01$, $\varepsilon = 0.01$ per year.)

et al. 2002). First, there are clear oscillations in the fraction of infecteds (summed over all strains) driven by the strain structure and rate of mutation; in practice, any oscillations in the prevalence of influenza are reinforced by seasonal effects (however, see Dushoff et al. 2004 and Viboud et al. 2006). Second, due to the cross-immunity invoked towards nearby strains, the next epidemic strain must be sufficiently distinct from previous strains; the time taken for these new distant mutants to arise sets the duration between epidemics.

Although such models are a powerful tool for providing insights into the evolutionary dynamics of diseases, the assumed structure of strain space is somewhat naive. A better approximation would be to assume that the space is high-dimensional (Gupta and Maiden 2001), and that, although in general most mutations give rise to nearby strains, some mutations (or recombination events) may cause a departure to more distant regions of strain space where there is little cross-immunity. These large jumps, known as *antigenic shift*, have profound public health implications and are likely to lead to far larger epidemics than normal. Genetic evidence suggests that the 1918 influenza pandemic that killed 10–20 million people was due to the recombination of co-circulating strains (Gibbs et al. 2001). It remains a considerable challenge to determine the precise structure of this strain space and the rate of mutation between various points. The current concerns over H5N1 influenza in poultry emphasizes the need for a detailed genetic understanding of immunity, virulence, and the factors affecting the ability to transmit between humans.

4.1.4.2. Oscillations Driven by Cross-Immunity

Persistent, large amplitude epidemic cycles are generally considered to be a signature of underlying seasonal effects or temporal forcing (Chapter 5), with the majority of unforced models asymptoting to a fixed prevalence. The dynamics of multi-strain models are an exception to this rule. Consider the behavior seen in Figure 4.8. If the strains were not positioned along a line, but the ends were joined together to make a circle, then an everlasting wave of epidemics could propagate around the circle. By the time one complete revolution of the circle has been made, there will have been sufficient births to increase susceptibility to the initial strain.

Slight variants of equations (4.8) and (4.9) can produce self-sustained oscillations with as few as four interacting strains (Andreasen et al. 1997; Gupta et al. 1998; Gomes et al. 2002; Dawes and Gog 2001). Following the work of Gupta et al. (1998), we consider the dynamics of four interacting strains in a circular arrangement where recovery from one strain offers partial protection to neighboring strains (e.g., recovery from strain 1 offers partial protection against strains 2 and 4, recovery from strain 2 offers partial protection to strains 1 and 3, etc.). Using the notation of Gupta et al. (1998), we have:

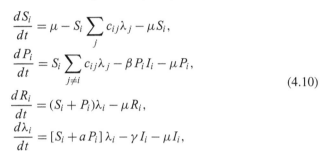

This is online program 4.2

$$\frac{dS_i}{dt} = \mu - S_i \sum_j c_{ij}\lambda_j - \mu S_i,$$

$$\frac{dP_i}{dt} = S_i \sum_{j\neq i} c_{ij}\lambda_j - \beta P_i I_i - \mu P_i,$$

$$\frac{dR_i}{dt} = (S_i + P_i)\lambda_i - \mu R_i,$$

$$\frac{d\lambda_i}{dt} = [S_i + aP_i]\lambda_i - \gamma I_i - \mu I_i,$$

(4.10)

where S_i, P_i, and R_i are those totally susceptible to, partially susceptible to, and fully immune to (either recovered from or infected with) strain i, and λ_i is the force of infection associated with strain i. (c_{ij} is one if i and j are neighboring strains, but zero otherwise.) Compared to the more complete notation given in equation (4.6), the above model assumes that partial immunity decreases an individual's transmissibility ($a < 1$) acting homogeneously, but does not affect its susceptibility ($\alpha = 1$); all strains have identical characteristics. In particular, we can map the terms in equation (4.10) to those in equation (4.6); for strain 1 we have:

$$S_1 = \sum_{Q\in\{S,I,R\}} N_{SSQS},$$

$$P_1 = \sum_{Q\in\{S,I,R\}} \left[\sum_{A\in\{S,I,R\}} \sum_{B\in\{S,I,R\}} N_{SAQB} - N_{SSQS} \right],$$

$$R_1 = \sum_{Q,A,B\in\{S,I,R\}} N_{R,A,Q,B} + N_{I,A,Q,B},$$

$$\lambda_1 = \beta \sum_{Q\in\{S,I,R\}} N_{ISQS} + a\beta \sum_{Q\in\{S,I,R\}} \left[\sum_{A\in\{S,I,R\}} \sum_{B\in\{S,I,R\}} N_{IAQB} - N_{SSQS} \right],$$

so that those individuals totally susceptible to strain 1 must not have encountered strains 1, 2, or 4, but could be in any state with respect to strain 3 because strains 1 and 3 do not interact. However, those who are partially resistant to strain 1 must not have encountered strain 1, but must have encountered either strain 2 or 4. Similar arguments can be made to construct the other terms.

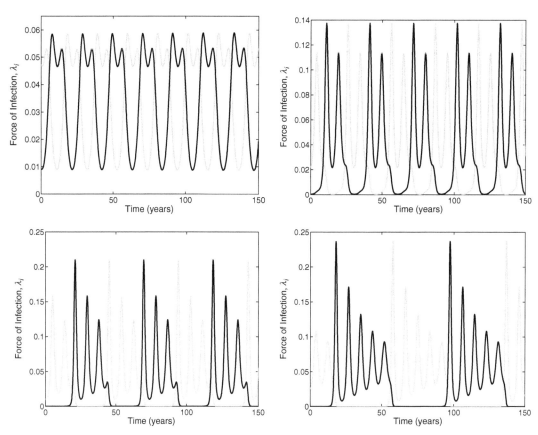

Figure 4.9. The dynamics of strains 1 and 2 of a four-strain system as typified by the force of infection for each strain, λ_i. The level of cross-immunity, a, increases from top left to bottom right ($a = 0.55, 0.6, 0.65, 0.7$). ($\mu = 0.02$ per year, $\gamma = 10$ per year ($1/\gamma = 36.5$ days), $\beta = 40$ per year, hence $R_0 = 4$.)

Figure 4.9 shows that the dynamics of strains 1 and 2 (strains 3 and 4) behave identically as the level of cross-immunity, a, varies. Increasing the level of cross-immunity increases the period of the dynamics and the complexity of the epidemic cycles; values of a less than about 0.53 lead to coexisting equilibrium dynamics where all strains asymptote to the same constant level. It is important to note that sustained regular oscillations are not always the signature of seasonal forcing. When $a = 0.55$ the period is around 21 years, whereas when $a = 0.7$ the period has increased to about 79 years. Although the existence of such cycles is inherently interesting, it is questionable whether such long-term oscillations play any meaningful role in the dynamics of influenza. However, recent work on dengue fever (Wearing and Rohani 2006), respiratory syncytial virus (White et al. 2005), and cholera (Koelle et al. 2005) all show the propensity for multi-strain diseases to exhibit complex cycles. More long-term strain-structured data and a more detailed understanding of the levels of partial immunity are required before the implications of these large-amplitude fluctuations can be practically assessed or their dynamics predicted.

Strain structure and partial cross-immunity between nearby strains can lead to long-period oscillatory dynamics without the need for external forcing.

The interaction between partially cross-immune strains can therefore lead to epidemic cycles in the total infection prevalence (summed across all strains). Two other long-term behaviors of cross-reactive multi-strain models exist, and both are readily achievable using equations (4.8) and (4.9). The first, and most simple, is homogeneous equilibria where all strains asymptote to the equilibrium level of abundance in the population. The second, and more interesting, is heterogeneous equilibria where some strains persist at a much higher prevalence than others; which strains are most abundant depends on the initial conditions. These three scenarios (temporal oscillations, homogeneous abundance, and heterogeneous equilibria) have interesting parallel with Turing patterns (Turing 1952; Murray 1982). Strain-structured models possess local suppression (in terms of local immunity), activation (in terms of susceptibles), and diffusion (in terms of mutation); it is therefore not surprising that Turing-like dynamics can occur in strain space.

Models of strain structure with local immunity and mutation can lead to traveling waves (observed as dynamic oscillations) or large amplitude stationary patterns in strain space, parallelling Turing patterns.

4.1.5. A General Framework

Finally, we present a comprehensive, flexible mathematical framework that incorporates various possible interactions between pathogens (Rohani et al. 2006). The framework is presented for two infectious agents (labeled disease 1 and 2), though extending it to include multiple pathogens is straightforward although lengthy. In developing the model, we envisage a simplified natural history of infection for each disease:

- All newborns are fully susceptible to both infections.
- Upon infection, a susceptible individual enters the exposed (infected but not yet infectious) class, and has a probability of contracting the "competing" infection simultaneously (represented by the cross-immunity parameter ϕ_i, where $i = 1, 2$).
- After the latent period, the individual becomes infectious but is not yet symptomatic and still has a reduced risk (ϕ_i) of becoming co-infected with the other disease.
- Typically, when symptoms appear, the disease is diagnosed and the individual is sent home to convalesce for an average period given by $1/\delta_i$. During convalescence, the competing infection may be contracted, with the transmission rate additionally modulated by the parameter ξ_i, which may represent quarantine or temporary cross-immunity (if less than one) or temporary immuno-suppression (if greater than one).
- Depending upon the disease, host age, and host condition (typically nutritional status), infection may be fatal owing to complications (such as pneumonia and encephalitis, in the case of measles and pertussis). This is represented by per capita infection-induced mortality probabilities ρ_i. It is assumed that mortality occurs at the end of the convalescent period, so that the effects of mortality can be separated from the effects of the infectious and convalescent period. This is a very different assumption to that used in Chapter 2, although it is equivalent under a (complex) change of variables.
- Upon complete recovery, the individual is assumed immune to the infection (disease 1) and reactivates susceptibility to disease 2, if previously not exposed to it. At this stage, we introduce the term α_i to explore the implications of long-lasting immuno-suppression ($\alpha_i > 1$) or cross-immunity ($\alpha_i < 1$) for the susceptibility to disease j following infection with disease i.

The model incorporates a large number of possible mechanisms for interaction among infections. At the immunological level, the parameter α_i represents the long-term cross-immunity or immuno-suppression resulting from becoming infected with infection i. The term ϕ_i represents the extent of cross-immunity to infection j while individuals are experiencing infection i. At the ecological level, the convalescent class and possible subsequent death following infection give rise to competition among infections for susceptibles. Hence, depending on the system of interest, the model can be adapted accordingly.

Mathematically, these assumptions lead to the following set of ordinary differential equations:

This is online program 4.3

$$\frac{dS}{dt} = \nu N - (\lambda_1 + \lambda_2)S - \mu S,$$

$$\frac{dE_1}{dt} = \lambda_1 S - \phi_2 \lambda_2 E_1 - (\sigma_1 + \mu)E_1,$$

$$\frac{dE_2}{dt} = \lambda_2 S - \phi_1 \lambda_1 E_2 - (\sigma_2 + \mu)E_2,$$

$$\frac{dI_1}{dt} = \sigma_1 E_1 - \phi_2 \lambda_2 I_1 - (\gamma_1 + \mu)I_1,$$

$$\frac{dI_2}{dt} = \sigma_2 E_2 - \phi_1 \lambda_1 I_2 - (\gamma_2 + \mu)I_2,$$

$$\frac{dC_1}{dt} = \gamma_1 I_1 - \xi_2 \phi_2 \lambda_2 C_1 - (\delta_1 + \mu)C_1,$$

$$\frac{dC_2}{dt} = \gamma_2 I_2 - \xi_1 \phi_1 \lambda_1 C_2 - (\delta_2 + \mu)C_2,$$

$$\frac{dR_1}{dt} = (1 - \rho_1)\delta_1 C_1 - \alpha_2 \lambda_2 R_1 - \mu R_1,$$

$$\frac{dR_2}{dt} = (1 - \rho_2)\delta_2 C_2 - \alpha_1 \lambda_1 R_2 - \mu R_2,$$

$$\frac{dR_{12}}{dt} = (1 - \rho_1)(1 - \rho_2)(\lambda_2 \phi_2 E_1 + \phi_2 I_1 + \xi_2 \phi_2 C_1 + \lambda_1 \phi_1 E_2 + \phi_1 I_2 + \xi_1 \phi_1 C_2),$$
$$+ (1 - \psi_2 \rho_2)\alpha_2 \lambda_2 R_1 + (1 - \psi_1 \rho_1)\alpha_1 \lambda_1 R_2 - \mu R_{12},$$

$$\frac{d\epsilon_1}{dt} = \lambda_1 S + \phi_1 \lambda_1 E_2 + \phi_1 \lambda_1 I_2 + \xi_1 \phi_1 \lambda_1 C_2 + \alpha_1 \lambda_1 R_2 - (\sigma_1 + \mu)\epsilon_1,$$

$$\frac{d\epsilon_2}{dt} = \lambda_2 S + \phi_2 \lambda_2 E_1 + \phi_2 \lambda_2 I_1 + \xi_2 \phi_2 \lambda_2 C_1 + \alpha_2 \lambda_2 R_1 - (\sigma_2 + \mu)\epsilon_2,$$

$$\frac{d\lambda_1}{dt} = \beta_1 \sigma_1 \epsilon_1 - (\gamma_1 + \mu)\lambda_1,$$

$$\frac{d\lambda_2}{dt} = \beta_2 \sigma_2 \epsilon_2 - (\gamma_2 + \mu)\lambda_2 .$$

This model represents an example of *history*-based formulation, rather than *status*-based. The variables require some explanation. All those susceptible to both infections are denoted by S. The variables E_i, I_i, and C_i ($i = 1, 2$) represent those currently exposed, infectious, or convalescing (respectively) after infection with disease i, with no previous exposure to the other pathogen. The term R_i ($i = 1, 2$) represents all individuals who have previously experienced infection i and as a result are now only (partially) susceptible to infection j ($j \neq i$). For bookkeeping purposes, we let ϵ_i and λ_i/β_i represent individuals latent and infectious with disease i ($i = 1, 2$), irrespective of their status for the other disease. Additionally, R_{12} are all those no longer susceptible to either infection, and may include those who are still exposed or infectious with one or both diseases and expected to fully recover. Thus, in terms of the parameters used earlier: $S = N_{SS}$, $E_1 = N_{ES}$, $I_1 = N_{IS}$, and so forth, $E_2 = N_{SE}$, and so forth, $\epsilon_1 = \sum_Q N_{EQ}$, $\epsilon_2 = \sum_Q N_{QE}$, $\lambda_1 = \beta_1 \sum_Q N_{IQ}$, $\lambda_2 = \beta_2 \sum_Q N_{QI}$. However, this new notation has the distinct advantage that the parameters ϵ_i and λ_i provide a useful shorthand. The total population size (N) is the sum of the first ten variables only ($N = S + \sum_{i=1}^{2}(E_i + I_i + C_i + R_i) + R_{12}$). The full derivation of this model is presented in detail by Vasco et al. (2007). The model's parameters are explained in Table 4.1.

TABLE 4.1.
Description of model parameters. Subscripts refer to disease i ($i = 1, 2$).

Parameter	Epidemiological Description	Typical Range
ν	*Per capita* birth rate	0.01–0.5 per year
μ	*Per capita* death rate	0.01–0.5 per year
$1/\sigma_i$	Latent period	1–2 weeks
$1/\gamma_i$	Infectious period	1–3 weeks
$1/\delta_i$	Quarantine period	1–4 weeks
ρ_i	Probability of infection-induced mortality	0–1
ϕ_i	Co-infection probability	0–1
ξ_i	Temporary immuno-suppression/cross-immunity	≥ 0
α_i	Permanent immuno-suppression/cross-immunity	≥ 0
ψ_i	Differential infection-induced mortality	0–1

One intuitively obvious possible consequence of interaction among infections is reduced abundance. Surprisingly, however, detailed analyses have demonstrated that disease interference does not manifest itself by significantly altering infection prevalence; changes in model parameters such as the convalescence period translate into negligible changes in the number of infectives of either infection (Huang and Rohani 2005). Perhaps more surprisingly, epidemiological interference exerts little influence on the coexistence likelihood of pathogens. Defining the basic reproductive ratio of each infection as $R_0^j = \beta_j \sigma_j / (\sigma_j + \mu)(\gamma_j + \mu)$ ($j = 1, 2$), it is straightforward to show that coexistence requires $R_0^j > 1$ and

$$R_0^j > \frac{R_0^i}{1 + a_i(R_0^i - 1)}, \tag{4.11}$$

where

$$a_i = \frac{1}{\sigma_i + \mu}\left\{\phi\mu + \frac{\sigma_i}{\gamma_i + \mu}\left(\phi\mu + \frac{\gamma_i}{\delta_i + \mu}(\xi\phi\mu + \alpha(1 - \rho_i)\delta_i)\right)\right\}, \tag{4.12}$$

where $i, j = 1, 2$, $j \neq i$, and the diseases are assumed to have symmetric values of ϕ, α, and ξ (such that, for example, $\phi_1 = \phi_2 = \phi$; details provided in Vasco et al. 2007).

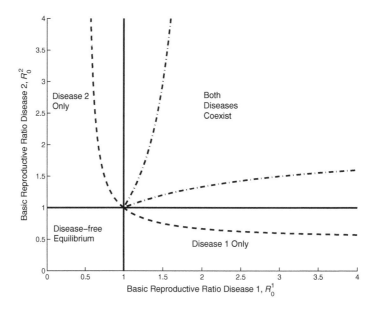

Figure 4.10. The figure demonstrates that coexistence of the two infections can be affected by immuno-suppression and disease interference. In the absence of immune-mediated interactions ($\phi_i = \alpha_i = 1$, $i = 1, 2$), large levels of disease-induced mortality ($\rho_i = 50\%$: dot-dashed line), can cause the region of two-disease coexistence to shrink somewhat. In contrast, strong levels of permanent immuno-suppression ($\phi_i = 1$, $\alpha_i = 2$, $\rho_i = 0$, $i = 1, 2$: dashed line) can expand the coexistence domain. Taken from Vasco et al. 2007. Model parameters were $\mu = 0.02$, $1/\sigma_1 = 1/\sigma_2 = 8$ days, $1/\gamma_1 = 5$ days, $1/\gamma_2 = 14$ days, $\xi = 1$, $1/\delta_1 = 7$ days, and $1/\delta_2 = 14$ days.

In Figure 4.10, we explore the conditions for disease coexistence in this model. In the absence of pathogen-induced mortality ($\rho_1 = \rho_2 = 0$) and with no long-term immunological interactions ($\alpha = 1$), the lack of coinfection alone has little effect on the stable two-disease equilibrium, with the coexistence criterion effectively reducing to $R_0^1, R_0^2 > 1$. It is only after we assume a 50% (dash-dotted line) probability of death following infection that the region of endemic coexistence of both diseases shrinks slightly. On the other hand, if we ignore ecological factors (such as pathogen virulence), immuno-suppression resulting from one infection can facilitate the invasion and persistence of the competing disease even if the invading infection has R_0 lower than one (dashed line).

4.2. MULTIPLE HOSTS

Although many diseases are host-specific, many others can infect multiple and often highly diverse species (Woolhouse et al. 2001b). The models for these types of disease mirror the risk-structured framework developed in Chapter 3, with species being the important risk factor. However, in contrast to risk-structured models, different species may have very different epidemiological and physiological responses to the same infection. In one species a infection may be short-lived and highly virulent, whereas in another species long-term chronic infection may be the norm; the spread of infection between two such host species is a complex problem that can only be understood with mathematical models. We focus on three distinct scenarios that cover a wide spectrum of infections.

First we consider two (or more) host species and a single disease that can be transmitted both within and between the species. This kind of interaction is especially of interest because, in some cases, the host species do not directly interact. The presence of a shared natural enemy—the infection—gives rise to an indirect or "apparent" competition (Holt 1977). A classic high-profile example of such a system is bovine-tuberculosis, where great attention has been focused on the spread of infection between badgers and cattle. Bovine-tuberculosis also exemplifies the parameterization difficulties that are encountered. In modeling terms, the extent of disease spread from badgers to cattle simply relates to one term in the transmission matrix and yet it is the subject of continual controversy despite many years of research (Krebs 1997; Bourne et al. 2000). Another example of a multi-host-pathogen system is foot-and-mouth virus, which can infect a large variety of livestock species such as cattle, sheep, and pigs, despite their physiological differences. Foot-and-mouth is a major problem for farmers in many areas of the world, and understanding the role that different species play in its transmission and persistence is vital for effective, and often species-specific, control measures.

The second type of multi-host systems are vector-transmitted diseases, such as malaria or dengue fever, which require a secondary "host" to spread infection between primary hosts. For both malaria and dengue fever this secondary host is the female mosquito, which spreads the infecting pathogen as it takes blood-meals from humans or other primary hosts. Vectors are almost always arthropods, and include a range of blood-sucking parasites such as fleas, ticks, lice, and mosquitoes. Unlike the range of infectious diseases considered so far, close contact between infected and susceptible humans is not a requisite of transmission; instead, vectors can spread the infection over a wide range. Infectious vectors can even be carried for thousands of miles in aircraft, promoting public health fears of transporting these vector-born diseases to naive, previously disease-free populations. Models for vector-transmitted diseases follow a similar pattern to standard multi-species models, but the parameterization tends to be simpler due to the absence of within-species transmission; mathematically, the diagonal terms of the transmission matrix are zero. Additionally, the rapid life cycle of the vector compared to epidemic timescales can be used to further simplify the modeling.

Finally we focus on zoonoses; these are infections of animals that can also be transmitted to humans, and therefore form a special class of multi-host model. In general, the animal population is the main reservoir for the infection and human cases are sporadic. Often these infections are also vector-transmitted, so that direct contact between humans and the reservoir species is not required. Examples of zoonoses include such high-profile diseases as bubonic plague, West Nile virus, and Ebola. The existence of an animal reservoir and the sporadic nature of transmission to humans has several implications. Without detailed surveillance of the reservoir species, the pattern of human cases may appear confusing and spontaneous. Additionally, by the time human cases arise and public health agencies are aware of the disease's presence, an epidemic with the reservoir species may be difficult to control. Finally, zoonotic diseases are often (re-)emerging pathogens, so outbreaks are unexpected, and experience of control measures is limited.

In the work that follows, due to the fact that we are dealing with different species whose populations may fluctuate independently, it will be prudent to utilize numbers rather than proportions. Hence we shall work with X for the number of susceptible individuals, rather than the proportion S.

4.2.1. Shared Hosts

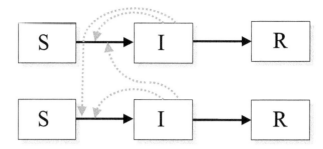

The approach to modeling a single infectious disease that can be transmitted within and between two different species has strong parallels with the work on risk-structured models (Chapter 3) where there is transmission within and between two different risk groups. Thus for two species (A and B), the SIR-type dynamics will be given by

$$\frac{dX_A}{dt} = \nu_A - X_A(\beta_{AA}Y_A + \beta_{AB}Y_B) - \mu_A X_A,$$
$$\frac{dY_A}{dt} = X_A(\beta_{AA}Y_A + \beta_{AB}Y_B) - \gamma_A Y_A - \mu_A Y_A,$$
$$\frac{dX_B}{dt} = \nu_B - X_B(\beta_{BA}Y_A + \beta_{BB}Y_B) - \mu_B X_B,$$
$$\frac{dY_B}{dt} = X_B(\beta_{BA}Y_A + \beta_{BB}Y_B) - \gamma_B Y_B - \mu_B Y_B.$$

(4.13)

The birth rates for the two species, ν_A and ν_B, may include complex density-dependent terms, such that in the absence of infection the population levels tend to a carrying capacity. The transmission term is not divided by the population size because we are dealing with two separate populations and the interaction is likely to depend on the density of the two species—we have assumed pseudo mass-action transmission. The main distinction from risk-structured models of Chapter 3 is that different species are likely to have different responses to infection and thus differing transmission rates and recovery periods, as well as differing demographic parameters. The effect of the different transmission parameters is to break the symmetry that is usually associated with the transmission matrix, β, of risk-structured models. This is because although the mixing between species A and species B is the same as the mixing between B and A, the transmission may be far stronger in one direction because one species may shed more pathogen than the other, or may have different physiological responses to infection.

In multi-host models, the transmission matrix is no longer expected to be symmetric, due to species differences.

From the perspective of wildlife species, it is useful to incorporate well-documented density-dependent population regulation into the modeling framework. This is done by Dobson (2004), who derived analytical expressions for R_0 under alternative assumptions of disease transmission (also see Diekmann et al. 1990). He found that host species diversity has an amplifying effect on outbreaks when transmission is density-dependent (pseudo mass-action; see Chapter 4), as might be the case for pathogens with a free living stage or transmitted by aerosol. On the other hand, for vector-borne diseases, where transmission is frequency-dependent (mass-action), increasing the number of host species has a detrimental effect on pathogen prevalence (see Section 4.2.2.1). Dobson (2004) also demonstrated that pathogen persistence is influenced by the relative strength of between- and within-species transmission, with greater heterospecific transmission leading to greater likelihood of long-term extinction.

4.2.1.1. Application: Transmission of Foot-and-Mouth Disease

The cause and implications of the asymmetry in the transmission matrix can be more readily seen by example. We focus on the spread of foot-and-mouth disease (FMD), which is a highly contagious infection with SIR-type dynamics that is rapidly transmitted between a variety a livestock, especially cattle, sheep, and pigs. Estimates from the 2001 epidemic within the United Kingdom (Keeling et al. 2001b) showed that FMD is transmitted slightly better by cattle than sheep (ratio cattle:sheep = 1.8 : 1) and that cattle are far more susceptible (ratio 15 : 1), although these factors are somewhat offset by the higher numbers of sheep within the United Kingdom (ratio 1 : 4.2). Hence, sheep and cattle respond very differently to this infection, so any attempt at prediction must recognize these differences and model the two species separately. Here, the susceptible and transmission ratios not only incorporate innate differences between the species, but also differences in farming practices and therefore the likelihood of infection being moved on and off a farm. Despite forming the index case, pigs played only a very minor role in the subsequent epidemic and can therefore be ignored.

If we naively assume random mixing between cattle and sheep, then the transmission matrix becomes:

$$\beta_{AB} = b s_A \tau_B \qquad \beta = b \begin{pmatrix} 27 & 15 \\ 1.8 & 1 \end{pmatrix}, \qquad (4.14)$$

where s and τ give the susceptibility and transmissibility for the two species, and the parameter b scales the transmission matrix to obtain the observed growth rate of reported cases such that $R_0 \approx 2.5$ (Woolhouse et al. 2001a). In truth, cattle and sheep are aggregated at both the farm and regional level, but to capture these effects requires detailed spatial models (see Chapter 7; Keeling et al. 2001b). Finally, although there is some speculation that sheep may be infected for slightly longer than cattle, partly due to the difficulty with diagnosing infected sheep, we shall assume that both species have an equal infected period $(1/\gamma)$ of around 11 days. This period is composed of around 4 days of latent period, a further 5 days of infectivity before symptoms emerge, and an average of around 2 days before the animals are slaughtered to prevent further transmission (Ferguson et al. 2001a).

Figure 4.11 shows the predicted epidemic dynamics of the foot-and-mouth model (equation (4.13) with the transmission matrix given by (4.14)), starting with 100 infected sheep and the approximate number of susceptible animals in the Cumbria at the start of the 2001 UK epidemic (number of cattle, $N_c \approx 5.12 \times 10^5$; number of sheep, $N_s \approx 2.64 \times 10^6 \Rightarrow b = 1.38 \times 10^{-8}$). The model results are highly reminiscent of those for risk-structured models where the lower-risk group (sheep) is more prevalent. We notice (inset to top graph) that initially the number of infected sheep decreases because the infection cannot be sustained in the sheep population alone and cattle are needed to maintain the transmission. In the absence of cattle, the basic reproductive ratio of this infection in the sheep population is just 0.4; even with cattle included, each sheep infects

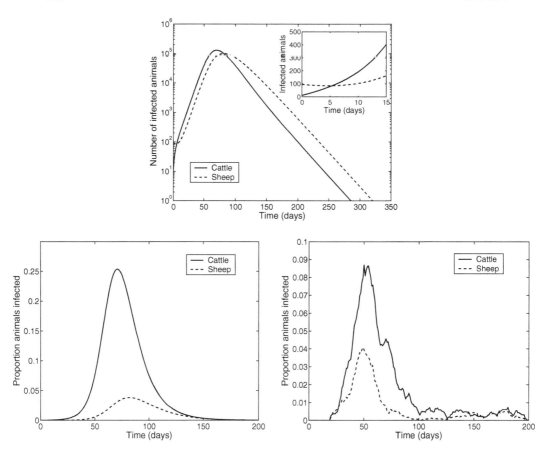

Figure 4.11. Dynamics of a multiple-host model of foot-and-mouth disease, based on the character-istic transmission and susceptibility parameters from the 2001 UK outbreak (Keeling et al. 2001b), and using the animal populations from Cumbria. Despite the much larger number of sheep in the UK, our model predicts that cattle are the main driving force in agreement with more complex simulations. The top graph gives the number of infected cattle (solid line) and sheep (dashed line), respectively, on a logarithmic scale; the inset shows the early epidemic behavior on a linear scale. The bottom graphs show the proportion of animals infected with the virus from the model (left) and the actual reported data from Cumbria in 2001 (right); the Cumbrian data assumes that all animals on a farm are infected. (Birth and deaths are not included in these model results.)

only 1.5 other animals. This strongly indicates that a control measure (such as vaccination) should be primarily focused toward the cattle industry (Tildesley et al. 2006) because without susceptible cattle the disease would soon die out.

The importance of cattle is further illustrated in the lower left-hand graph, showing that throughout much of the epidemic the prevalence of infection in cattle was far higher than it was in sheep; however, toward the end of the epidemic the levels became comparable. The actual data from the 2001 epidemic in Cumbria (lower right-hand graph) supports this basic result, although the difference between cattle and sheep is less pronounced.

Many elements are missing from this simple model, most obviously that livestock do not randomly mix but are aggregated into farms. It may therefore be more appropriate to formulate the "multi-host" model at the farm level, partitioning the population into different farm types, rather than different species.

$$\frac{dX_F}{dt} = v_F - X_F \sum_f \beta_{Ff} Y_f,$$

$$\frac{dY_F}{dt} = X_F \sum_f \beta_{Ff} Y_f - \gamma_F Y_F,$$

where F and f are farm types, X_F and Y_F refer to the number of farms of a particular type, and again $1/\gamma_F \approx 11$ days is the time from infection to slaughter. A plausible partitioning of farms would distinguish between large and small numbers of livestock as well as predominantly sheep, predominantly cattle, or mixed (Ferguson et al. 2001a), leading to six distinct "species" of farms. Although such a high-dimensional model can be parameterized from our knowledge of the 2001 epidemic, a realistic model would also need to account for the complex temporally varying control measures and the intense local spread of infection (Keeling et al. 2001b). Such a data-intensive model is beyond the scope of this book, although see Chapter 8, Box 8.1.

4.2.1.2. Application: Parapoxvirus and the Decline of the Red Squirrel

Since its introduction from America at the start of the twentieth century, the gray squirrel (*Sciurus carolinensis*) has displaced the red squirrel (*S. vulgaris*) from much of its home range in the United Kingdom and mainland Europe (Lloyd 1983; Reynolds 1985). Although gray squirrels have an innate competitive advantage over reds (MacKinnon 1978), this advantage is not sufficient to explain the gray's rapid expansion and the reds' decline (Rushton et al. 1997). The action of a disease, parapoxvirus, has therefore been postulated as a likely cause of red squirrel decline—two-species disease models are therefore needed to understand and predict the likely competitive outcome. This infection has a negligible effect on gray squirrels but causes high mortality in reds, and therefore further decreases the red's competitive ability. Following the work of Tompkins et al. (2003), the following set of equations form a suitable model for the two populations:

$$\frac{dX_G}{dt} = \left[r_G - \frac{N_G + c_R N_R}{K_G} \right] N_G - \mu_G X_G - (\beta_{GG} Y_G + \beta_{GR} Y_R) X_G,$$

$$\frac{dY_G}{dt} = (\beta_{GG} Y_G + \beta_{GR} Y_R) X_G - \gamma_G Y_G - \mu_G Y_G,$$

$$\frac{dZ_G}{dt} = \gamma_G Y_G - \mu_G Z_G, \tag{4.15}$$

$$\frac{dX_R}{dt} = \left[r_R - \frac{N_R + c_G N_G}{K_R} \right] N_R - \mu_R X_R - (\beta_{RG} Y_G + \beta_{RR} Y_R) X_R,$$

$$\frac{dY_R}{dt} = (\beta_{RG} Y_G + \beta_{RR} Y_R) X_R - m_R Y_R - \mu_R Y_R.$$

The first term in both of the susceptible equations leads to a density-dependent birth rate, with the parameters c_G and $c_R = 1/c_G$ measuring the competitive effect of gray squirrels on red and visa versa (Begon et al. 1996). The remaining terms are the familiar

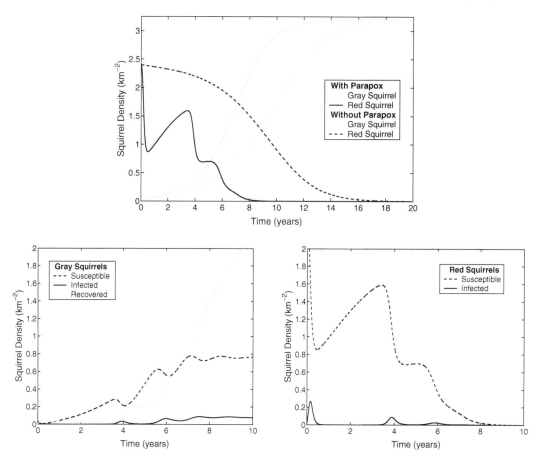

Figure 4.12. Invasion dynamics of gray squirrels into a population of red squirrels. The top graph shows the competitive effects of gray squirrels on the red population in the presence (solid line) and absence (dashed line) of the parapoxvirus. The lower left- and right-hand graphs give the levels of susceptible, infected, and recovered squirrels in the gray and red populations, respectively. The simulations are initialized with the invasion of one gray squirrel (either infected or susceptible) into a 5×5 kilometer area. (Parameters taken from Tomplins et al. (2003) are $r_G = 1.2$, $r_R = 1.0$, $K_G = 4$, $K_R = 4$, $c_R = 0.61$, $c_G = 1.65$, $\mu_G = \mu_R = 0.4$, $\beta_{GG} = \beta_{GR} = \beta_{RR} = \beta_{RG} = 17.5$, $\gamma_G = 13$, $m_R = 26$.)

epidemiological ones, although whereas gray squirrels recover from infection (at rate γ_G), the disease is always fatal to red squirrels (with mortality rate m_R). We have again assumed density-dependent (pseudo mass-action) transmission, in line with the standard paradigm on how directly transmissible wildlife diseases spread.

Figure 4.12 (top graph) shows the competition between red and gray squirrels, following the release of a low number of grays, both when parapox disease is present (solid lines) and absent (dashed lines). With parapoxvirus, during the first year there is a dramatic early decline in the red squirrel population; this is predominantly due to the epidemic dynamics rather than interspecific competition. In general, the addition of this infection into the model leads to the localized extinction of red squirrels within about 8 years, approximately twice as fast as without the infection and more in keeping with field observations (Rushton et al. 1997; Reynolds 1985).

Although the assumption of random mixing between red and gray squirrels can be justifiable at the scale of an individual wood, when contemplating invasion and extinction at a national scale spatial factors play a more dominant role—with invasion moving in a wave-like fashion across the country (Reynolds 1985). A fully predictive model would need to account for such localized movements, as well as the variablity in habitat quality between regions (see Chapter 7). In addition, when dealing with invasions and extinctions—both of which involve low numbers of individuals—a stochastic element to the model becomes vital (see Chapter 6). Despite these shortcomings, this relatively simple model highlights the importance of pathogens that may be introduced along with an invading and competitive species.

4.2.2. Vectored Transmission

Many infections are transmitted via blood-sucking arthropods known as vectors. Malaria, yellow fever, dengue fever, trypanosomiases, and leishmania are all highly prevalent diseases of tropical and subtropical regions that are spread by this mechanism. Malaria alone is responsible for one million deaths and 300 million acute illnesses per year worldwide, making it one of the most devastating of all diseases. From a public health perspective, models may be crucial in determining which strategies or combinations of strategies are likely to be most successful against these devastating infections. Given the vast scale of these diseases, in some of the poorest areas of the world, cost-effective and optimally targeted controls are vital—well parameterized models, informed by good epidemiology and entomology, can play an important role in assessing the likely success of any policy.

In general, vector-borne diseases cannot be passed between primary hosts (person to person or animal to animal) but only through an intermediate insect host or vector. The natural history of vector-borne diseases therefore follows a standard pattern. An insect vector takes a blood-meal from an infected primary host (human or animal); therefore, with a given probability it becomes infected and is soon infectious. When the insect next feeds on a susceptible (and hence different) host, the pathogen enters the host's bloodstream and infection can occur, again with a given probability. In this way the 2×2 transmission matrix has zero on the diagonal elements, with all the transmission operating through the off-diagonal terms.

For vector-borne diseases, because no transmission occurs between humans (or animals) or between vectors, the diagonal elements of the transmission matrix are zero.

As mentioned in Chapter 1, malaria is caused by a single-celled protozoan, generally *Plasmodium falciparum*, but also *Plasmodium vivax*, *Plasmodium malariae*, and *Plasmodium ovale*. The life cycle associated with malaria is more complex than we have considered thus far, largely because there is sexually reproducing phase within humans and an asexually reproducing phase within *Anopheles* mosquitoes. However, despite these complexities, dynamics within humans are sufficiently fast that the SIR modling paradigm represents a reasonable description.

4.2.2.1. Mosquito Vectors

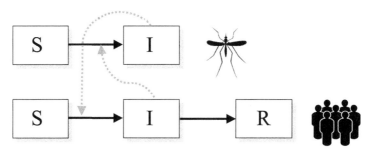

We now consider a simple model for the spread of infection between humans (or other primary hosts) via mosquitoes following the framework founded by MacDonald (1957). This model is also applicable to other vectors, such as tsetse flies, midges, or ticks, that take a single blood-meal from a host and then move on. First, we focus on the distinguishing feature of these models—the rate, r, at which a particular human is bitten by a particular mosquito:

$$r = \frac{b}{N_H},$$

where b is the bite rate of mosquitoes (the number of bites per unit time) and N_H is the number of humans. This formula therefore assumes that each mosquito bites at a constant rate b and that this is shared among all the human hosts within an area. The equations for the disease dynamics (dealing with numbers of individuals) now become:

$$\frac{dX_H}{dt} = \nu_H - r T_{HM} Y_M X_H - \mu_H X_H,$$
$$\frac{dY_H}{dt} = r T_{HM} Y_M X_H - \mu_H Y_H - \gamma_H Y_H,$$
$$\frac{dX_M}{dt} = \nu_M - r T_{MH} Y_H X_M - \mu_M X_M,$$
$$\frac{dY_M}{dt} = r T_{MH} Y_H X_M - \mu_M Y_M,$$

(4.16)

This is online program 4.4

where T_{HM} (≤ 1) is the probability that an infected mosquito biting a susceptible human transmits the infection, with T_{MH} being the probability of transmission in the reverse direction. Although at first this transmission mechanism appears to be density-dependent, the inclusion of the parameter r means that transmission is actually frequency-dependent with respect to the human population. This is because it is assumed that each mosquito bites at a constant rate (irrespective of the number of available humans), whereas the rate at which humans are bitten will increase proportionally to the number (or density) of mosquitoes (Box 4.1). Finally, we note that this formulation assumes that the mosquitoes (or other appropriate vector) can feed only on humans; when other species are part of the menu, a three- or more species model is required, accounting for the different epidemiological parameters associated with each host.

Box 4.1 Minimum Infected Ratio

Measuring the proportion of infected mosquitoes, $I_M = Y_M/N_M$, for wild populations is a difficult and time-consuming task. In general, the fraction of infecteds is low, so testing each individual mosquito would realize a vast number of uninfected mosquitoes for every positive one. Instead, groups of mosquitoes are tested in batches (or pools), thereby increasing the chance that a batch is infected. This leads to the Minimum Infected Ratio (MIR) per thousand mosquitoes, which is defined as:

$$\text{MIR} = 1000\frac{\text{Number of infected batches}}{\text{Total number of mosquitoes tested}}$$

In the ideal scenario, every batch would contain exactly M mosquitoes, which provides a direct link between MIR and the proportion of infected mosquitoes, I_M. Suppose that b batches are tested, then:

$$\text{MIR} = 1000\frac{b \times [\text{proportion of infected batches}]}{b \times M} = 1000\frac{[1 - (1 - I_M)^M]}{M}.$$

Thus, when I_M is small, the Minimum Infected Ratio scales almost linearly with increasing prevalence:

$$\text{MIR} = 1000 I_M - 500(N - 1)I_M^2 + \dots$$

but as I_M and the batch size, M, increase, MIR becomes an underestimate hence—the name *Minimum* Infected Ratio.

The relationship between the actual prevalence of infection in mosquitoes and that estimated by MIR. Clearly, using smaller batch sizes produces more reliable results, but with the disadvantage that more batches need to be tested for the same total number of mosquitoes.

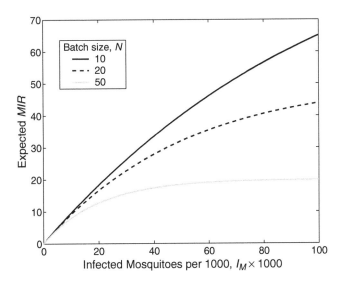

Simulations show that (assuming a plentiful supply of mosquitoes), the number per batch M should be chosen such that $1 - (1 - I_M)^M \approx \frac{1}{2}I_M M$. This ensures that I_M can be estimated from MIR with the greatest accuracy; if M is too large or too small, then either most of the batches will be infected or most will be uninfected, reducing the sensitivity of the results.

The disease transmission dynamics are specified in terms of the bite rate of mosquitoes and the probabilities of transmission following a bite.

In terms of a traditional multi-host model, the transmission matrix is given by:

$$\beta = \begin{pmatrix} 0 & rT_{HM} \\ rT_{MH} & 0 \end{pmatrix}.$$

The set of equations (4.16) is for the number of individuals or the density within a given area (and not the proportion); often we can interchange proportions and numbers at will (Chapter 3), but because human and mosquito populations may fluctuate independently, this interchange is no longer viable.

The mosquito parameters, ν_M and μ_M, are likely to vary with climatic conditions. Thus, in regions of the world where there are pronounced climatic variations, strong seasonal effects may dominate the dynamics (Chapter 5). The extreme case of this is temperate regions, where only a low number of mosquitoes successfully overwinter, and hence transmission may be negligible for a significant fraction of the year.

To get a better understanding of the range of dynamics of these vector-transmitted diseases, we shall calculate the basic reproductive ratio, R_0. This can be done from first principles, which provides a more intuitive understanding of the early dynamics. Let us start with one mosquito that has just become infected, then R_0 is the number of secondary infections in mosquitoes that will be generated. First, we calculate the expected number of infected humans from this primary mosquito assuming all humans are susceptible:

$$\text{infected humans} = \frac{rT_{HM}N_H}{\mu_M} = \frac{bT_{HM}}{\mu_M}.$$

Now we calculate the number of mosquitoes infected by an infectious human:

$$\text{infected mosquitoes} = \frac{rT_{MH}N_M}{\gamma_H + \mu_H} = \frac{bT_{MH}N_M}{(\gamma_H + \mu_H)N_H}.$$

Thus, R_0 is given by the product of these two terms:

$$R_0 = \frac{b^2 T_{HM} T_{MH} N_M}{\mu_M(\gamma_H + \mu_H)N_H}. \tag{4.17}$$

Note that each mosquito could infect less than one human on average, and yet R_0 could still be more than one. This definition of R_0, which is used throughout vector-borne epidemiology, does not correspond exactly with the definition from two-species or risk-structured models (Chapter 3). Although both methods agree at the critical point when $R_0 = 1$, the vector approach is the square of the two-species approach because the vector approach includes the multiplication of two transmission steps.

The ratio of mosquitoes to humans is vital in determining both R_0 and the dynamics of infection. When there are many more humans compared to mosquitoes, sustained transmission may be impossible.

R_0 increases with the number (or density) of mosquitoes, but surprisingly decreases with the number (or density) of humans. This is because when there are many humans (and relatively few mosquitoes), the chance of someone being bitten twice in quick succession—once to catch the infection and once to pass it on before recovery—is very small. Therefore, for the infection to successfully spread and invade, the ratio of mosquitoes to humans has

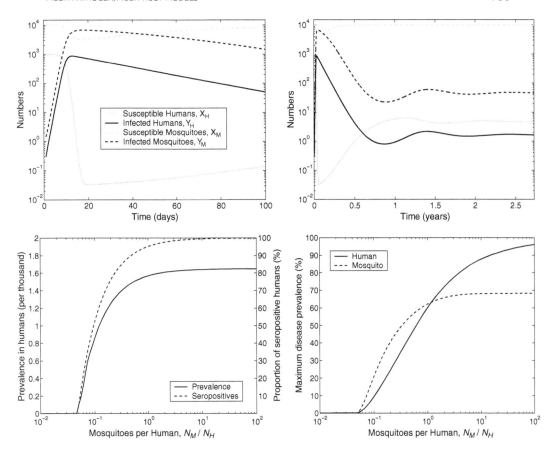

Figure 4.13. The top two graphs show typical dynamics of infected and susceptible humans and mosquitoes against time, starting with a single infected mosquito. The left-hand graph focuses on the early epidemic dynamics, whereas the right-hand graph shows the eventual equilibrium distribution. The lower graphs give the prevalence of infection in humans as the ratio of mosquitoes to humans ($\frac{N_M}{N_H}$) varies. The left-hand graph shows the final equilibrium prevalence in humans (solid line) and the proportion that have recovered from the infection and are seropositive (dashed line). The right-hand graph shows the maximum prevalence of infection in both humans and mosquitoes which occurs during the peak of the initial epidemic. ($N_H = 1000$, $\mu_H = 5.5 \times 10^{-5}$ (50-year human life span), $\nu_H = N_H\mu_H$ (constant human population size), $\mu_M = 0.143$ (1-week mosquito life span), $\nu_M = N_M\mu_M$ (constant mosquito population size), $\gamma_H = 0.033$ (infectious period of one month), $T_{HM} = 0.5$, $T_{MH} = 0.8$, $b = 0.5$. In the top graphs $N_M = 10^4$, which means that $R_0 \approx 200$.)

to be sufficiently large that double bites are common:

$$\text{mosquitoes per human needed for invasion,} \quad \frac{N_M}{N_H} > \frac{\mu_M(\gamma_H + \mu_H)}{b^2 T_{HM} T_{MH}}.$$

Figure 4.13 gives an example of the typical dynamics, starting with a single infectious mosquito and susceptible mosquitoes and humans at equilibrium. Again, as expected from this type of two-species model, the rates of increase during the early epidemic are slaved (Chapter 3). Due to the high value of R_0, a large epidemic occurs on the timescale of a few weeks, after which the prevalence settles toward their equilibrium values, with the vast majority of humans having experienced infection. The lower two graphs consider the dynamics because the ratio of mosquitoes to humans varies. The critical ratio of around 0.048 mosquitoes per human, when $R_0 = 1$, is clear in both graphs as the point where an epidemic is just possible. Above this level, the equilibrium prevalence and proportion of seropositive individuals (left-hand graph) rapidly increase to their asymptotic values. In contrast, the peak human prevalence during the initial epidemic shows much weaker saturation as the ratio of mosquitoes to humans increases. Therefore, although ratios in excess of 1:1 have little impact on the equilibrium prevalence, when faced with the invasion of a infection the precise ratio has a significant impact of the scale of the human epidemic.

In most situations, we expect the number of mosquitoes to far exceed the number of humans (or other hosts). The only notable exception is in subtropical or temperate regions where the number of vectors may be low during the colder winter months and hence disease transmission may be negligible. Such regions often form the boundary to areas where these infections are endemic, and thus understanding their dynamics is crucial if we wish to predict the spread of infection and the epidemiological implications of global warming. Including temporal (climatic) forcing into these vector-based models requires a detailed understanding of the entomology and vector ecology of these species and the models will rely on the techniques developed in Chapter 5. The example of West Nile Virus in Section 4.2.3.2 illustrates how this climatic forcing could be included.

Given that the life cycle of mosquitoes is much faster than both the epidemic and human timescales, each mosquito effectively experiences a constant level of human infection during its lifetime. Mathematically, we can use this fact to produce quasi-equilibrium calculations (Box 4.2) which assume that the mosquito population rapidly converges to an equilibrium state that depends exclusively on the current host population levels.

Due to the rapid life cycle of mosquitoes, a quasi-equilibrium approach can be used wherein mosquito populations are assumed to rapidly converge to equilibrium levels that are functions of the human population.

The quasi-equilibrium solution shows that the force of infection to humans rapidly saturates with increasing levels of human infection. This contrasts with the linear behavior of directly transmitted infections.

The quasi-equilibrium relationship between infection prevalence in mosquitoes and its prevalence in humans provides a deeper understanding of the behavior of vector-borne diseases (Figure 4.14). We observe that the quasi-equilibrium prevalence in mosquitoes begins to saturate with increasing prevalence in the host (solid line). This allows a comparison between the dynamics of vector-born and directly transmitted infections. The dashed line shows the expected linear behavior for a directly transmitted infec-tion with a similar basic reproductive ratio, $R_0 \approx 200$; in contrast, the saturation of the vector-based infection curve, Y_M^*, shows that the force of infection (to humans) saturates. Therefore, relatively less transmission occurs when the prevalence in the host is high, and hence the equilibrium level of seropositives is lower in a vector-borne disease compared to a directly transmitted infection with the same R_0. The converse argument is also true; for a given level of seropositives in the human population, R_0 is larger for a vector-borne infection and the infection is more difficult to erad-icate than results based on directly transmitted pathogens would suggest.

Box 4.2 Fast Vector Dynamics

One difficulty with such vector-based transmission of diseases is that there are double the number of equations compared to standard single-species models: a set for both the host and the vector. However, we can take advantage of the rapid vector life cycle (often 1 to 2 weeks) to simplify the equations. By assuming that the dynamics of the vector are fast compared to those of the host, we can find (quasi-)equilibrium vector abundances for any host population levels by setting the vector rates of change equal to zero:

$$\frac{dX_M}{dt} = \nu_M - rT_{MH}Y_H X_M - \mu_M X_M = 0,$$

$$\frac{dY_M}{dt} = rT_{MH}Y_H X_M - \mu_M Y_M = 0,$$

which implies that

$$X_M^*(X_H, Y_H) = \frac{\nu_M}{rT_{MH}Y_H + \mu_M}, \qquad Y_M^*(X_H, Y_H) = \frac{rT_{MH}\nu_M Y_H}{(rT_{MH}Y_H + \mu_M)\mu_M}.$$

These (quasi-)equilibrium vector population levels, which depend on the current host population, can then be substituted into the host equations to give a smaller, but more complex, set of differential equations:

$$\frac{dX_H}{dt} = \nu_H - T_{HM}b\frac{bT_{MH}\nu_M Y_H}{(bT_{MH}Y_H + \mu_M N_H)\mu_M N_H}X_H - \mu_H X_H,$$

$$\frac{dY_H}{dt} = bT_{HM}\frac{bT_{MH}\nu_M Y_H}{(bT_{MH}Y_H + \mu_M N_H)\mu_M N_H} - \mu_H Y_H - \gamma_H Y_H.$$

These equations give *exact* equilibrium solutions; however, the dynamic approach to equilibrium is an approximation and will be affected by the quasi-equilibrium assumption for the vector.

These results emphasize the crucial point that although the vector-borne infections generally spend the vast majority of their time in the primary (human or animal) host, the role of the vector cannot be simply ignored. The nonlinear behavior due to the obligatory role of the vector in transmission between hosts can have a pronounced effect on our understanding and parameterization of such disease models.

4.2.2.2. Sessile Vectors

Some blood-sucking arthropods, such as fleas and lice, tend to remain with a host for several generations. In such cases, the bite rate of the vector has little relevance to infection transmission because the vector is unlikely to have left the host. A more plausible model is therefore to consider transmission through a pool of free-living infected vectors (Y_V) that are in search of a new host:

$$\frac{dX_H}{dt} = \nu_H - rT_{HV}N_H Y_V \frac{X_H}{N_H} - \mu_H X_H,$$

$$\frac{dY_H}{dt} = rT_{HV}N_H Y_V \frac{X_H}{N_H} - \mu_H Y_H - \gamma_H Y_H - m_H Y_H, \qquad (4.18)$$

$$\frac{dY_V}{dt} = T_{VH}(\mu_H + m_H + l)Y_H K_V - rN_H Y_V - \mu_V Y_V,$$

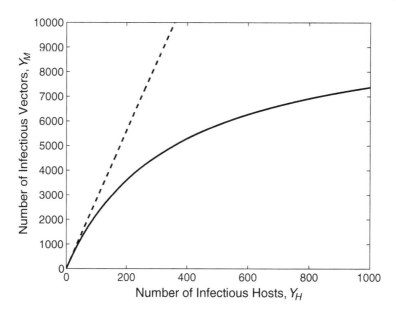

Figure 4.14. The number of infected mosquitoes (solid line) as calculated from the quasi-equilibrium equations. The dashed line shows the linear function that would be needed for the dynamics to approximate host-to-host transmission. All parameters are the same as Figure 4.13, with $N_M = 10,000$.

where the subscripts V and H refer to vector and host. Whenever an infected host dies (either naturally at rate μ_H or due to disease-induced mortality at rate m_H), the vectors leave the dead host in search of another live one at rate r. It is assumed that the vector life cycle is rapid so that each host on average supports a population of K_V vectors, and that a proportion T_{VH} of these are infected. Additionally, vectors may leave a living infected host at rate l, thus increasing transmission. The free-living infected vectors then encounter a new host at rate rN_H, and if this host is susceptible ($\frac{X_H}{N_H}$), they may transmit the infection with the probability T_{HV}. Finally, free-living vectors have a natural death rate, μ_V.

Figure 4.15 gives typical equilibrium-level dynamics for this type of vector. Not surprisingly, as the number of vectors per host increases, so does prevalence in the population (left-hand graph)—this increase in the number of vectors will have a linear effect on the overall transmission. For the parameters used in the model, each host must support more than approximately 25 vectors before transmission is possible. The dynamics with respect to mortality is more complex (right-hand graph). With directly transmitted infections, an increase in the mortality rate, m_H, decreases the infectious period and therefore decreases R_0. However, for this class of vector-borne diseases, the death of an infected host actually releases infected vectors into the environment. Thus as the mortality increases, the number of new cases (gray line), and the number of infected vectors in the environment (dashed lines), also increases.

- For infections spread by ticks, fleas, or lice, a high disease mortality may lead to greater transmission (despite the shorter infectious period) because it increases the rate at which vectors leave the host.

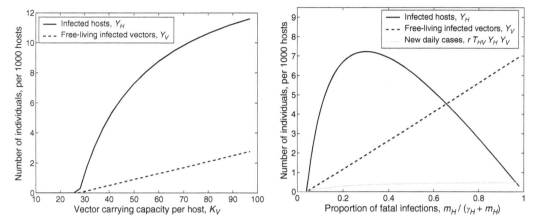

Figure 4.15. The left-hand figure shows the equilibrium number of infected hosts and free-living infected vectors, because the average number of vectors per host, K_V, is varied. The right-hand figure considers the effects of changes in the disease mortality; it also shows the equilibrium number of infected hosts and free-living infected vectors, but additionally gives the number of daily cases. ($N_H = 1000$, $\mu_H = 5.5 \times 10^{-4}$ (5-year host life span), $\nu_H = N_H \mu_H$ (constant host population size), $\mu_V = 0.071$ (2-week life span of free-living vectors without a host), $\gamma_H = 0.033$ (infectious period of one month), $T_{HM} = 0.5$, $T_{MH} = 0.8$, $r = 10^{-3}$, $l = 2.7 \times 10^{-3}$ (once per year). Left-hand graph $m_H = 0$, right-hand graph $K_V = 10$ (an average of 10 vectors per host).)

Again, some understanding is gained by calculating R_0 from the first principles starting with one free-living infected vector:

$$R_0 = \text{Probability that a vector finds a host} \times \text{Probability of infection}$$

$$\times \text{Number of infected vectors released during host's infectious period}$$

$$= \frac{rN_H}{rN_H + \mu_V} \times T_{HV} \times T_{VH} \frac{\mu_H + m_H + l}{\mu_H + m_H + \gamma_H} K_V. \tag{4.19}$$

Although complex, this formula explains the model behavior because K_V and m_H are varied.

4.2.3. Zoonoses

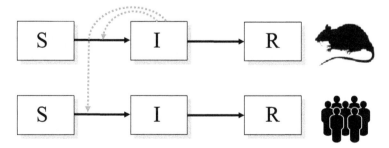

Zoonotic diseases are defined as those that can be passed from animals to humans. In general, the animal host is the main reservoir for the infection, and humans contribute little to the overall transmission. From the disease's perspective, human cases are an irrelevance, but are obviously the focus of public health interest. This dichotomy is also present in the models, where the animal population controls the infection dynamics but the human population determines the disease's impact. In this way, models of directly transmitted zoonoses are much simpler than standard multi-species models, because the 2×2 transmission matrix has zeros in one column (representing the lack of transmission from humans). Models of vector-borne zoonoses follow the standard template for all vector-borne diseases but include the additional transmission to humans. These two distinct classes are dealt with separately.

4.2.3.1. Directly Transmitted Zoonoses

Zoonoses are a ubiquitous challenge to human health, and are associated with a wide range of reservoir species (Acha and Szyfres 1989; Frank and Jeffrey 2001). It has been estimated that three-quarters of emerging human pathogens are zoonotic (Woolhouse 2002), thus a better understanding of their dynamics is likely to play an important role in public health planning. Many prominent zoonoses are associated with reservoirs in household pets (e.g., toxoplasmosis in cats) and livestock (e.g., brucellosis in cattle); this prominence reflects the greater mixing, and therefore transition, between these species and humans, rather than any epidemiological characteristics of the infections. Although the model given below (equation (4.20)) is generic, and could be applied to many directly transmitted zoonoses with appropriate parameterization, several zoonotic diseases deserve special mention due to their epidemiological importance, and scientific and public interest:

- Anthrax is a bacterial infection that affects a wide range of species, especially herbivores. Whereas early modeling focused on the transmission of infection between animals (Furniss and Hahn 1981; Hahn and Furniss 1983), more recent attention has concerned its use as a bioterrorism weapon (Webb and Blaser 2002; Wein et al. 2003), although the risk of subsequent transmission (and therefore R_0) in such situations is very low.

- Brucellosis is a coccobacilli that can be transmitted to humans from cattle, pigs, sheep, and dogs (Corbel 1997). It was once a major public health concern, and although veterinary efforts have dramatically reduced the number of cases in the United States and Europe, there are still several hundred thousand cases per year in humans worldwide. Modeling interest has primarily focused on the effect of the disease on the natural bison population, due to conservation issues, rather than the implications for human health (Peterson et al. 1991; Dobson and Meagher 1996).

- Ebola is one of the most notorious zoonotic diseases, causing a rapid onset hemorrhagic fever and very high mortality. The need for very close contact, and the severity of the symptoms soon after infection, means that outbreaks have been locally isolated. The animal reservoir species for Ebola is still unknown, despite much research.

- Hantavirus is primarily associated with rodent hosts, with each viral type having its only preferred host species. The "Four-Corners" outbreaks in Arizona, Colorado, New Mexico, and Utah in the 1990s have been linked to an infectious reservoir in the deer mouse (*Peromyscus maniculatus*) (Mills et al. 1999). Infection leading to

Hantavirus Pulmonary Syndrome is frequently fatal in humans, unless early treatment is given. Modeling of hantavirus has focused on the role of fluctuating rodent reservoir populations in an attempt to understand the observed large amplitude spatial and temporal variability (Abramson and Kenkre 2002; Abramson et al. 2003; Sauvage et al. 2003; Buceta et al. 2004).

- Rabies has been the focus of much mathematical modeling, due to its public health importance and the long-term spatiotemporal data that has been available. Many species can act as a reservoir for rabies, with foxes (in Europe), raccoons (in United States), and dogs (worldwide) being the primary sources for human infection. Early characteristics of rabies in humans occur 1 to 3 months after infection and are nonspecific and flu-like, with rapid progression to neurological symptoms including anxiety, confusion, slight or partial paralysis, excitation, hallucinations, agitation, hypersalivation, and hydrophobia. Rabies is almost inevitably fatal once symptoms emerge. Modeling efforts can be partitioned into two overlapping groups: Following the lead of Murray and co-workers (Kallen et al. 1985; Murray et al. 1986), many models consider the spatial spread of rabies in a wavelike manner (Moore 1999; Smith et al. 2002), whereas more applied models focus on the impact of specific control mechanisms for either preventing epidemic invasion (Smith and Harris 1991) or reducing the impact where the disease is endemic (Tischendorf et al. 1998; Rhodes et al. 1998; Suppo et al. 2000; Bohrer et al. 2002; Kitala et al. 2002; Smith and Wilkinson 2003). All of these models focus on the infection dynamics within the host reservoir, with little quantitative consideration given to the number of human cases.

- Toxoplasmosis is one of the most well-known zoonotic infections in the developed world. Its natural reservoir is the domestic cat (although sheep and other livestock are often infected), which explains its high prevalence in humans. Generally the symptoms of toxoplasmosis are mild and flu-like, but if caught during pregnancy the effects on the unborn child may be severe (Dubey 1988). Due to its usually benign nature, little quantitative informative is known about the epidemiology and transmission rates of this infection, which limits the amount of predictive modeling that is feasible (Ades and Nokes 1993).

We now consider a very general model for the dynamics of a zoonotic disease in both its animal reservoir and in humans. Assuming SIR-type dynamics in both the human (H) and animal (A) reservoir populations, the equations for directly transmitted zoonoses are:

$$\frac{dX_A}{dt} = \nu_A - \beta X_A Y_A - \mu_A X_A,$$

$$\frac{dY_A}{dt} = \beta X_A Y_A - \mu_A Y_A - \gamma_A Y_A - m_A Y_A,$$

$$\frac{dX_H}{dt} = B_H - \varepsilon \beta X_H Y_A - \mu_H X_H,$$

$$\frac{dY_H}{dt} = \varepsilon \beta X_H Y_A - \mu_H Y_H - \gamma_H Y_H - m_H Y_H,$$

$$(4.20)$$

where ε is generally small and measures the trickle of infection from the animal population into the human one. The birth and death rates for the animal population (ν_A and μ_A) may be quite complex; seasonal factors, density dependence, and stochastic variation may all impact the dynamics. In fact, it is often our lack of quantitative knowledge of the basic ecology of the reservoir species that limits our modeling of the zoonoses. Because all transmission events are assumed to be from infectious animals, we have again adopted a density-dependent approach; this means that large fluctuations in the wildlife population can greatly increase the risk of an epidemic. Outbreaks of hantavirus are thought to arise via such a mechanism.

In a purely deterministic setting, the number of human cases parallels the number of animal cases ($\beta X_A Y_A \propto \varepsilon \beta X_H Y_A$), although there will be far fewer human cases. However, due to the low numbers involved, it is often far better to use a stochastic approach (Chapter 6). As such, $\varepsilon \beta X_H Y_A$ is the probabilistic rate of new cases in humans, and the probability of detecting at least one human case within time-interval t_1 to t_2 is:

$$P(t_1, t_2) = 1 - \exp\left(-D\varepsilon\beta X_H \int_{t_1}^{t_2} Y_A dt\right), \tag{4.21}$$

where D is the probability of successful diagnosis. For many zoonoses, it is difficult (and time consuming) to monitor the infection within the animal population, therefore the onset of human cases is usually the only indicator of a major epidemic within the animal population and therefore an elevated risk to humans. The problem is then a statistical one, determining whether the increase in human cases is merely a statistical fluctuation or the signature of an underlying epidemic.

For zoonotic diseases when human cases are rare, it may be difficult to separate the observation of a few chance human cases and the start of a larger-scale outbreak.

The typical dynamics of a zoonotic infection where humans play a negligible role in transmission are shown in Figure 4.16. The epidemic in the animal population has the characteristic shape that we have come to expect from such simple epidemics, and is unaffected by the behavior of the human population. As seen in the left-hand figure, the chance of observing human cases increases with number of animals infected so far, the relative transmission rate to humans ε, the size of the susceptible human population X_H, and the detection rate D. In particular, the size of the detected human epidemic can be estimated as:

Number suscept. humans $\times$ prob. of infection $= X_H(0)\left[1 - \exp(-\varepsilon D R_0 R_\infty)\right]$

where, using standard notation, R_∞ is the proportion of animals infected. Hence, if R_0 is significantly larger than 1, and therefore R_∞ is close to one, we expect to see human cases if $\varepsilon D R_0 X_H > 1$.

The right-hand graph of Figure 4.16 considers the probabilistic nature of human infection in more detail, assuming that at least one human case is diagnosed and hence the epidemic is identified. The gray line shows the expected number of observed human cases after the initial diagnosis; clearly, this increases rapidly with the scaling factor, $\varepsilon D X_H$, and shows remarkably little stochastic variation. In contrast, the number of infected animals when the first human case is observed shows much more variation and, in general,

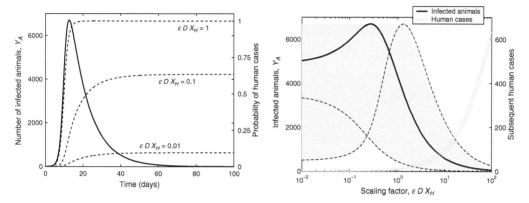

Figure 4.16. The left-hand figure shows a typical animal epidemic (solid line) ($N_A = X_A + Y_A + Z_A = 10^4$, $\gamma_A = 0.1$, $m_A = 0$, $R_0 = 10$, $\mu_A = 2.74 \times 10^{-4}$ (life expectancy of around 10 years), $\nu_A = \mu_A N_A$), and the probability of there being at least one human case detected (dashed line), (scaling factor $\varepsilon D X_H = 0.01, 0.1, 1$). The right-hand graph uses the same animal epidemic, but is conditional on there being at least one human case detected. The black line shows the expected number of infected animals, $Y_A(t)$, when the first human case is detected. The dashed lines correspond to values of Y_A with early and late detection (based on 95% confidence intervals for the timing of the first human case), and the shaded area shows the corresponding range of Y_A. The gray lines give the number of detected human cases (and 95% confidence intervals), assuming no control and $X_H = 1000$.

decreases as the scaling factor increases. Given that there is at least one human case, the first case is expected to occur at time T_1, such that:

$$\frac{P(0, T_1)}{P(0, \infty)} = \frac{1 - \exp\left(-\varepsilon D\beta X_H \int_0^{T_1} Y_A dt\right)}{1 - \exp\left(-\varepsilon D X_H R_0 R_\infty\right)} = \frac{1}{2},$$

where P is defined in equation (4.21) as the probability of identifying at least one case within a given time interval. The solid black line gives the number of infected animals at this time, $Y_A(T_1)$. Similarly, times can be found when the conditional probability is 0.05 and 0.95; the number of infected animals at these times are shown as dashed lines and the range incorporated is shaded.

Two opposing public health implications are associated with the results of this simple model. Although a low scaling factor ($\varepsilon D X_H$) means that few human cases will arise, it also implies that the animal epidemic is likely to be large before cases are detected and therefore difficult to control. Conversely, a high scaling factor should mean that the epidemic is detected far sooner allowing for easier control, but the cost to human health for not controlling the disease is more severe. Intermediate values of the scaling may present the greatest challenge; there is a potential for many human cases so the epidemic must be controlled, but detection is often delayed, making control much more difficult. All these problems become exacerbated if early cases are not quickly diagnosed, as tends to be the case with emerging zoonoses.

When a zoonoses is identified only by rare human cases, an epidemic within the animal hosts can be large before it is discovered. In such cases the epidemic may be difficult to control.

4.2.3.2. *Vector-Borne Zoonoses: West Nile Virus*

There are many infectious diseases that have a primary animal host, but that can be spread to humans via an insect vector. Examples include Chagas' disease (which infects dogs and is spread by the *Triatoma* and *Rhodnius* species of the True Bugs or Heteroptera family), Lyme disease (which infects rodents and dogs and is spread by *Ixodes* ticks), Q fever (which infects birds, rodents, and a range of household pets and is spread by ticks), leishmaniasis (which infects dogs and is spread by mosquitoes), and bubonic plague (which infects rats and other rodents and is spread by fleas). In recent years, West Nile virus (WNV, which infects birds and is spread by mosquitoes) has hit the headlines due to a significant number of deaths in the United States. Here we will concentrate on developing a model of WNV as an illustration of the general methods and complexities involved with understanding vector-borne zoonoses.

West Nile virus (WNV) provides an encompassing example of all that has been discussed in this chapter. It is a vector-borne zoonoses that has multiple host reservoirs, and during the 1990s a new strain (Lineage 1) emerged that has been associated with increased virulence and a range expansion. These elements make WNV a major health concern (especially in the United States but increasing in Europe), negate standard epidemiological rules-of-thumb which are based on experience from directly transmitted single-species pathogens, and make the formulation and parameterization of a detailed model extremely complex.

West Nile virus was first identified in 1937 in the West Nile region of Uganda, hence its name. It is a flavivirus commonly found in Africa, West Asia, and the Middle East. The natural host reservoir for WNV is birds (of many different species), with infection vectored by mosquitoes; occasionally an infected mosquito will bite a human, leading to infection. For the vast majority of human cases, symptoms are mild and flu-like with most individuals not even realizing that they have been infected. However, in a small proportion of cases, the infected person can develop meningoencephalitis, which can be fatal.

West Nile virus made international headlines in 1999 when it was responsible for a number of deaths in New York state, echoing an increasing trend for severe human cases and a high rate of avian mortality (Hubalek and Halouzka 1999; Petersen and Roehrig 2001; Campbell et al. 2002). In subsequent years this infection has spread to cover the majority of the United States, has invaded Canada and the Caribbean, and the death toll has continued to rise (see Figure 4.17). In the United States in 2002 there were, 3,873 clinical cases and 246 deaths, and data from New York City suggests that around 80% of the cases are subclinical (asymptomatic). Despite the shocking number of severe cases and fatalities, the actual incidence in the human population is very low. Levels of seroprevalence in Queens (New York) after the 1999 outbreak were estimated at only 3%. This is in direct contrast to the data from areas of Africa where the infection is endemic (and probably Lineage 2), where seroprevalence levels are about 50% in children and 90% in adults.

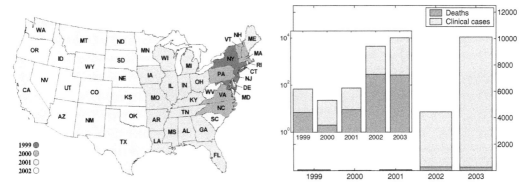

Figure 4.17. Data from the spread of West Nile Virus in the USA from 1999 to 2003. From its initial focus in New York State, the left-hand graph shows the dramatic range expansion that has occurred as the invading wave spreads west. The right-hand graph gives the number of reported clinical cases (light gray) and deaths (dark gray) due to WNV over the same period; the inset graph is plotted on a logarithmic scale to improve the clarity of the early data.

Modeling of West Nile virus within the United States (and its possible spread to other areas) is complicated by a variety of factors:

1. WNV has been found in 138 bird species within the United States, with susceptibility, transmissibility, and infectious period varying between species. House sparrows (*Passer domesticus*) may be a major reservoir due to their long infectious period and high-level of exposure to the virus—up to 60% (Komar et al. 2001). In contrast, the American crow (*Corvus brachyrhynchos*) is considered a sentinel species, due to its high level of mortality—crows comprised over 70% all the dead antibody positive birds reported. Sentinel species may be pivotal in providing an early warning of increasing incidence within the bird population (Eidson et al. 2001).
2. Multiple mosquito vectors may be responsible for transmission. Some, such as *Culex restuans*, feed predominantly on birds (ornithophilic) and therefore are responsible for amplification of the infection within the bird population but cause few human cases. In contrast, more opportunistic mosquito species (such as *Cx pipiens*) that feed on both birds and mammals may generate more human cases.
3. The temperate climate in the United States means that mosquito (and bird) populations fluctuate throughout the year—introducing temporal forcing into the model. The persistence of WNV from one year to the next relies on the successful overwintering of infected mosquitoes.

We have a very limited quantitative knowledge of the basic ecology of the species concerned and the epidemiological parameters and characteristics of their infection. However, a plausible attempt can be made at defining the basic structure of a full model for West Nile virus, after which simplifications can be made that will allow us to parameterize and simulate its dynamics. There are four basic components to the full dynamics of West Nile virus: birds, ornithophilic mosquitoes, opportunistic mosquitoes, and humans, with each being subdivided into a number of species. We start with the transmission matrix β, which can be partitioned into a number of nonzero components:

$$\beta = \begin{pmatrix} 0 & \text{to birds from ornithophilic mosquitoes} & \text{to birds from opportunistic mosquitoes} & 0 \\ \text{to ornithophilic mosquitoes from birds} & 0 & 0 & 0 \\ \text{to opportunistic mosquitoes from birds} & 0 & 0 & 0 \\ 0 & 0 & \text{to humans from opportunistic mosquitoes} & 0 \end{pmatrix}.$$

Given the 29 mosquito species and 138 bird species that are known to have been infected with WNV within North America, the transmission matrix is 168×168 with over 8,000 parameters to be estimated. This is clearly impractical; we therefore focus on a much reduced model, which contains five basic elements: house sparrows (as the reservoir bird species), crows (as a sentinel bird species), ornithophilic and generalist (opportunistic) mosquitoes, and finally humans. With five interacting groups, parameterization will still be difficult, although some progress can now be made. The formulation of the equations follows the same mechanisms as elsewhere in this chapter, just with a greater number of components. We subdivide the populations into X, W, Y, and Z corresponding to susceptible, exposed, infectious, and recovered, and use the subscripts S, C, O, G, and H, to refer to sparrows, crows, ornithophilic mosquitoes, generalist mosquitoes, and humans:

$$\frac{dX_b}{dt} = \nu_b - (r_O T_{bO} Y_O + r_G T_{bG} Y_G) X_b - \mu_b X_b,$$

$$\frac{dW_b}{dt} = (r_O T_{bO} Y_O + r_G T_{bG} Y_G) X_b - \sigma_b W_b - \mu_b W_b,$$

$$\frac{dY_b}{dt} = \sigma_b W_b - \gamma_b Y_b - m_b Y_b - \mu_b Y_b,$$

$$\frac{dZ_b}{dt} = \gamma_b Y_b - \mu_b Z_b \qquad \text{where } b \in \{S, C\},$$

$$\frac{dX_m}{dt} = \nu_m - (r_m T_{mS} Y_S + r_m T_{mC} Y_C) X_m - \mu_m X_m,$$

$$\frac{dW_m}{dt} = (r_m T_{mS} Y_S + r_m T_{mC} Y_C) X_m - \sigma_m W_m - \mu_m W_m,$$

$$\frac{dY_m}{dt} = \sigma_m W_m - \mu_m Y_m \qquad \text{where } m \in \{O, G\},$$

$$\frac{dX_H}{dt} = \nu_H - r_G T_{HG} Y_G X_H - \mu_H X_H,$$

$$\frac{dW_H}{dt} = r_G T_{HG} Y_G X_H - \sigma_H W_H - \mu_H W_H,$$

$$\frac{dY_H}{dt} = \sigma_H W_H - \gamma_H Y_H - m_H Y_H - \mu_H Y_H, \tag{4.22}$$

$$\frac{dZ_H}{dt} = g_H Y_H - \mu_H Z_H,$$

$$r_O = \frac{b_O}{N_S + N_C} \qquad r_G = \frac{b_G}{N_S + N_C + N_H}.$$

Here we explicitly assume that sparrows always recover from infection ($m_S = 0$), crows always die of the disease ($\gamma_C = 0$), mosquitoes can catch WNV only from birds, and humans can catch WNV only from generalist (opportunistic) mosquitoes. As with many wildlife diseases and vector-borne infections, the birth and death rate of the birds and mosquitoes may well be seasonal and density dependent.

The results of the model for West Nile virus given by equation (4.22) are shown in Figure 4.18. Although this model is much reduced in complexity from one that includes all species and all possible interactions, many of the parameters are largely a matter of speculation, and a rich variety of dynamics are possible. The parameters we have chosen mean that after the initial epidemic the level of seroprevalence in the sparrow population is around 50%, which is in agreement with observations (Komar et al. 2001). We also find from this model that the peak numbers of infectious generalist mosquitoes occur in late August early September, which is slighter later than the peak in mosquito numbers and agrees with the times when humans are most at risk.

This model demonstrates that using the methodology developed within this chapter, we can readily create models for a large number of interacting species. The primary difficulty comes from parameterizing such models, because data on the individual constituent mechanisms is difficult and time consuming to obtain—our knowledge of the basic ecology of both birds and mosquitoes is still too poor to allow a detailed parameterization of this model. However, such models can still be used to consider a variety of control measures (such as the use of insecticides) in order to limit the disease dynamics, with the ultimate aim of minimizing human cases. However, great care must be taken to ensure that the results are robust to the uncertainties in parameter values. To fully predict the complete behavior of West Nile virus however, would necessitate a model that can capture the heterogeneities at a variety of scales, from the local patchiness of mosquito breeding grounds to the national-scale spread of infection. Such spatial models are explored in Chapter 7.

4.3. FUTURE DIRECTIONS

In the coming decades it is likely that far more genetic, molecular, and immunological data will become available. The challenge will be to integrate this knowledge with the evolutionary disease models developed earlier in this chapter. Currently, models of disease evolution are in their infancy—far more work is required to integrate the type of models outlined in this chapter with the novel immunological models that are currently being developed (Nowak and May 2005). A comprehensive understanding of disease evolution would

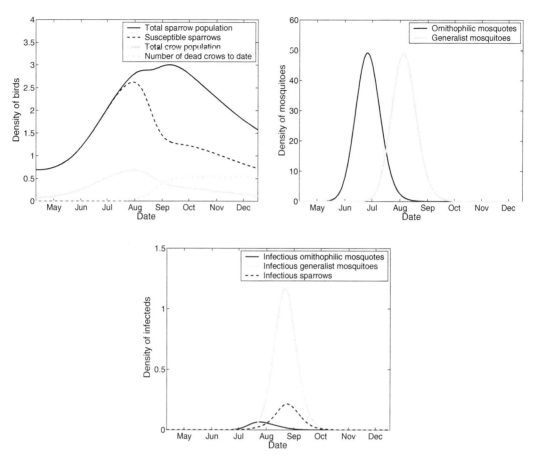

Figure 4.18. Model results for the spread of West Nile virus between species, using the formulation given in equation (4.22). Some parameters are easily found from the literature, $1/\mu_G \approx 1/\mu_O \approx 7$ days, $T_{SO} \approx T_{CO} \approx T_{SG} \approx T_{CG} \approx 0.4$ (Goddard et al. 2002), $T_{OS} \approx T_{OC} \approx T_{GS} \approx T_{GC} \approx 0.5$ (range 0.3–0.8), $1/\sigma_S \approx 1/\sigma_C \approx 1/\gamma_S \approx 1/m_C \approx 3.5$ days (Komar et al. 2001), $1/\sigma_H \approx 1/\gamma_H \approx 5$ days, $m_H \approx 1.3 \times 10^{-3}$ (Campbell et al. 2002), $1/b_O \approx 1/b_G \approx 3$ days. Other parameters can be found by matching to the observed proportion of seropositive birds, the minimum infectious ratio of mosquitoes, and the number of human cases. In addition, the number of bird and mosquito births are assumed to have a Gaussian distribution $\nu_Q = \hat{\nu}_Q \exp(-\frac{1}{2}(t - t_Q)^2/V_Q)$, where $Q \in \{S, C, O, G\}$. ($1/\hat{\nu}_S = 20$ days, $1/\hat{\nu}_C = 60$ days, $1/\hat{\nu}_O = 1/\hat{\nu}_G = 0.127$ days, $t_S = t_C = 190$, $t_O = 170, t_G = 210, V_S = V_C = 1800, V_O = V_G = 150$.)

allow us predict the short-term behavior of influenza, including the probable strains for the next season as well as the likelihood of a pandemic. In addition, a more complete knowledge of viral and bacterial genetics may allow us to predict with greater accuracy methods of preventing the evolution of drug-resistant strains and their spread through the population.

A second area where substantial advances are required is in the parameterization of multi-species models, where the number of parameters usually grows quadratically with the number of species. However, it is often our understanding of the basic ecology of the host species that is lacking, and only detailed field work can resolve many of the issues.

From a modeling perspective, it is important that we ascertain how sensitive models are to these unknown ecological factors, so that field work can be directed toward the key factors that can shape an epidemic.

4.4. SUMMARY

This chapter addressed two contemporary but very different issues, the competition and evolution of infections/strains in a single host population and the spread of a single infection between multiple host species. Both of these modeling issues have important applied implications to public health. Understanding disease evolution would enable us to predict and prepare for future epidemics. The study of infections in multiple hosts has a more immediate impact, because it concerns a range of high-profile diseases (such as malaria) that are responsible for millions of deaths worldwide every year.

The main findings about competing and evolving infections can be summarized as follows:

➤ When competing strains provide complete protection for each other, the strain with the largest R_0 will force the other strain to extinction, although a rapid life cycle may allow short-term dominance.

➤ Evolution will favor mutants with higher R_0, leading to higher transmission, life-long infections with low mortality. However, trade-offs between transmission rates and duration of infection mean that R_0 is maximized for intermediate values and runaway evolution is prevented.

➤ Application of antibiotic treatments requires a careful balance between combating infection and not providing suitable conditions for resistant mutants to outcompete the wild type.

➤ Even when there is no cross-immunity, the absence of multiply infected individuals is epidemiologically plausible, reflecting the reduced number of contacts when ill. This is believed to be why cases for measles and whooping cough are often out of phase.

➤ Research into malaria strains shows that when there is limited cross-immunity, the individual values of R_0 for each strain are lower than estimated from seropositive levels that ignore strain structure, reducing R_0 for each malaria strain to as low as 6 or 7.

➤ Having one sexually transmitted infection can often increase the susceptibility to others, promoting coinfection. In such circumstances the Allee effect may operate, and reducing the levels of one infection may lead to a reduction of the other.

➤ Coexistence of competing strains is possible when their respective R_0 values are close and the level of cross-immunity is weak.

➤ Models of strain structure with immunity to genetically close strains and mutations can lead to both traveling waves or large amplitude patterns in strain space.

Multiple-host models have much in common with the risk-structured models of Chapter 3, with species playing the role of risk. A number of general issues of importance are:

➤ In multi-host models (unlike risk-structured models), the transmission matrix is nolonger expected to be symmetric due to species differences. However, we still expect to see very early dynamics determined by the initial conditions before the behavior of infection in all the hosts becomes slaved, increasing with an exponent determined by R_0.

➤ For vector-borne diseases, such as malaria, because there is no transmission between humans (or animals) and no transmission between vectors, the diagonal elements of the transmission matrix are zero—which dramatically simplifies the calculation of R_0.

➤ The ratio of mosquitoes to humans is vital in determining both R_0 and pathogen dynamics. When there are many more humans compared to mosquitoes, sustained transmission may be impossible because humans rarely experience two bites—one to infect the human and one to infect subsequent mosquitoes.

➤ Due to the rapid life cycle of mosquitoes, a quasi-equilibrium approach can be used where mosquito populations are assumed to rapidly converge to equilibrium levels that are functions of the human population. The quasi-equilibrium solution shows that the force of infection to humans rapidly saturates with increasing levels of human infection. This contrasts with the linear behavior of directly transmitted infections.

➤ For infections spread by ticks, fleas, or lice, a high disease mortality may lead to greater transmission (despite the shorter infectious period) because it increases the rate at which vectors leave the host.

➤ For zoonotic diseases (those spread from animals to humans) when human cases are rare, it may be difficult to separate the observation of a few chance human cases and the start of a larger scale outbreak. When the zoonoses is identified only by rare human cases, an epidemic within the animal hosts can be large before it is discovered, making the epidemic difficult to control.

➤ For zoonotic diseases such as West Nile virus, the vast number of animal hosts and mosquito vectors makes parameterization of even the simplest model very difficult—a greater understanding of host and vector ecology is needed.

Chapter Five

Temporally Forced Models

In this chapter, we consider how seasonally varying parameters act as a forcing mechanism and examine their dynamical consequences. For the most part, we will use measles as a prototypical directly transmitted infectious disease. We demonstrate how such temporally forced models allow us to better capture the observed pattern of recurrent epidemics in contrast to unforced models, which predict oscillations that are damped toward equilibrium (see Chapter 2). We will follow the historical progress of work in this field, because it provides a natural progression from simple models to their more complex and realistic refinements.

5.1. HISTORICAL BACKGROUND

Understanding the mechanisms that generate periodic outbreaks of childhood infectious diseases had been the subject of much debate among Victorian epidemiologists (e.g., Farr 1840; Ransome 1880, 1882; Hamer 1897). In 1880, for example, Arthur Ransome systematically considered numerous "plausible" mechanisms that may generate regular epidemics of measles, whooping cough, and smallpox. Having dismissed factors including meteorological elements (for example, sunspots), isoclinal magnetic lines, the "age-theory" of disease (where only specific ages may be prone to infection), reduced virulence following successive transmission events, Ransome settled on changes in the density of susceptibles as the most likely explanation. He argued that exanthematous diseases wipe out nearly all susceptibles and, as a consequence, "must necessarily wait a number of years before the requisite nearness of susceptible individuals has been again secured." This is essentially a verbal version of the threshold theorem, which, as discussed in Chapter 2, gives rise to damped oscillations. The next important conceptual breakthrough came as a result of classic work by H. E. Soper in 1929. He noticed that in a large population, case report data for measles, which conforms well to the assumptions of the SIR model, show large amplitude recurrent epidemics with very dramatic peaks and troughs. This is in direct contrast to the equilibrium dynamics predicted by simple models, with a steady incidence of disease (Chapter 2). This pattern of pronounced fluctuations in incidence has, since then, been documented for a number of other human infectious diseases such as chickenpox, whooping cough, mumps, and rubella (see, for example, Figure 5.1). When data and models disagree in such a stark manner, there is invariably an important opportunity to re-examine the key assumptions of the model and explore ways in which it can be made more realistic. Focusing on the monthly case reports for measles in Glasgow from 1905–1916, Soper (1929) proceeded to estimate relative transmission rates per month.

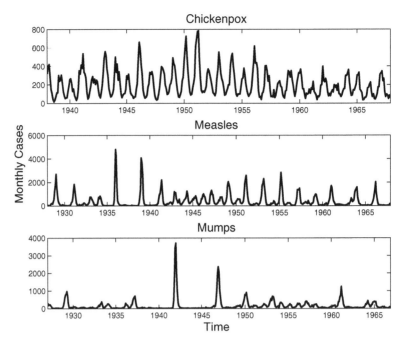

Figure 5.1. Monthly case reports of chickenpox, measles, and mumps in Copenhagen during the twentieth century, demonstrating dramatic patterns of recurrent outbreaks.

His methodology centered on the argument that

$$\frac{\text{Cases this interval}}{\text{Cases last interval}} \sim \frac{\text{Number of susceptibles now}}{\text{Equilibrium number of susceptibles}},$$

which can be expressed as the following equation

$$\left(\frac{C_{t+1}}{C_t}\right)^\alpha = k_\theta \frac{X_{t+1}}{X^*}. \tag{5.1}$$

The parameter α relates the realistic infection "time interval" (the sum of the infectious and latent periods) to the time scale of the data. For measles, this is approximately two weeks; therefore for monthly data, we set $\alpha = 1/2$. The term k_θ is "the factor representing the influence of season θ" (Soper 1929). To estimate X^*, Soper followed Hamer's (1906) calculation that the mean number of susceptibles (X^*) is approximately equivalent to 70 weeks' case reports. Then, once we take into account the fact that at the peak of an epidemic, $C_{t+1} \sim C_t$, we have an initial estimate for X_{t+1}, which can be updated by adding the documented births and subtracting the number of cases. All that remains now is to fit the seasonality parameter k_θ.

Soper's findings, averaged over the 12-year period of the data, are presented in Figure 5.2 and clearly demonstrate that estimated transmission was very low in the summer months, and peaked dramatically in the early autumn (October). Soper argued that based on his results, a key missing ingredient in the SIR model proposed originally by Hamer (1906) was seasonal change in "perturbing influences, such as might be brought about by school breakup and reassembling, or other annual recurrences."

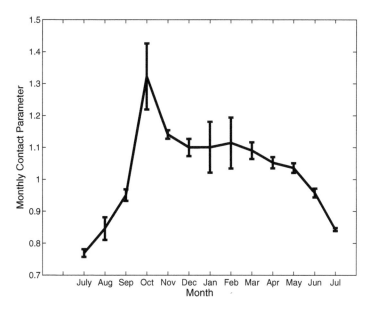

Figure 5.2. Average monthly contact coefficient for measles in Glasgow 1905–1916, as estimated by Soper (1929). The graph clearly highlights the non-constant nature of transmission, with the highest intensity observed during the fall/winter months. Error bars represent standard errors.

The important, though too often ignored, work of Soper was followed by the highly influential studies of London and Yorke (1973) and Yorke and London (1973). These authors were also interested in exploring seasonal influences on transmission, and estimated the mean monthly transmission rates for measles, mumps, and chickenpox in New York City from 1935 to 1972. The key concept in their analysis was based on an earlier empirical observation by Hedrich (1933) that the number of susceptibles has the same value, X_p, at the peak of every outbreak. Mathematically, we can see this is true because at the epidemic peak, the number of infectious individuals, Y, has reached its maximum; therefore, $\frac{dY}{dt} = 0$:

$$\frac{dY}{dt} = 0 \;\Rightarrow\; \beta XY/N - \gamma Y = 0 \;\Rightarrow\; X = X_p = \frac{\gamma N}{\beta}.$$

Hence, at the start of the epidemic year (which they defined to be from the beginning of September to the end of August), the number of susceptibles is X_p plus the cumulative number of reported cases for that year. Then, by using a discrete-time model (see Section 2.7), they were able to explore the pattern of transmission rates that provided model exposures consistent with the observed case reports. Although there are subtle differences in the details of London and Yorke's results compared to those of Soper, they also found a clearly seasonal pattern of transmission for all three diseases, with a peak that coincided with the start of school terms in the autumn and a trough that occurred during the summer months. Since then, more mechanistic approaches for the estimation of transmission rates have been developed, which involve a more detailed "reconstruction" of the number of susceptibles in the population. First Fine and Clarkson (1982) and later Finkenstädt and Grenfell (2000) used case report data for measles in England and Wales, together with information on the population size and birth rates to estimate transmission rates

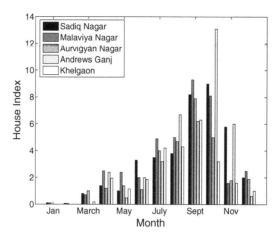

Figure 5.3. House index for *Aedes aegypti* larvae in five different localities in Delhi, India during 1996–1998. The chart clearly demonstrates the seasonal nature of fluctuations in mosquito numbers. Data from Ansari and Razdan (1998).The house index is defined as the percentage of houses infected with mosquito larvae or pupae.

through time. The overall signature detected in these data is consistent with the work of Soper (1929) and London and Yorke (1973).

A range of statistical approaches have revealed that transmission of childhood infections varies seasonally, peaking at the start of the school year and declining significantly in the summer months.

Although epidemiologists have been aware of the importance of seasonal factors in the transmission of childhood diseases, the modeling of such phenomena has been facilitated by the advent of accessible computational power. Analytical methods for dealing with forced models are woefully lacking, and therefore detailed computer integration of forced equations is often the only practical means of understanding or predicting the dynamics.

5.1.1. Seasonality in Other Systems

Changes in transmission rates through time is increasingly recognized as important in a range of infectious diseases (for a review, see Altizer et al. 2006; Grassley and Fraser 2006). For human infectious diseases that are vector transmitted (such as malaria or dengue), seasonality plays an important dynamical role. In these instances, however, the time dependency in transmission is brought about by the biology of the vector population. Hence, models need to capture the fact that mosquito numbers in the tropics, for example, are substantially higher during the rainy season (Figure 5.3). As a result, the forcing required in these models would essentially represent environmental trends, such as precipitation levels through the year (vectored populations are covered in detail in Chapter 4).

In food or waterborne infections, such as cholera, the role played by temporal forcing is more subtle and interesting. There is strong evidence, for example, that the multi-annual dynamics of cholera are interlinked with long-term environmental factors. Studying historical records of cholera outbreaks in Bangladesh, Pascual et al. (2000) have established a correlation with the El Niño Southern Oscillation. However, cholera data also contain a pronounced annual signature, which is thought to be due to an increase in transmission during the monsoon seasons (Koelle and Pascual 2004).

Interesting seasonal components are also found in wildlife diseases. These most frequently arise from changes in host behavior throughout the year. Increased transmission may result from increased contact arising from flocking behavior (e.g., housefinches; Hosseini et al. 2004), seasonal migration (e.g., Monarch butterflies; Altizer 2002), or congregations during the breeding and molting season (e.g., harbor seals; Swinton et al. 1998).

These well-cited studies have established seasonal changes in the contact rates between susceptible and infectious individuals as an important feature of the dynamics of many infectious diseases. This chapter, reviews the various methods used to model time-dependent transmission in human and animal systems.

5.2. MODELING FORCING IN CHILDHOOD INFECTIOUS DISEASES: MEASLES

The last section reviewed the historical studies of seasonality in case reports of childhood infections. A large body of theoretical work has also examined the dynamical consequences of temporal changes in transmission. These studies started with the work of Soper (1929), Bartlett (1956), and Bailey (1975), who incorporated seasonality in SIR models with the primary aim of establishing the amplitude of variation in contact rates necessary to produce the observed 80% fluctuation in epidemics. (It is often difficult to consider the forcing of childhood infections without considering age structured models (Chapter 3). Throughout this chapter, age structure is ignored for simplicity; however, toward the end we highlight that a true mechanistic description of any childhood disease must take into account the interaction of forcing and age structure.) Bailey (1975) explored a simplified SIR model:

$$\frac{dX}{dt} = \mu N - \beta(t) X Y N, \tag{5.2}$$

$$\frac{dY}{dt} = \beta(t) X Y / N - \gamma Y. \tag{5.3}$$

As usual, μ is the per capita birth rate and γ is the recovery rate from the infection. These equations ignore the death of susceptible and infectious individuals; it therefore assumes that all individuals contract the infection during their lifetime, which is a reasonable approximation for measles. The transmission rate is a function of time, $\beta(t)$, and was taken by Bailey to be a sinusoid:

$$\beta(t) = \beta_0 (1 + \beta_1 \cos(\omega t)). \tag{5.4}$$

The parameter β_0 denotes the baseline or average transmission rate, ω is the period of the forcing, and β_1 is the amplitude of seasonality which is restricted to the unit interval. For this form of forcing, the basic reproductive ratio, R_0, is given by $\frac{\beta_0}{\gamma}$; this value represents a yearly average and at certain times of the year (when $\cos(\omega t) \approx 1$), instantaneous growth rates may be much larger than predicted by this average value. Bailey (1975) proceeded to explore the dynamics of small perturbations to the unforced equilibrium assuming a small amplitude of seasonality ($\beta_1 \ll 1$). This was achieved by making the substitutions $X = X^*(1 + x)$ and $Y = Y^*(1 + y)$, which, after omitting some intermediate steps, gives a second order differential equation in the small infectious perturbation y:

$$\frac{d^2 y}{dt^2} + \mu R_0 \frac{dy}{dt} + \mu \beta_0 y = -\beta_1 \omega \gamma \sin(\omega t).$$

The particular integral of this equation (see, for example, Strang 1986) gives the period and amplitude of oscillations as driven by the seasonal term. Although the period of the oscillations is the same as the period of the forcing, the amplitude, M, of the oscillations is given by:

$$M = \beta_1 \omega \gamma \left\{ \left(\mu \beta_0 - \omega^2 \right)^2 + \left(\omega \mu R_0 \right)^2 \right\}^{-\frac{1}{2}}. \qquad (5.5)$$

Now, making the appropriate substitutions for measles, we set $1/\gamma = 2$, $\mu R_0 \sim 0.014$, and $\omega = \frac{\pi}{26}$ (taking the week as our basic time unit), which gives $M \sim 7.76\beta_1$. The implication of this result is that a 10% variation in the transmission parameter translates into seasonal variations of 78% in case notifications, as envisaged by Soper (1929).

Relatively modest levels of variation in the transmission rate can translate into large amplitude fluctuations in the observed disease incidence.

5.2.1. Dynamical Consequences of Seasonality: Harmonic and Subharmonic Resonance

The first systematic examination of seasonality affecting the dynamical pattern of epidemics was made, as far as we are aware, by Klaus Dietz in his seminal 1976 paper. Dietz carried out a stability analysis of the familiar SIR model:

$$\frac{dX}{dt} = \mu N - \left(\beta(t) \frac{Y}{N} + \mu \right) X, \qquad (5.6)$$

$$\frac{dY}{dt} = \beta(t) X \frac{Y}{N} - (\mu + \gamma) Y, \qquad (5.7)$$

This is online program 5.1

where $\beta(t) = \beta_0(1 - \beta_1 \cos(\omega t))$ (note that he used a minus sign in his formulation in order to ensure that contact rates were at their lowest at the start of the epidemic year). He demonstrated that in the absence of seasonal forcing, the system fluctuated with frequency F (c.f. Chapter 2, Box 2.4), where

$$F^2 = \mu(\gamma + \mu)(R_0 - 1) - \left(\frac{\mu R_0}{2} \right)^2. \qquad (5.8)$$

In many realistic situations, $\mu R_0 \ll 1$, hence we may ignore the final term in equation (5.8). Dietz pointed out that for cases in which the natural period of oscillations in the SIR model are approximately the same as that of the seasonal forcing (i.e., $F \approx \omega$), we observe *harmonic resonance*, where model dynamics mimic those of the forcing, although the amplitude of oscillations may be greatly increased.

When the forcing is relatively small, this result can be made more precise. Looking at equation (5.5), the amplification of sinusoidal forcing (M/β_1) is largest whenever the forcing frequency, ω, is close to the natural frequency of oscillations, $F \approx \sqrt{\mu \beta_0}$.

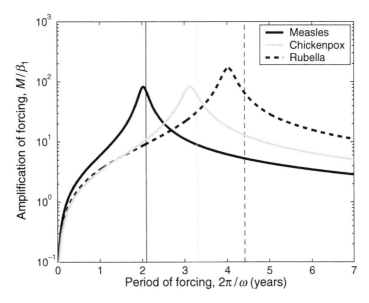

Figure 5.4. The predicted amplification of small amounts of forcing for three childhood infections using equation (5.5). Clear peaks of amplification exist for all three diseases, and these are reasonably close to the natural period of oscillations, $2\pi/F$, shown by the vertical lines. ($\mu = 0.02$ per year. Measles: $1/\gamma = 13$ days, $R_0 = 17$. Chickenpox: $1/\gamma = 20$ days, $R_0 = 11$. Rubella: $1/\gamma = 18$ days, $R_0 = 6$)

Essentially, by forcing the transmission in synchrony with the natural period, we ensure that sequential forcing effects accumulate rather than cancel (Dushoff et al. 2004). By and large, we all have firsthand experience of the effects of resonance in a forced oscillator—on the playground. When pushing (the forcing) a child on a swing (the oscillator), we aim to optimize our effort by timing each push to coincide with the natural period (i.e., the high point) of each oscillation. In Figure 5.4, we show how the amplification of sinusoidal forcing varies with both the forcing frequency and the natural frequency of oscillations (vertical lines) for three childhood diseases.

Forcing is most greatly amplified when the forcing period is close to the natural oscillatory frequency of the unforced dynamics.

For different ratios of $\omega{:}F$, however, it is possible for forcing to excite *sub-harmonic resonance* that gives rise to oscillations with a longer period than the period of the forcing. This phenomenon can occur whenever the natural period of the oscillations $1/F$ is close to an integer multiple of the period of the forcing $1/\omega$. Subharmonic resonance is dependent on the nonlinearities within the transmission process dynamics and requires substantial levels of forcing. As a result, subharmonic resonance is generally studied numerically (see, for example, Greenman et al. 2004; Choisy et al. 2006). Dietz (1976) showed that for β_1 small, some analytical understanding can be gained by setting $T = \omega/F$, and noting that the subharmonic resonance occurs only when T is an integer. We can rearrange equation (5.8) to obtain the following equation that links the observed period of oscillations (T) to the infectious period of the infection ($1/(\mu + \gamma)$) and its mean age at

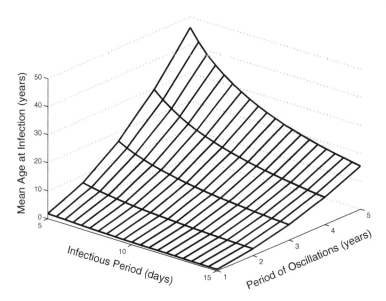

Figure 5.5. The subharmonics of a seasonally forced SIR model, as given by equation (5.9). The surface depicts the relationship between the infectious period, the mean age at infection and the observed period of oscillations. The period of forcing is assumed to be one year ($\omega = 2\pi$).

infection ($A = 1/[\mu(R_0 - 1)]$):

$$A = (\gamma + \mu)\frac{T^2}{\omega^2}. \tag{5.9}$$

For childhood infections, we can assume that the period of forcing is one year ($\omega = 2\pi$) and then examine the predicted period of oscillations as the infectious period and mean age at infection are varied (Figure 5.5). This permits a relationship to be established between the natural period of oscillations resulting from the low-level forcing of the system (T), pathogen transmissibility (in terms of the transmission rate β), and host demography (birth rate μ).

To explore in more general terms how the *amplitude* of seasonality affects dynamics, Dietz (1976) resorted to numerical integration of the underlying equations. In this way, he was able to demonstrate that changes in either R_0 or β_1 can lead to qualitatively different epidemic patterns. For example, when R_0 is large and the level of seasonal forcing is small, the fraction of infecteds shows harmonic oscillations with small-amplitude annual epidemics (Figure 5.6, top left). As β_1 increases to 0.1, we observe subharmonic resonance (as $\omega \approx 2F$) giving rise to biennial dynamics (Figure 5.6, middle left). A further increase in β_1 gives rise to four-year cycles that have a noticeable and pronounced biennial as well as annual component (Figure 5.6, bottom left). However, Dietz (1976) noted that the precise sequence of dynamical transitions that is observed depends on R_0. The second and third columns in Figure 5.6 exhibit qualitatively different dynamics in response to changes in the level of forcing. In these cases, when R_0 is smaller, increases in seasonal amplitude do not influence the period of epidemics that remain annual, but do substantially alter the magnitude of oscillations.

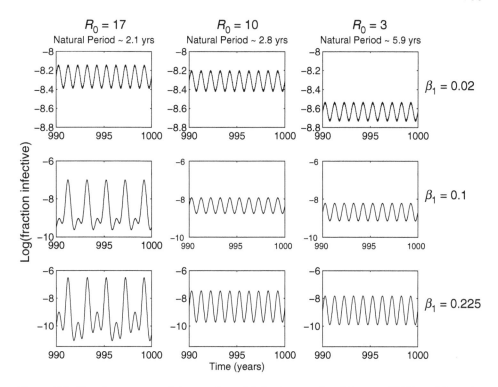

Figure 5.6. The time course of the number of infectives predicted by an *SEIR* model as the amplitude of seasonal forcing (β_1) and R_0 are varied. The parameters used to generate these panels were $\mu = 0.02$ per year, $1/\sigma = 8$ days, $1/\gamma = 5$ days. All simulations were started with $S(0) = 6 \times 10^{-2}$ and $E(0) = I(0) = 10^{-3}$. The logarithm of the proportion of infectives is plotted for clarity. The low troughs predicted between epidemics could be sustained only by large population sizes.

In the absence of seasonal forcing, the *SIR* family of models exhibit a stable equilibrium (see Chapter 2). The introduction of time-dependent transmission rates can generate a variety of dynamical patterns—depending on parameter values— ranging from simple annual epidemics to multiennial outbreaks and eventually chaos.

5.2.2. Mechanisms of Multi-Annual Cycles

It is straightforward and intuitive to understand how making the transmission rate oscillate with a period of one year may generate annual epidemics. As described above, this phenomenon is referred to as simple harmonic resonance, whereby the dynamical system (in this case, the $S[E]IR$ model) simply tracks the temporal changes in the forcing (i.e., $F = \omega$). There is also the possibility (as shown in Figure 5.6) of obtaining subharmonic resonance, where oscillatory dynamics have a period that is an integer multiple of the forcing. This phenomenon can be understood by thinking about dynamics within a more ecological framework.

Consider the equation giving the rate of change of infectives in the SIR system (equation (5.7)). As demonstrated by Kermack and McKendrick (1927) and discussed in Chapter 2, the condition for growth of disease incidence is:

$$\frac{dY}{dt} = \beta X \frac{Y}{N} - (\gamma + \mu)Y > 0,$$

$$= Y(\mu + \gamma)(R_0 \frac{X}{N} - 1) > 0,$$

$$\implies \frac{X}{N} > \frac{1}{R_0} \approx \frac{\gamma}{\beta}.$$

Therefore, the spread of the pathogen can occur only if there is a sufficient fraction of susceptibles in the population, with the critical value determined by the reproductive ratio (R_0). Although this concept is typically thought of in terms of the introduction of a pathogen into a population, it is also informative when thinking about seasonal systems. In such systems we can use the seasonally varying value of $\beta(t)$ to inform whether the current level of susceptibles is sufficient for the number of cases to increase. We demonstrate this concept in Figure 5.7, with the top panel depicting two sinusoidal transmission rates and two lower panels showing the relationship between the fraction of susceptibles (thick lines) and the threshold for infection spread (thin black lines) when epidemics are annual (middle graph) and biennial (bottom graph).

In the case of annual epidemics, there is a straightforward sequence of events. The peak in disease incidence coincides with the point at which the fraction of susceptibles ($\frac{X}{N} = S$) falls below $\gamma/\beta(t)$ (labeled point 1 in Figure 5.7). The fractions of susceptibles and infectives continue to decline until the rate of transmission is less than births, at which point S begins to increase. Once $S > \gamma/\beta(t)$ (labeled point 2 in Figure 5.7) disease incidence rises. This pattern is repeated once transmission has depleted susceptibles below the threshold (labeled point 3 in Figure 5.7). In this instance, the seasonal changes in the transmission rate are intimately associated with the dynamics of susceptibles in driving harmonic resonance.

For a slightly higher amplitude of seasonality, however, the picture changes in important ways (Figure 5.7, bottom graph). As before, the peak in the fraction of infectives coincides with $S = \gamma/\beta(t)$ (point 4). In this instance, however, the peak in the infectives is substantially larger than in the middle graph (due to the greater transmission rate), and as a result the fraction of susceptibles falls to much lower levels than before. Consequently, it takes much longer for births to replenish the susceptibles above the critical threshold (point 5). While at this point $S > \gamma/\beta(t)$, the transmission rate is very near its annual maximum (the threshold is near its minimum), and as a result the susceptibles do not remain above the threshold long enough to produce a large epidemic. It is only when the level of susceptibles exceeds the threshold for a second time (point 6) that a large epidemic begins. The entire process from point 4 to point 7 takes two years, representing subharmonic resonance. What these results show is that as the amplitude of seasonality increases, larger epidemics are generated that lower the level of susceptibles such that recovery to above the threshold takes far longer, resulting in longer period cycles.

5.2.3. Bifurcation Diagrams

How do we visually summarize the dynamics of seasonally forced models without needing to resort to figures containing numerous panels (such as those in Figure 5.6)? This may be achieved by constructing *bifurcation diagrams*, where a bifurcation refers to a qualitative

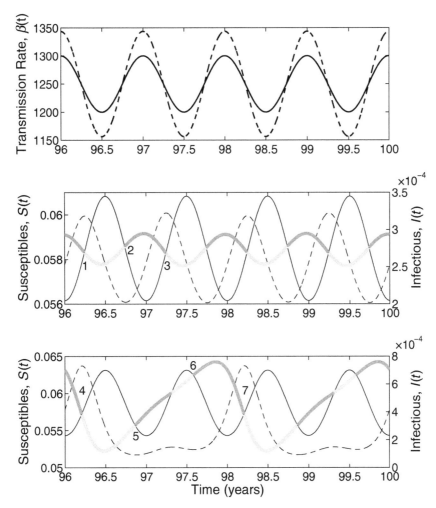

Figure 5.7. The anatomy of seasonally forced epidemics. In the top panel the seasonally varying transmission rate ($\beta(t)$ per year) is plotted, with the solid line representing the weaker forcing used in the middle graph and the dashed line representing the stronger forcing used in the bottom graph. The lower two panels demonstrate outbreak dynamics in annual (middle) and biennial (bottom) regions of parameter space (with $\beta_1 = 0.04$ and $\beta_1 = 0.075$, respectively). In addition to the fraction of susceptibles (thick line) and infectives (dashed line), we have plotted the threshold level of S required for instantaneous epidemic growth (thin line). The level of susceptibles is color coded indicating when it is above or below the threshold. The parameters used to generate these panels were $\mu = 0.02$ per year, $1/\sigma = 8$ days, and $1/\gamma = 5$ days. All simulations were started with $S(0) = 6 \times 10^{-2}$ and $E(0) = I(0) = 10^{-3}$.

change in model dynamics as a *control parameter* is altered. This is most painlessly and commonly achieved by the numerical integration of the equations as the parameter of interest (in this case, β_1) is systematically varied (Figure 5.8). For any specific parameter value, the model is started according to some specified initial conditions and integrated for a "reasonable" period of time, after which it is assumed the dynamics have reached their long-term (or asymptotic) state. Then, some measure of the population is plotted at one

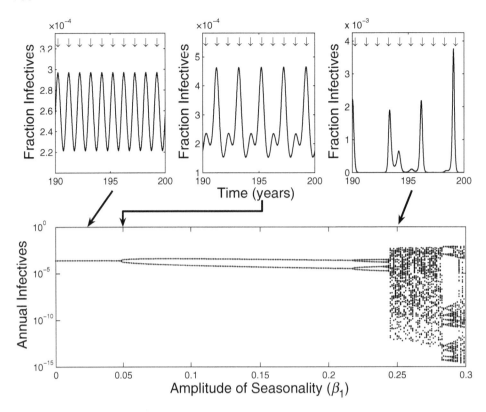

Figure 5.8. Constructing a bifurcation diagram. The top three panels depict time-series data for the $SEIR$ model with different levels of seasonality ($\beta_1 = 0.025, 0.05$, and 0.25, respectively). The arrows at the top of the panels indicate the points when the time series are sampled in order to construct the bifurcation diagram below. The parameters used to generate these panels were $\mu = 0.02$ per year, $\beta_0 = 1250$, $1/\sigma = 8$ days, and $1/\gamma = 5$ days. All simulations were started with $S(0) = 6 \times 10^{-2}$ and $E(0) = I(0) = 10^{-3}$.

particular time-point each year for the subsequent n years; the precise value of n depends on the timescale of the system but for epidemiological systems 50–100 years represent good rule-of-thumb values. Often, we produce a graph with different values of β_1 along the x-axis and the prevalence of infection (at one time each year) on the y-axis (Figure 5.8).

A typical bifurcation diagram for parameter values representative of measles is shown in Figure 5.8. To interpret the figure, we need to consider a value of β_1 and count the corresponding number of points found on the graph. For example, for $\beta = 0.1$ we observe two dots—this informs us that the dynamics are biennial and repeat every two years (see Figure 5.6, middle left, for an example of the dynamics). The bifurcation diagram shows the increasing dynamical complexity as seasonality becomes stronger. For modest levels of forcing, the dynamics mimic those of the forcing function and are rigidly annual. For β_1 greater than approximately 0.0455, the dynamics are biennial, which give way to multiennial and then aperiodic dynamics when the amplitude of seasonality exceeds 0.2. For measles, attempts to fit β_1 from time-series data have provided estimates of around 0.1–0.2 (Keeling and Grenfell 2002), therefore we have focused our attention to a restricted region of parameter space: $\beta_1 \in [0, 0.3]$. For values of β_1 greater than 0.3, the dynamics are largely chaotic with occasional "windows" (regions of β_1 that give qualitatively similar dynamics) of multiennial cycles.

However, many elements can complicate the formulation of a bifurcation diagram, which is why several different bifurcation diagrams for the $SEIR$ model with measles parameters are given throughout this chapter. For example, although Figure 5.8 uses the amplitude of seasonality, β_1, as the control parameter, many other parameters or combinations of parameters could be used. Additionally, in Figure 5.8 a fixed set of initial conditions were used to generate each point on the bifurcation diagram and, as we show below, different initial conditions can lead to very different bifurcation patterns. Therefore, the richness of the dynamics and the large number of possible scenarios that can be considered means that a single bifurcation diagram can never fully capture the entire range of behavior.

5.2.4. Multiple Attractors and Their Basins

In Chapter 2, the solutions we obtained were globally attracting—as long as we started with some infecteds we eventually settled to the same equilibrium level. However, for seasonally forced models this simple property does not always hold; for some parameter values (generally higher values of β_1 associated with more complex behavior) the qualitative dynamics are sensitive to the initial conditions. Therefore, more than one possible dynamical behavior may exist at each point on the bifurcation diagram. In dynamical systems terminology, there are *multiple stable attractors* and which attractor is observed depends on the initial conditions—we no longer have a single globally attracting solution. Each attractor has an associated *basin of attraction*, such that whenever we start from a specific combination of variables (e.g., S, E, and I) within the basin, we eventually observe the same dynamics after transients. To demonstrate this idea, in the top graph of Figure 5.9 the number of infectives is plotted for $\beta_1 = 0.19$, depicting clear biennial dynamics as predicted by the bifurcation diagram in Figure 5.8. However, the bottom graph is also for $\beta_1 = 0.19$—the only difference between these figures is the initial conditions. The dynamics of the lower graph are qualitatively different, showing a pronounced six-year cycle.

The basin of attraction for the biennial and six-year cycles can be determined by extensive simulations examining different combinations of initial conditions. To achieve this, we fix all parameter values and systematically explore a grid of initial conditions and the resulting long-term dynamics. In Figure 5.10, we plot the results of such an exploration for $\beta_1 = 0.19$. The dark regions represent the basin of attraction for the six-year cycle and the light regions represent the combination of initial conditions for which the biennial dynamics are observed. For some combinations of initial conditions, there is considerable structure (top-right quadrant of Figure 5.10), whereas in other regions there is extreme sensitivity to the initial conditions and very small deviations can flip the long-term dynamics between attractors.

In seasonally forced systems, qualitatively different dynamical patterns can be stable for any specific combination of parameter values. Which attractor is observed depends on whether initial conditions are within the basin of attraction.

Given the possibility of many coexisting attractors for the same parameters, it may be impossible to produce a complete bifurcation diagram that displays and differentiates between the various multiple attractors (see Figure 5.11). This is especially true if some of the attractors have very small basins such that finding the appropriate set of initial

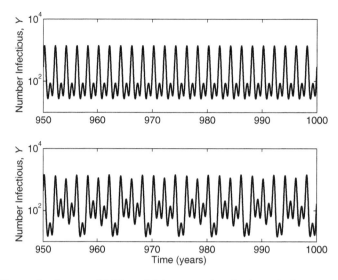

Figure 5.9. Time series from an $SEIR$ model demonstrating the dynamical consequences of using different initial conditions. The two panels share identical levels of seasonality ($\beta_1 = 0.19$), but their initial conditions vary in the number with the exposed class (top graph $W(0) = 10^{-2} \times N$, whereas for the bottom graph $W(0) = 10^{-3} \times N$). All other initial variables are the same between the two graphs ($X(0) = 6 \times 10^{-2} \times N$ and $Y(0) = 10^{-3} \times N$). The parameters used to generate these panels were $\mu = 0.02$ per year, $1/\sigma = 8$ days, $1/\gamma = 5$ days, $R_0 = 17$, and $N = 5 \times 10^6$. Note that the number of infectives is plotted on a logarithmic scale.

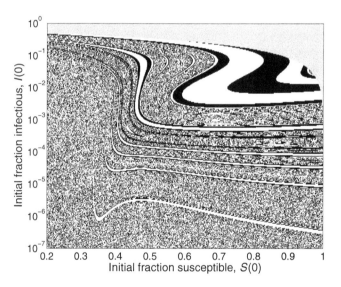

Figure 5.10. The basins of attraction for biennial and six-year dynamics. The black regions represent the combination of the initial fraction of susceptibles ($S(0)$) and infectives ($I(0)$) that give rise to six-year dynamics, and the white regions lead to the biennial attractor. Gray regions show initial conditions where $S(0) + E(0) + I(0) > 1$. Note that we set $E(0) = I(0)\gamma/\sigma$ corresponding to an equal rate of movement through the two classes. A grid of 1,000 susceptible and 460 infectious initial values was used to construct this picture. Model parameters were $\mu = 0.02$ per year, $1/\sigma = 8$ days, $1/\gamma = 5$ days, $\beta_1 = 0.19$, $R_0 = 17$.

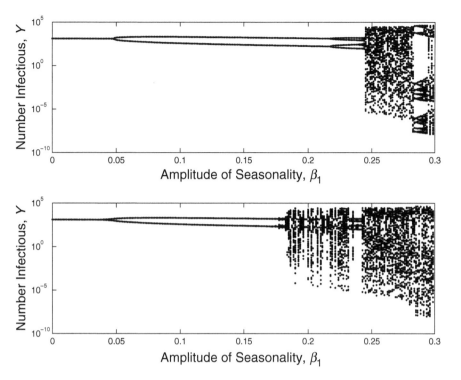

Figure 5.11. Bifurcation diagrams demonstrating the dynamical consequences of using different methods of generating initial conditions. The top graph is constructed using extrapolated initial conditions starting with $\beta_1 = 0$, where the lower graph is plotted starting with $\beta_1 = 0.3$. The parameters used to generate these panels were $\mu = 0.02$ per year, $1/\sigma = 8$ days, $1/\gamma = 5$ days, and $N = 5 \times 10^6$. Note that the number of infectives is again plotted on a logarithmic scale.

conditions is very difficult. In contrast, using a single fixed set of initial conditions can produce a rather disjoint bifurcation diagram because the changing shape of the basins of attraction means that we may suddenly jump between different attractors. An alternative is to use "extrapolated" initial conditions whereby the numbers of susceptibles, exposed, and infectives at the end of one simulation (when $\beta_1 = 0.19$, for example) are used to start the next simulation (for $\beta_1 = 0.2$). This approach ensures that in general our initial conditions start near an attractor, leading to more continuous behavior. Hence, any structural changes in the bifurcation diagram as the control parameter is varied can be confidently attributed to bifurcations (the loss of stability of one attractor giving way to a new stable regime), rather than resulting from the effects of crossing the "basin of attraction."

The use of "extrapolated" initial conditions clearly means that we could obtain different plots depending on whether we start the bifurcation diagram from the right or from the left (Figure 5.11). The top graph shows a bifurcation diagram starting from $\beta_1 = 0$ and increasing β_1 with each increment, whereas the lower graph starts with $\beta_1 = 0.3$ and decreases β_1. The overall qualitative patterns in the two graphs are similar, though they clearly differ in much of the dynamical detail.

As usual in mathematical modeling, there is a trade-off between obtaining a speedy understanding of model dynamics and mathematical rigour. Some of the issues we have

---

Box 5.1 Bifurcation Methods

The bifurcation diagrams plotted in, for example, Figures 5.8 and 5.11 were obtained numerically, by integrating the model equations and documenting the observed patterns. Ideally, however, we would explore model dynamics using the same kinds of principles employed for unforced models, where we used the Jacobian matrix to evaluate eigenvalues and establish the stability or otherwise of equilibria. In the forced model, we also need to establish whether perturbations made to a known trajectory are likely to die out or grow. This is achieved by studying the stability of the map $P^{(k)}$, which iterates the dynamics forward by k years:

$$P^{(k)} : (X(0), W(0), Y(0)) \mapsto (X(k), W(k), Y(k)).$$

In particular, we are interested in orbits of period k, which corresponds to $P^{(k)}$ having a fixed point such that after waiting for a period of k years we recover the state variables that we started with. These periodic solutions are referred to as *period k fixed points*. To examine their stability, we need to study the fate of perturbations to the fixed point.

Specifically, we define $X_k^*(t)$, $Y_k^*(t)$, $Z_k^*(t)$ as a k-period trajectory and make the substitutions $X(t) = X_k^*(t) + x(t)$, $Y(t) = Y_k^*(t) + y(t)$, and $Z(t) = Z_k^*(t) + z(t)$, then form equations for the dynamics of perturbations $(x(t), y(t), z(t))$ (ignoring terms that are of order $x^2(t)$ or higher). This leads to the following differential equation:

$$\begin{pmatrix} dx/dt \\ dy/dt \\ dz/dt \end{pmatrix} = J_{X_k^*(t), Y_k^*(t), Z_k^*(t)} \begin{pmatrix} x(t) \\ y(t) \\ z(t) \end{pmatrix}.$$

Whether the perturbations eventually die out is determined by the dominant eigenvalue $\Lambda^{(k)}$ of the Jacobian matrix J_P of the map $P^{(k)}$ evaluated at its fixed point. (The Jacobian J_P of the map can be formed by integrating the Jacobian $J_{X_k^*(t), Y_k^*(t), Z_k^*(t)}$ over a k-year period). The dominant eigenvalue is called the "floquet multiplier" of the period k solution. A fixed point is stable if and only if it has no multipliers with $\|\Lambda^{(k)}\| \geq 1$. In practical terms, this means that if we are interested in exploring the stability of the annual trajectory, for example, then we need to integrate both the $SEIR$ equations and their Jacobian for one year and then examine the resultant Jacobian's dominant eigenvalue. Note that the initial conditions here are crucial. We need to integrate the model equations starting *on* the annual attractor, whereas the initial conditions for the Jacobian (also referred to as the relational equations) are the identity matrix. It is possible to use relatively simple root-finding schemes (such as the Newton-Raphson method) to work out the period k solutions before we establish the stability.

The main advantage of this dynamical-systems approach is that abrupt changes in the floquet multipliers are very informative about the dynamics we may expect. For example, if at some value of the control parameter the multiplier $\Lambda^{(k)}$ becomes -1, then we know that we expect a period doubling bifurcation, leading to solutions with a period of $2k$. The major disadvantage of this method is that it perhaps requires much more experience with dynamical systems theory. There are, however, at least two well-established freeware programs that can be used to carry out numerical bifurcation analyses: AUTO (Doedel et al. 1998) and Matcont (Dhooge et al. 2003).

---

come across concerning multiple stable states and the resulting numerical bifurcation diagrams may be overcome by using more mathematically sophisticated methods that allow us to establish the stability of different solutions from basic principles (see Box 5.1). Such approaches, though substantially more technically involved, will allow us to establish *a priori* the range of dynamics we might expect to observe (see, for example, Kuznetsov 1994; Kuznetsov and Piccardi 1994; Seydel 1994).

TABLE 5.1.
Timings of the major school holidays when $Term = -1$; during all other times $Term = +1$. Note that the autumn half-term break is included because this is the only short holiday that has an identifiable signature in the England and Wales data.

Holiday	Model Days	Calendar Dates
Christmas	356–6	December 21–January 6
Easter	100–115	April 10–25
Summer	200–251	July 19–September 8
Autumn Half Term	300–307	October 27–November 3

5.2.5. Which Forcing Function?

So far, we have explored seasonality by assuming that the transmission rate is time-dependent and specifically is determined by a simple sinusoidal function. The precise shape of seasonality was historically thought to be dynamically unimportant, though this view has changed in recent years. Starting with the influential work of Schenzle (1984), seasonally forced models of childhood infections now more often use a square wave, attempting to capture the aggregation of children in schools. Specifically, the transmission rate is assumed to be high during school terms and low at other times (Bolker and Grenfell 1993; Keeling and Grenfell 1997a; Rohani et al. 1999; Earn et al. 2000; Keeling et al. 2001a). In this manner, equation (5.4) is rewritten as follows:

$$\beta(t) = \beta_0(1 + b_1\, Term(t)), \tag{5.10}$$

where $Term(t)$ is $+1$ during the school term and -1 at other times. We now use the parameter b_1 to represent the amplitude of seasonality: a slightly different notation in order to distinguish between sinusoidal and term time forcing. The historical dates of school holidays in England and Wales are presented in Table 5.1, and the resulting transmission rate throughout the year is plotted in Figure 5.12 (top graph). An important point to note is that if we sum the number of school holidays given in Table 5.1, we obtain 92, leaving 273 days of school. Because there are many more "$+1$" days than "-1" days, adopting the basic equation (5.10) is going to give rise to a mean transmission rate—averaged over the year—that exceeds β_0, with the level of excess depending on b_1. As a result, in order to ensure R_0 is constant irrespective of the precise forcing function used and the amplitude of seasonality, we need to implement a correction to equation (5.10). In general terms, if there are D_+ days of school and D_- holidays, then our forcing function would be:

$$\beta(t) = \frac{\beta_0}{\frac{1}{365}\big((1 + b_1)D_+ + (1 - b_1)D_-\big)}\Big(1 + b_1\, Term(t)\Big). \tag{5.11}$$

This is online program 5.2

The denominator in equation (5.11) is the mean of the forcing term, division by which ensures $\overline{\beta(t)} = \beta_0$. The changes in the transmission rate resulting from the above correction are depicted in Figure 5.12 (top graph), and the dynamical consequences of this correction are illustrated in the lower two panels. It is evident that dynamics of the basic term-time forced model (equation (5.10)) are more complex than when the corrected forcing is applied, as demonstrated by the period-doubling bifurcation occurring at a smaller amplitude of seasonality ($b_1 \sim 0.0975$ as opposed to $b_1 \sim 0.1285$). Also, without the correction term, quadrennial and higher-period epidemics ensue once $b_1 > 0.5$, whereas

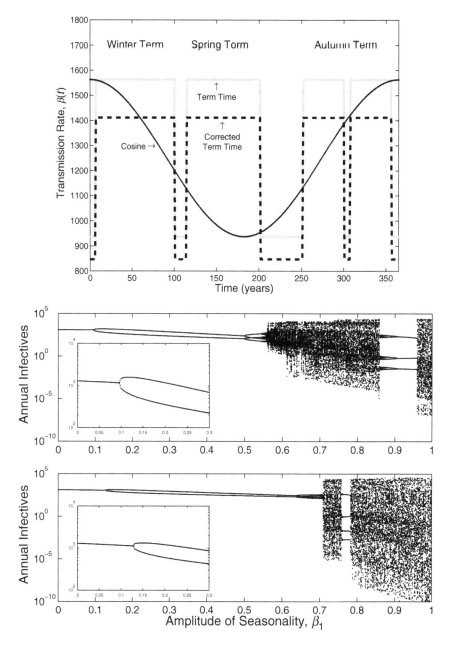

Figure 5.12. The top graph gives the transmission rate plotted through the calendar year for three different forcing functions: the sinusoidal function (solid line), the basic term time (gray line), and the corrected term time (dashed line). The figure was plotted assuming $\beta_0 = 1250$, $\beta_1 = b_1 = 0.25$, and $N = 5 \times 10^6$. The mean transmission rates ($\bar{\beta}$) are 1,250 for the sinusoidal and corrected term time ($R_0 = 17$) compared with 1,384 for the basic term time ($R_0 = 19.35$). The school term dates are given in Table 5.1. The lower two graphs show measles bifurcation dynamics with the basic (middle panel) and corrected term-time forcing function (bottom panel). The insets depict the region of parameter space around the first period-doubling bifurcation.

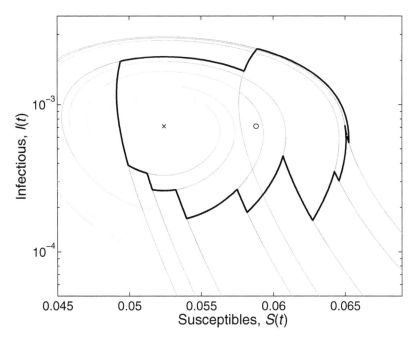

Figure 5.13. The biennial dynamics of a seasonally forced measles-like $SEIR$ model depicted in state space: the time-evolution of susceptibles versus infectious individuals as school terms change through the year. The black line demonstrates the actual biennial cycle whereas the gray lines represent continuation of the trajectory had there not been a switch at those specific points. The cross represents the position of the term-time fixed point (the holiday-time fixed point is off the right-hand side of the graph), whereas the circle represents the fixed-point of the unforced model. The figure was plotted assuming measles parameters: $\beta_0 = 1250$, $1/\sigma = 8$ days, $1/\gamma = 5$ days, $\mu = 0.02$, and $b_1 = 0.25$.

if equation (5.11) is used, complex dynamics are observed only after $b_1 > 0.65$. However, term-time forcing in general produces a much simpler bifurcation picture than sinusoidal forcing (compare Figures 5.11 and 5.12, noting the very different ranges of forcing), with complex/chaotic dynamics only occurring for relatively high values of b_1.

One way to conceptualize the behavior of a disease subject to such binary term-time forcing is as switching between two stable points or spiral sinks (one for term-time when β is high, another for holidays when β is low). Therefore, during term-times β remains constant at the higher value and the trajectories spiral toward the fixed point given by this transmission rate in exactly the same manner as predicted by an unforced model. When holidays start, a new fixed point exists and the trajectories spiral toward that. We therefore view the changing values of β as switching the model between two attracting fixed points. This idea is demonstrated in Figure 5.13 for measles parameter values; by "extending" the orbits (gray lines), the switching between the two spiral-sink attractors is clearly visible. The orbits are traced out counter-clockwise, with an abrupt change in direction every time a switch from term-time to holidays (or vice versa) occurs.

Naturally, we need to establish qualitative and the quantitative consequences of different forcing functions on model dynamics. One way of assessing this is to explore comparable bifurcation diagrams. In Figure 5.14, we present two-dimensional bifurcation figures for

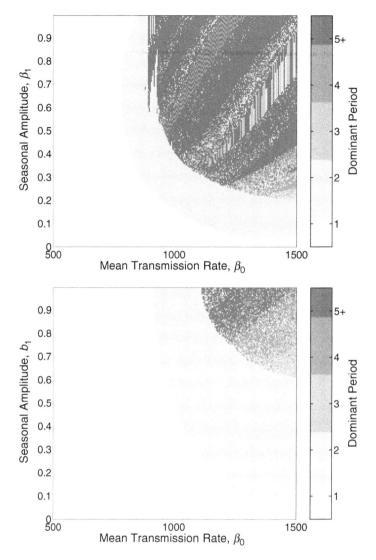

Figure 5.14. Bifurcation diagrams for the $SEIR$ model as the mean transmission rate (per year) and amplitude of seasonality are altered for sinusoidal (top graph) and term-time forcing (bottom graph). The diagrams were constructed using extrapolated initial conditions and as a result are much "cleaner" than equivalent diagrams with fixed initial conditions because these figures do not exhibit multiple stable attractors, especially for small β_1 and b_1. The period plotted is the dominant multiennial period of the Fourier spectrum (Box 5.2). The figure was plotted assuming $1/\sigma = 8$ days, $1/\gamma = 5$ days, and $\mu = 0.02$.

the sinusoidal and corrected term-time forced $SEIR$ model, using extrapolated initial conditions. (Results using a fixed set of initial conditions tend to show more multiennial cycles; extrapolation allows the annual attractor to be tracked through the parameter space.) A number of observations need to be made here. In comparison with the sinusoidally forced model (top graph), term-time forcing (bottom graph) is in some general sense more "stable"

(Bolker and Grenfell 1993). For example, with the mean transmission rate fixed at 1,250 per year, the bifurcation from annual to biennial epidemics occurs at a larger amplitude of seasonality for the term-time forcing ($b_1 \sim 0.1285$ compared with $\beta_1 \sim 0.0455$). In addition, the term-time forced models exhibit biennial epidemics for a far larger region of the parameter space, with irregular (quasiperiodic or chaotic) outbreaks observed only once b_1 exceeds approximately 0.6 (compared to $\beta_1 > 0.2$). Additionally, the sinusoidal-forced model generates very large amplitude dynamics with the proportion of infectives often falling below 10^{-20} when β_1 is large; in contrast, term-time forcing the proportion of infectives in the troughs of epidemics in Figure 5.12 always exceeds 10^{-10}. These figures have been produced using extrapolated initial conditions and as a result do not document the full dynamical complexity of these systems, especially when mean transmission rates are small (explained in detail below). Finally, if we consider the dominant period rather than the dominant multiennial period, then the region of annual behavior is extended further in both models as expected.

The choice of functional form used to represent seasonality in the transmission term can have a substantial qualitative, as well as quantitative, dynamical effect.

The introduction of time-dependence in transmission has introduced a wide array of interesting dynamics to the model. The unforced models discussed in Chapter 2 focused on equilibrium properties, whereas the models here deal with periodic epidemics. The exact period of these oscillations is determined by the characteristics of the disease, such as its mean transmission rate (β_0) and infectious period ($1/\gamma$), as well as host characteristics, such as the per capita birth rate (μ) or the amplitude of seasonality (β_1 or b_1). In order to relate models to case report data from a specific population we therefore need to be able to establish the appropriate amplitude of seasonality, as well as determine the more usual demographic and disease parameters. Unfortunately, a straightforward approach to this problem does not exist, although a number of authors have proposed different methods.

Bolker and Grenfell (1993), for example, studied the dynamics of measles in England and Wales during the 1950s and 1960s, when epidemics were clearly biennial. They constructed the average biennium (gray lines in Figure 5.15) and explored the relative goodness of fit of different models. This kind of exercise allows a clear visual inspection of the comparison between model output and data, but is not statistically rigorous. An alternative approach, adopted by Keeling and Grenfell (2002), has been to attempt to fit, in a rigorous sense, the amplitude of seasonality that is most consistent with the data. Keeling and Grenfell's findings are presented in Table 5.2. The best-fit model with sinusoidal forcing results in a lower error because it more accurately captures the timing of the epidemic peak; in contrast, the best-fit model with term-time forcing generates an epidemic peak that is slightly delayed, even though it captures more of the qualitative properties of the biennial cycle. This discrepancy highlights the need for extra biological detail: A more realistic distribution for the latent and infectious periods or including age structure (see Chapter 3) greatly reduces the error associated with term-time forcing.

For other childhood diseases, such as pertussis and rubella, where there is no systematic epidemic pattern, determining the correct level of seasonality is more complex. One approach is to adapt the time-series methods used for measles (Finkenstädt and Grenfell 2000, Bjørnstad et al. 2002), where the weekly transmission rate is estimated from a statistical viewpoint. Alternatively, we can attempt to match more generic features of

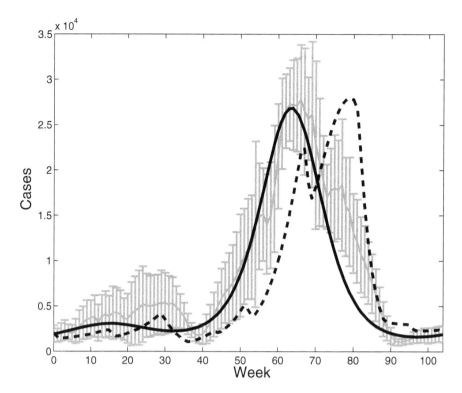

Figure 5.15. Comparing model dynamics and measles data from England and Wales. The gray lines represent the number of weekly cases averaged over the nine biennia from 1950 to 1968, with the error bars representing the standard error. The black solid line is the best fit $SEIR$ model with sinusoidal transmission ($\beta_1 = 0.09$), whereas the dotted line is the fit from the corrected term-time forced model ($b_1 = 0.29$). (The results from the basic (uncorrected) term-time model are very similar.) The figure was plotted assuming $\beta_0 = 1,250$ per year, $1/\sigma = 8$ days, $1/\gamma = 5$ days, $\mu = 0.02$ per year, $\beta_1 = 0.09$, and $b_1 = 0.29$. The estimates for β_1 and b_1 were taken from Keeling and Grenfell 2002.

TABLE 5.2.
Optimum level of seasonality estimated from measles
case reports in England and Wales data and the
goodness of fit (associated error, E_V) for the $SEIR$
model with the two different seasonal forcing
functions (Keeling and Grenfell 2002).

Seasonal Forcing	Term-time	Sinusoidal
Best Fit	$b_1 = 0.29$	$\beta_1 = 0.11$
Associated Error	$E_V = 1.18$	$E_V = 0.64$

the data, such as the strength of annual, biennial, and multiennial signals in the data (found by taking the Fourier transform). However, if low troughs exist between the major epidemics, a stochastic framework might be required (Chapter 6) that can account for chance extinctions—with stochastic models there is the possibility of obtaining likelihood estimates which provides an alternative means of parameterization.

Box 5.2 Determining Periodicity

Here we outline two basic methods to characterize two distinct but related elements of periodicity. In particular, we look at (i) the period of the deterministic attractor and (ii) the dominant period of the epidemic cycle.

Period of Attractor

If we have an attractor with period n years, then once the dynamics are on the attractor we have the relationship:

$$S(t+n) = S(t) \quad \text{and} \quad I(t+n) = I(t) \quad \forall t,$$

where the time, t, is measured in years. Despite this obvious definition, the computational reality is more far complex due to the time taken for a set of initial conditions to converge close to the attractor. We therefore stipulate that the dynamics are of period n if:

$$\begin{aligned} |\log(S(t+n)) - \log(S(t))| < \varepsilon \text{ and} \\ |\log(I(t+n)) - \log(I(t))| < \varepsilon, \end{aligned} \qquad \forall t > T_0 \qquad (5.12)$$

where the time T_0 allows for convergence and ε is a small numerical tolerance. The difficulty comes in deciding on appropriate convergence times and tolerances—long times and small tolerances are more accurate, but more computationally intensive.

A further complication is that multiples of the true period will also meet criteria (5.12); in particular if the dynamics are period n, then, once close to the attractor:

$$|\log(S(t+2n)) - \log(S(t))| < |\log(S(t+n)) - \log(S(t))| \qquad \forall t > T_0,$$

and similarly for I. We must therefore insist that our dynamics are period n, if n is the smallest value for which criteria (5.12) holds.

Dominant Epidemic Period

The difficulty with the above definition is that it cannot be readily applied to real observational data or the results from stochastic models (Chapter 6). In addition, it is possible for an attractor to be of very high period and yet display annual epidemics that may take many years before they precisely repeat. An alternative is therefore to look for the *dominant* period of the epidemic cycle using Fourier spectra. We define the strength, Q_m, of the m-year cycle to be:

$$Q_m = \frac{1}{T} \left\| \int_{T_0}^{T_0+T} Y(t) \exp\left(\frac{2\pi t}{m} i\right) dt \right\|,$$

$$Q_m = \frac{1}{T} \left[\left(\int_{T_0}^{T_0+T} Y(t) \cos\left(\frac{2\pi t}{m}\right) dt \right)^2 + \left(\int_{T_0}^{T_0+T} Y(t) \sin\left(\frac{2\pi t}{m}\right) dt \right)^2 \right]^{\frac{1}{2}}$$

where T_0 is again a convergence time, and time is measured in years.

For observational data that is collected or collated at discrete time points, the integrals in the above equations are replaced by sums. To make a fair comparison, the length of the time-series used, T, must be a multiple of the period, m, being investigated. We define the dominant period as that which gives rise to the largest Q value.

One difficulty with this Fourier Spectra approach is that many multiennial cycles have a strong annual signature, due to the annual pattern of seasonal forcing. For this reason we often consider the dominant multiennial period ($m \geq 2$) and define the dynamics as annual only if the strength of all other periods is zero or insignificant. A similar result can be achieved by calculating the Fourier Spectra of annual data Y_t (such as disease incidence at a particular time each year, or the total annual cases).

$$Q_m = \frac{1}{T} \left\| \sum_{t=T_0}^{T_0+T-1} Y_t \exp\left(\frac{2\pi t}{m} i\right) \right\|$$

For a deterministic model that has converged onto the attractor, the largest m for which Q_m is nonzero generally corresponds to the period of the attractor n defined earlier.

When dealing with observational data, or models where the periodicity appears to change, a refinement to the Fourier spectra is possible. *Morlet wavelet analysis* provides a means of determining the dominant period at any given time. At time T, we define the strength $Q_m(T)$ of the m-year cycle to be:

$$Q_m(T) = \left\| \int_{-\infty}^{\infty} \log(I(T+t)+1) \exp\left(\frac{2\pi(T+t)}{m} i\right) \exp\left(-\frac{t^2}{2m^2 V}\right) dt \right\|.$$

In effect this provides a moving average at each time, weighted by a normal distribution. Such methodology has been used to great effect in interpreting the dynamical effects of changing birth rates for measles in England and Wales (Grenfell et al. 2001).

5.2.6. Dynamical Transitions in Seasonally Forced Systems

Now that we have a best-fit estimate for the amplitude of forcing for the average biennial pattern of measles in England and Wales during the 1950s and 1960s, we wish to better understand why case notification data exhibit a more annual pattern of epidemics from 1944–1950 and irregular 2–3-year fluctuations from 1968 onward (Figure 5.16). This kind of dynamical variability has been noticed in other childhood disease incidence data (such as measles in Baltimore or rubella in Copenhagen) and has generated a substantial body of work in the search for explanations. During the 1980s, the general consensus among theoretical epidemiologists was that the epidemics of measles were chaotic, determined by a strange attractor (see, for example, Olsen and Schaffer 1990). In more recent years, however, authors have increasingly focused on more biological explanations of childhood disease epidemics. Finkenstädt et al. (1998), for example, demonstrated strong dependence of measles epidemics on population birth rates with relatively high rates associated with annual patterns, whereas modest birth rates coincided with biennial cycles.

The interaction between factors affecting host population demography and infection dynamics was clarified by Earn et al. (2000). As explained in Chapter 8, Earn et al. pointed out that changes in the recruitment rate of susceptibles—either via systematic trends in the birth rate or resulting from vaccination—can be dynamically expressed as an effective change in the mean transmission rate (β_0) of the disease. For example, by carrying out a simple change of variables, it can be demonstrated that vaccination at level p will induce epidemic patterns identical to those in an unvaccinated population with mean transmission rate $\beta(1-p)$. Similarly, changes in the birth rate μ by a given factor should produce exactly the same dynamical transitions as changing β by the same factor (see Chapter 8). Although a simple observation, this has powerful implications because it allows a single bifurcation diagram to be constructed in order to examine the dynamical consequences of varying host demography.

Such a summary diagram for the term-time forced $SEIR$ model, with epidemiological parameters chosen to correspond to measles, is shown in Figure 5.17. The control

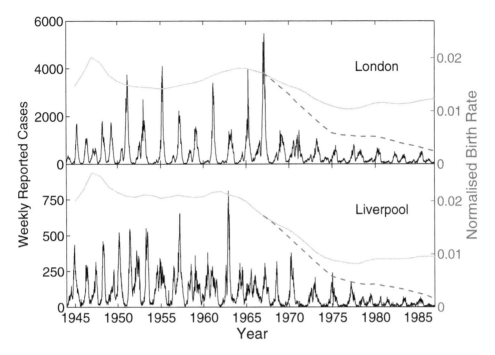

Figure 5.16. Case reports for measles in London and Liverpool from 1944 until 1988. The black line demonstrates weekly reported cases (around 60% of the true cases), with the gray line depicting the per capita birth rate. The darker gray dashed lines demonstrate the effective birth rate, correcting for vaccination that started in 1968 (see Earn et al. 2000).

parameter (x-axis) is the mean transmission rate $\overline{\beta}$ (per year), although this is representative of changes in the vaccination level or birth rate as well. The y-axis shows measles incidence on January 1 of each year, so annual cycles are represented by a single point, biennial cycles by two points, and so on. Different shades correspond to different stable solutions of the model, which attract different sets of initial conditions (basins of attraction). For four values of the mean transmission rate ($\overline{\beta} = 500, 750, 1000, 1750$), basins of attraction of the various coexisting attractors are shown above the bifurcation diagram. Where multiple stable solutions coexist, stochasticity can induce complicated dynamics due to shifts among attractors (Chapter 6). The upper panels of Figure 5.17 show that the basins of coexisting attractors are more intermixed if β is smaller, so we expect the effects of stochasticity to be greater for smaller β (or, equivalently, when the effective β is reduced by vaccination or a decrease in birth rate).

How can we use this diagram to understand measles epidemics in London and Liverpool (Figure 5.16)? The estimated mean transmission rate for this period is $\beta_0 \sim 1{,}240$ per year (Anderson and May 1991), corresponding to a biennial attractor (gray line, Figure 5.17). Before 1950, epidemics were roughly annual; over the same brief period the birth rate was much higher, which greatly increased the effective mean transmission rate, leading to an annual cycle becoming stable (black line, far right of Figure 5.17). After 1968, recruitment rates steadily decreased because of mass vaccination (for example, when vaccine uptake reached 60%, the effective mean transmission rate was reduced to $\beta_0 \sim 750 = 0.6 \times 1240$ per year); this brought the system into the parameter region where there are multiple

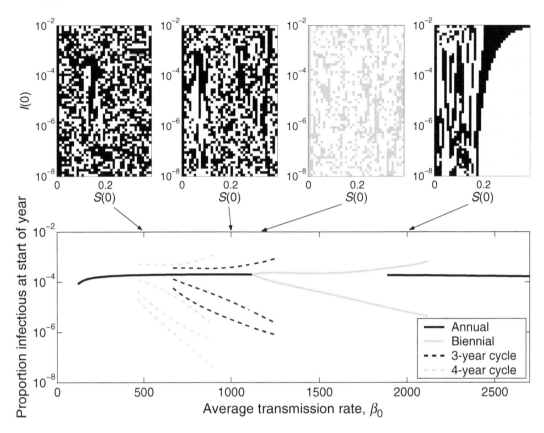

Figure 5.17. Exploring the consequences of changes in the mean transmission rate equivalent to changing the effective host birth rates (μ) for measles-like parameters. In the top panels, initial conditions that lead to annual epidemics are colored black, those that lead to biennial epidemics are dark gray, and those that lead to higher period cycles are light gray. Given the analyses of Keeling and Grenfell (2002), we assumed $b_1 = 0.29$ throughout the dynamic range. We also take a per capita birth rate of $\mu = 0.02$ per year, together with a latent period of $1/\sigma = 8$ days and an infectious period of $1/\gamma = 5$ days.

coexisting attractors with extremely intermixed basins. Stochastic effects then appear to cause frequent random jumps between these attractors (this can be confirmed using simulations—see Chapter 6), providing an explanation for the irregular epidemics in the vaccine era (Rohani et al. 1999). In Liverpool (Figure 5.16, lower graph), the birth rate was much higher than the mean in England and Wales from 1944 to 1968, leading to a higher effective transmission rate. This explains the roughly annual cycle of measles epidemics in this location over the same period. After 1968, the combination of vaccination and a lower birth rate brought Liverpool, like London, into the regime where irregular dynamics are predicted.

The bifurcation diagram in Figure 5.17 is plotted for a particular seasonal amplitude ($b_1 = 0.29$), but the qualitative conclusions of the above discussion are similar for a wide range of amplitudes. For much higher seasonality, the region with many attractors contains chaotic attractors as well. Such high seasonal amplitudes would not change our conclusion that measles dynamics will be irregular in this region. For lower seasonality, many of the attractor sequences cease to exist or end at higher β_0, but the "ghosts of departed attractors" influence the dynamics for low β_0. This means that often the attractor becomes weakly unstable, such that if the initial conditions are close to this attractor they may take a long time to converge to a stable attractor; this generates extremely long and erratic transient dynamics especially when the unstable attractor is chaotic such that many initial conditions are close by (Rand and Wilson 1991; Earn et al. 2000). Again, this supports the prediction of irregularity, so we would expect the same qualitative dynamical picture to emerge in places that have significantly higher or lower externally imposed seasonality.

5.3. SEASONALITY IN OTHER DISEASES

Although the dynamics of measles provides an ideal test case for our ideas about the effects of seasonality, it is important to extend these concepts to other diseases (with different parameters) or to other forms of seasonal forcing.

5.3.1. Other Childhood Infections

The approach outlined above allows us to understand primarily measles epidemics in large population centers in the modern era. We followed the analysis of Earn et al. (2000) to demonstrate how changes in birth rates and the onset of vaccination can give rise to different epidemic patterns, which may provide a qualitative explanation of observed case notifications in the big cities of the United Kingdom and the United States. However, this argument ignores a potentially very important issue concerning age structure (see Chapter 3). Whether thinking about measles epidemics under extensive vaccination regimes or different infections with smaller R_0 values, we need to consider the fact that a reduction in the basic reproductive ratio gives rise to increasing transmission among older age groups, for which the effects of seasonality may be substantially different. (From Chapter 2, we know that the average age of infection is $A \approx \frac{1}{\mu(R_0-1)}$.)

To understand these effects, we use some straightforward analysis to predict how the amplitude of seasonality might change with R_0. The value of b_1 or β_1 can be approximated by calculating the extent of mixing between susceptible and infectious school children of the same age. Based on the differential equations describing the SIR system, and ignoring heterogeneities in transmission and host vital dynamics, we can calculate (at equilibrium) the proportion of the population of age a that are susceptible by solving:

$$\frac{dX^*}{da} = -\beta(t)X^*(a)\frac{Y_T^*}{N} = -\frac{X^*(a)}{A}, \tag{5.13}$$

where Y_T is the total number of infecteds summing over all ages and A is the average age of infection (Anderson and May 1991). This gives a solution $X^*(a) = \exp(-\mu(R_0 - 1)a)$. We can now use equation (5.13) to obtain an explicit expression for the number of infectives of age a. Assuming $\frac{dY^*(a)}{dt} \sim 0$, we get $Y^*(a) = \frac{1}{\gamma}\frac{dX^*}{da}$, which yields

$$Y^*(a) = \frac{\mu(R_0 - 1)}{\gamma}e^{-\mu(R_0-1)a}. \tag{5.14}$$

Now, having established explicit equations describing $X(a)$ and $Y(a)$, we can estimate the relative importance of mixing at school by studying the ratio (ψ) of mixing within

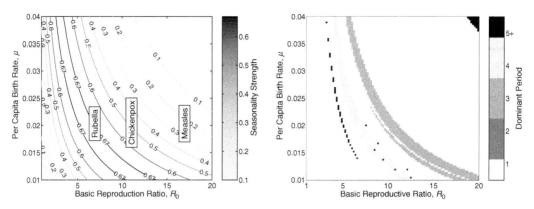

Figure 5.18. Exploring the consequences of changes in infection R_0 and host birth rates (μ) for the strength of seasonality in dynamics. Given the analyses of Keeling and Grenfell (2002), we assumed $b_1 = 0.29$ for measles during the 1950s and 1960s, when per capita birth rates were approximately 0.018 and R_0 is estimated to be 17. This allows us to use equation (5.15) to scale the amplitude of seasonality for other infections, such as chickenpox ($R_0 \sim 11$) and rubella ($R_0 \sim 7$). To produce this figure, we assumed children enter school at age $A_S = 5$ and leave at age $A_L = 16$. The precise shape of the contours (left-hand) giving the highest levels of seasonality varies slightly given different assumptions about A_S and A_L values, though the qualitative picture remains largely unaffected. The right-hand graph shows the dominant epidemic period as determined from the Fourier spectrum (Box 5.2) using an $SEIR$ with the average incubation and infection periods set to one week. All simulations are started at the unforced equilibrium solution.

school compared to random mixing:

$$\psi \propto \frac{\int_{\text{school ages}} X(a)Y(a)da}{\int X(a)da \int Y(a)da},$$

$$\propto \mu R_0 \{ e^{-2\mu R_0 A_S} - e^{-2\mu R_0 A_L} \}, \tag{5.15}$$

where A_S and A_L represent ages at which children start and leave school. Equation (5.15) allows us to establish the relative role of seasonality as host demographic rates (μ) and the basic reproductive ratio (R_0) vary (Figure 5.18). Intuitively, we would expect that for a given rate of births into the susceptible population, a range of R_0 values ensures that significant transmission occurs in the school-age classes and within this range the effects of seasonality in transmission will be most pronounced. For England and Wales data in the 1950–1960 era, with mean annual per capita births of approximately 0.018 per year, the window of R_0 values resulting in the largest seasonal variation in transmission is from 6–8, consistent with the estimated reproductive ratio for rubella, which is observed to have complex multiennial dynamics (Anderson and May 1982). As R_0 increases, however, the mean age at infection decreases and so does the variation in seasonal transmission. Hence, when attempting to explain the dynamics of childhood infectious diseases, it is crucial not only to obtain accurate estimates for the primary epidemiological parameters (such as transmission rates, the infectious and latent periods), but also to take into account the differential exposure to seasonality (Keeling et al. 2001a). This varying seasonality obviously has dynamic consequences for the infection within a population (Figure 5.18, right-hand graph). Multiennial epidemic cycles can be generated; the correspondence between the contours in the left-hand graph and the regions of particular periods in the right-hand graph (Figure 5.18) shows that the degree of mixing between school children (ϕ), and therefore the level of seasonality, plays the dominant role in determining the periodicity.

For a specified pattern of school openings and closures, differences in host demography (e.g., the per capita birth rate) and the basic reproductive ratio (R_0) result in different amplitudes of seasonality in disease transmission.

The methodology outlined here is only an approximation to using fully age-structured models (Chapter 3) in which the seasonal variability is naturally included in the mixing between schoolchildren. In addition, changes in the age distribution of infectious and susceptible individuals during the course of an epidemic may cause changes in the strength of seasonality experienced, leading to deviations away from the simple binary pattern.

5.3.2. Seasonality in Wildlife Populations

In wildlife populations, seasonal changes in flocking and social mixing could also generate pulses of high transmission rates for directly transmitted infections (Altizer et al. 2006). Indeed, seasonal changes in social grouping have been demonstrated for a wide range of species in response to variation in food resources and breeding behavior (e.g., Newton-Fisher et al. 2000). Fall and winter flocking behavior and aggregations at bird feeders have been suggested as increasing the transmission and prevalence of Mycoplasma gallisepticum in house finches (Altizer et al. 2004; Hosseini et al 2004). Regular increases in the incidence of rabies in skunks during the winter and spring could be driven by seasonal host crowding (Gremillion-Smith and Woolf 1988). Outbreaks of phocine distemper virus in seals have coincided with the breeding period when animals haul out and aggregate on beaches (Swinton et al. 1998). Seasonal changes in aggressive interactions (male:male) or courtship-related contacts during the breeding season could provide further opportunities for the transmission of directly transmitted pathogens. In many animals that breed in seasonal environments, annual aggregations coincide with long-distance movement events (Dingle 1996), and seasonal migration in insects, birds, and mammals could further drive variation in parasite pressure (Folstad et al. 1991; Loehle 1995). These seasonal changes to the transmission rate for wildlife infections can be dealt with in a similar manner to the seasonal changes experienced for human diseases because their root cause—the aggregation of hosts—is essentially the same.

5.3.2.1. Seasonal Births

An alternative source of seasonality can arise from concentrating host births into a period that is short relative to the full year. This will generate a pulse of hosts that are recruited into the population at approximately the same time each year, thus effectively expanding and contracting the base of susceptible hosts throughout the course of each year (Gremillion-Smith and Woolf 1988; Bolker and Grenfell 1995). Furthermore, juveniles recruited into the population are likely to be immunologically naive and more susceptible to a variety of pathogens. Levels of herd immunity could also decline when a pulse of new juvenile hosts enters the population, leading to greater risks of infection among susceptible adults.

A growing number of empirical studies point to seasonal births, possibly in combination with changes in social behavior, as a factor important to the dynamics of wildlife pathogens, with examples spanning cowpox virus and macroparasites infecting voles and wood mice (Montgomery and Montgomery 1988; Begon et al. 1999) to phocine distemper virus in seals (Swinton et al. 1998). Mathematical models based on the biology of both vertebrate and invertebrate systems further show that the seasonal timing of reproduction can influence the dynamics of host-pathogen systems (e.g., White et al. 1996; Dugaw et al. 2004; Ireland et al. 2004; Bolzoni et al. 2006).

One way to approach modeling seasonal births is simply to make the influx rate into the susceptible population time dependent, such that the susceptible equation becomes:

$$\frac{dX}{dt} = \alpha(t)N - \beta XY - \mu X, \tag{5.16}$$

This is online program 5.3

where $\alpha(t)$ represents the time-dependent per capita birth rate. Here we have assumed transmission is density dependent, in keeping with the standard models of wildlife diseases. Often, it is assumed that $\alpha(t) = \alpha_0(1 + \alpha_1 \cos(2\pi t))$, where as before α_0 is the baseline per capita birth rate and α_1 is the amplitude of seasonality. In this instance, because birth rates are seasonally varying, it is useful to add one extra equation to keep track of the total population size N:

$$\frac{dN}{dt} = \alpha(t)N - \mu N.$$

Using the same bifurcation analysis approach as described in previous sections, it is straightforward to demonstrate that dynamics in such a model are substantially simpler than when seasonality affects the transmission rate (Figure 5.19). For illustration purposes, if we assume model parameters for measles, we find that annual outbreaks dominate unless the variation in the birth rate is extreme. This is clearly in stark contrast with Figure 5.8, where oscillations with period 2 or higher are observed for $\beta_1 > 0.0455$. This would seem to suggest that incorporating environmental variability in births is perhaps less likely to give rise to complex dynamics than similar levels of variability in transmission.

The assumption of a sinusoidally varying birth rate is somewhat naive, we therefore consider an alternative extreme where births occur in an annual pulse. This parallels the work in measles where researchers have generally switched from sinusoidal to term-time seasonality. The per capita birth rate now becomes:

$$\alpha(t) = \begin{cases} \frac{\alpha_0}{T} & \text{if } 0 \leq mod(t, 1) < T \\ 0 & \text{otherwise.} \end{cases}$$

Therefore all births for the year are compressed into a portion T at the start of each year. As such, the dynamics represent the behavior of a large number of species in temperate regions where the breeding season may be relatively short. Additionally, we stipulate that $\alpha_0 = \mu$ such that births and deaths over the year cancel each other.

Figure 5.20 shows the period of attractors that exist for hosts with an annual pulse of births. For simplicity we have let $T \to 0$, such that all births happen at the same instant; however, further numerical studies have shown that the general pattern of periods remains unchanged even when T takes quite large values (such as $T = 0.5$ years). Several interesting features exist in this pattern of coexisting periods. First, low transmission and

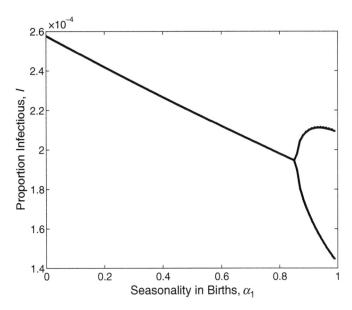

Figure 5.19. The bifurcation diagram showing the period of solutions when the birth rate is seasonally forced: The control-parameter, plotted on the x-axis, is the level of seasonality in the birth rate. Measles-like parameters are chosen so that comparison can be made with previous results: $\beta = 1240$ per year, $b_1 = 0$, $1/\sigma = 8$ days, $1/\gamma = 5$ days, $\alpha_0 = \mu = 0.02$ per year.

long life expectancy, or high transmission and short life expectancy, lead to an annual attractor (pale gray) being globally stable. Short life expectancy and low transmission can give rise to a four-year cycle (dark gray) which is again globally stable because no other attractors coexist for these parameters. Finally, we turn our attention to the bands of period 2 (diagonal hashing) and period 3 (vertical hashing). We observe that the 3-year cycle can coexist with both the annual and biennial attractor; in addition, the appearance and loss of this 3-year cycle as we move through parameter space is independent of the stability of the other periods. In contrast, the biennial cycle at its upper edge (long life expectancy) is formed by the annual cycle undergoing a period-doubling bifurcation, whereas at its lower edge both biennial and annual attractors coexist.

Despite the simplicity of the seasonality within this model, the emergent dynamics are relatively complex. This complexity increases as the infectious period becomes smaller such that longer-period attractors and even chaotic dynamics can be found.

5.3.2.2. Application: Rabbit Hemorrhagic Disease

In order to examine the relative dynamical consequences of different forms of seasonality, we now focus on wildlife diseases more closely and explore model dynamics using parameters estimated for Rabbit Hemorrhagic Disease in the United Kingdom (White et al. 2001, Ireland et al. 2004). In contrast to the above examples of seasonal variation in wildlife populations, we now allow *both* the transmission rate, β, and the per capita birth rate, ν, to be seasonally varying. In addition, following the generally observed behavior of

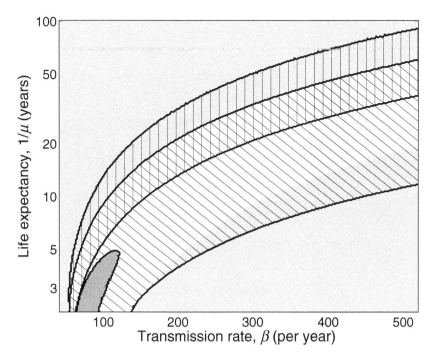

Figure 5.20. For an annual pulse of births, the parameter regimes for which various period attractors exist. Pale gray signifies annual dynamics, diagonal hashing shows the existence of biennial dynamics, vertical hashing represents three-year cycles, and dark gray signifies four-year dynamics. The diagram is computed using $1/\gamma \approx 13$ days, while the duration of the birth pulse T is reduced to zero. The period shown is the exact period of the attractor (Box 5.2), and the dominant period from the Fourier spectrum generally gives similar results with the exception that the region of exact period 4 is dominant by biennial epidemics.

natural populations, we include density-dependent regulation of the hosts (in terms of an increased death rate), which results in rewriting of the SIR system taking the basic host ecology into account:

$$\frac{dX}{dt} = \nu(t)N - \beta XY - \left(\mu + \frac{N}{K}\right)X, \tag{5.17}$$

This is online program 5.4

$$\frac{dY}{dt} = \beta XY - \left(\mu + m + \gamma + \frac{N}{K}\right)Y, \tag{5.18}$$

$$\frac{dZ}{dt} = \gamma Y - \left(\mu + \frac{N}{K}\right)Z, \tag{5.19}$$

$$\frac{dN}{dt} = \left(\nu(t) - \mu - \frac{N}{K}\right)N - \alpha Y. \tag{5.20}$$

Note that $(\nu - \mu)K$ represents the carrying capacity of the habitat, and competition for resources is assumed to affect all categories equally. In the absence of any seasonality, this system possesses a disease-free and an endemic equilibrium. The key difference between this system of equations (excluding seasonality) and the classic SIR equations is the presence of density-dependent host regulation (represented by the $\frac{N}{K}$ per capita mortality term), which makes this system analytically intractable. For the very high mortality levels associated with this disease ($m \gg \gamma$), the SIR model above approximates SI dynamics with very few rabbits entering the recovered class. Numerical exploration, however, reveals that as expected density dependence has a strongly stabilizing effect on the endemic equilibrium (Ireland et al. 2004).

We can explore the dynamics of incorporating a breeding season by making the birth rate seasonal (such that $v(t) = v_0(1 + \alpha_1 \cos(2\pi t))$). As shown in Figure 5.21 (top graph, black points), increasing levels of seasonality in the birth rate can generate a cascade of dynamical exotica with long-period multi-ennial cycles and eventually chaos. A qualitatively similar picture emerges when seasonality is incorporated solely into the transmission term, Figure 5.21 (top graph, gray points). An important aspect of these analyses is, however, that the bifurcation from annual to biennial cycles occurs for larger values of β_1 compared to the human versions of the SIR equations (without density dependence). Indeed, the results shown in Figure 5.21 (top graph) are rather sensitive to the choice of v_0. Higher values of this parameter result in a large influx of susceptibles, leading to annual epidemics irrespective of the amplitude of seasonality.

In some wildlife disease systems, such as conjunctivitis transmitted among house finches, two sources of seasonality arise from a breeding season which forces the birth rate and the winter/fall flocking behavior which forces the transmission rate (Hosseini et al. 2004). It is perhaps surprising that the incorporation of time-dependence in both host birth rates ($v(t)$) and transmission ($\beta(t)$) in equations (5.17)–(5.20) does not substantially change the dynamics, Figure 5.21 (bottom graph). Once again, we find that when the amplitude of seasonality is small, purely annual cycles ensue. As the total seasonality exceeds approximately 0.4, dynamics become extremely complex; we find a window of coexisting annual, biennial, and multiennial dynamics is found which gives way to long-period and eventually chaotic epidemics. However, when both breeding and transmission are very strongly seasonal, we once again recover more simple behavior with stable biennial cycles. In fact, if we examine the Fourier spectra (Box 5.2) in some detail we find that annual epidemics occur across the entire parameter regime, although their amplitude is modulated by multiennial cycles. This aanlysis although demonstrating some broad dynamical effects, is clearly both rather brief and somewhat ecologically naive. In reality, we may need to examine carefully the precise timing of the two seasonal mechanisms. For the house finch system, for example, the breeding season is during April to September whereas flock sizes peak during the winter months. To take these factors into account, any detailed study of the system would need to incorporate a fixed phase difference (ψ) between seasonality in breeding and transmission dynamics (i.e., $\beta(t) = \beta_0(1 + \cos(2\pi t + \psi))$ and $v(t) = v_0(1 + \cos(2\pi t + \psi))$). As shown by Hosseini et al. (2004), the addition of such features generates realistic double epidemics within a calendar year.

5.4. SUMMARY

In this chapter, we have explored epidemics in seasonally varying environments. The source of this environmental variability is varied, ranging from social aggregation of hosts (as seen in measles, phocine distemper, or mycoplasma), pulses of susceptible recruitment

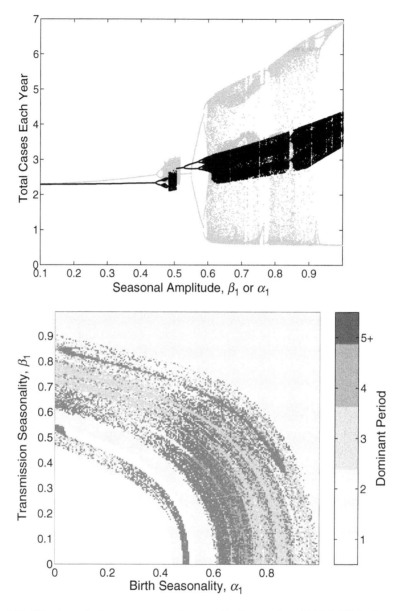

Figure 5.21. The dynamical consequences of seasonality in equations (5.17-5.20). In the top graph seasonality is implemented in either the *per capita* birth rate ($\nu = \nu_0(1 + \alpha_1\cos(2\pi t))$, $\beta = \beta_0$; black points) or the transmission rate ($\beta = \beta_0(1 + \beta_1\cos(2\pi t))$, $\nu = \nu_0$; gray points). In the bottom graph both disease transmission and population birth rates are assumed seasonal and in phase, although asynchronous forcing produces similar results. The period shown is the dominant multiennial period of the Fourier spectrum (Box 5.2). These bifurcation diagrams are computed using $\gamma = 0.025$ per day, $\mu = 0.01$ per day, $m = 0.475$ per day, $\nu_0 = 0.02$ per day, $\beta_0 = 0.936$ per day and $K = 10000 \Rightarrow N \approx 100$.

following breeding seasons (such as rabbit hemorrhagic disease), and regular disturbances (such as lettuce stuff). We have seen that a number of different dynamical changes can follow the introduction of seasonality. In some cases, we get harmonic oscillations, whereby epidemics simply track the periodic forcing (e.g., low levels of forcing in the measles model). As the amplitude of forcing increases, however, the phenomenon of subharmonic resonance is observed with epidemics that are integer-multiples of the forcing period (Greenman et al. 2004). In other cases, if the period of forcing matches the underlying natural period of the system, we obtain harmonic resonance.

By concentrating largely on the SIR paradigm, we have been able to demonstrate a number of important epidemiological principles:

➤ For childhood infections such as measles, chickenpox, and rubella, it is empirically established that rates of transmission peak at the start of the school year and steadily decline, reaching a trough during the summer months.

➤ Small levels of seasonal forcing in transmission can give rise to harmonic oscillations, with large-amplitude cycles in disease prevalence that have the same period as the forcing function.

➤ Moderate levels of seasonality can result in dynamics that have periods that are integer-multiples of the forcing function period—this phenomenon is known as subharmonic resonance.

➤ The effects of forcing are most pronounced when its period matches the natural period of the system—this is referred to as harmonic resonance.

➤ In systems that experience seasonal forcing, it is possible to observe qualitatively different dynamics (or multiple attractors) for the same combination of parameter values, depending on initial conditions.

➤ Whenever multiple attractors coexist, a full understanding of the system requires the basins of attraction to be determined.

➤ Seasonality may be mathematically implemented using alternative formulations, which qualitatively affect the predicted dynamics.

➤ The strength of seasonality in transmission depends on the mean age at infection, as determined by an infection's basic reproductive ratio, R_0.

➤ Seasonality in births is dynamically less destabilizing, with bifurcations from annual to multiennial dynamics occurring at larger seasonal amplitudes.

Chapter Six

Stochastic Dynamics

All the models considered thus far have been deterministic. This means they are essentially fixed "clockwork" systems; given the same starting conditions, exactly the same trajectory is always observed. Such a Newtonian view of the world does not apply to the dynamics of real pathogens. If it were possible to "re-run" a real-world epidemic, we would not expect to observe exactly the same people becoming infected at exactly the same times. Clearly, there is an important element of chance. Stochastic models are concerned with approximating or mimicking this random or probabilistic element. In general, the role played by chance will be most important whenever the number of infectious individuals is relatively small, which can be when the population size is small, when an infectious disease has just invaded, when control measures are successfully applied, or during the trough phase of an epidemic cycle. In such circumstances, it is especially important that stochasticity is taken into account and incorporated into models.

This chapter details three distinct methods of approximating the chance element in disease transmission and recovery: (1) introducing chance directly into the population variables, (2) by random parameter variation, and (3) individual-level, explicit modeling of the random events (Bartlett 1957). This third method is generally the most popular (Mollison et al. 1994; Levin and Durrett 1996) and will be the predominant focus of this chapter, illustrated by examples from the recent literature. All these examples have two elements in common. First, they predict different outcomes from the same initial conditions and, as such, multiple simulations are required to determine the expected range of behavior. Second, they all require the use of a random number-generating routine—this can be thought of as a rather clever computational trick that can deliver an apparently random sequence of numbers in some prescribed range (Box 6.1).

Box 6.1 Random Number Generation

The question of how we can use a computer algorithm to generate a truly random series of numbers has received considerable attention from mathematicians and computer scientists. Ultimately, because computers are deterministic devices, any random number generator (RNG) must also be deterministic. Hence, the set of results given by a RNG must at some point repeat and exhibit some form of pattern (albeit a very complex one) due to the underlying computer code. Some researchers have attempted to overcome these limitations by sampling and processing sources of entropy outside the computer, such as atmospheric noise, radioactive decay, or even lava lamps! In the absence of these more elaborate methods, others have focused on writing RNGs with cycles as long as possible and the patterns too complex to be detected or introduce spurious results. As such, RNGs are inherently quite complex and slow, often being the slowest step of a computer program; therefore, having an efficient routine is a must.

Fortunately, such routines already exist and are widely available. Many computer languages have RNGs built in as standard, or sample code can be found on the Web or in Numerical Recipes books (Press et al. 1988). This chapter assumes that the RNG, $RAND$, returns a high-precision number uniformly distributed and strictly between 0 and 1—values of exactly 0 or 1 should be discounted because they can produce spurious results. Some RNGs produce integers between zero and some maximum value, in which case the random number must be rescaled before use.

One element all RNGs share is their need for an initial seed value to start the processes. Due to their deterministic nature, given the same seed, the same sequence of "random" numbers will be generated. For this reason, the seed is usually chosen to be the current time (to the nearest second) because this is an ever-changing value.

Other routines exist that will generate random numbers from different distributions, such as normal, Poisson, or binomial (see, for example, Press et al. 1988). Because it is frequently used in the first two sections of this chapter, the following pseudo-code (Box-Mueller 1958) will pick random numbers from a normal distribution, with mean $MEAN$ and variance VAR.

1. $R1 = 2 \times RAND - 1$.
2. $R2 = 2 \times RAND - 1$.
3. $R3 = R1 \times R1 + R2 \times R2$.
4. if $R3 \geq 1$ or $R3 = 0$, go to line 1.
5. $R4 = \sqrt{-2 \times VAR \times \log(R3)/R3}$.
6. $NORM = MEAN + R1 \times R4$.

To enable a fair comparison of the different methods of implementing stochasticity, we primarily focus on the SIR model with births and deaths, which, in the deterministic framework, always converges to an equilibrium in an oscillatory manner (Chapter 2). We also utilize a constant set of parameters: $R_0 = 10$, $1/\gamma = 10$ days, and $\nu = \mu = 0.02$ per year, which corresponds to a mean life expectancy of 50 years and a constant population size. For stochastic models, we need to consider both the long-term behavior and the short-term "transient" dynamics, as well as study both the mean and variance of the number of infected cases. Because stochastic models have close relationships to integer-valued (or individual-based) models, we generally deal with the number (as opposed to the proportion) of susceptibles, infecteds, and recovereds, labeled X, Y, and Z.

Five key features distinguish stochastic models from their deterministic counterparts. Although not evident for all forms of stochastic models, these features can have a profound impact on epidemiological dynamics and will be a consistent theme throughout this chapter. In brief, these distinguishing features are:

1. *Variability between simulations*. The most obvious element of any stochastic model is that different simulations give rise to different outcomes. This implies that although the general statistical properties (such as the mean and the variance) may be accurately predicted, it is generally impossible to predetermine the precise disease prevalence at any given point in the future.
2. *Variances and covariances*. The continual perturbations caused by the random nature of stochastic equations leads to variation in the prevalence of disease and the number of susceptibles. Additionally, the interaction between stochasticity and deterministic dynamics generally causes negative covariance between the numbers of infectious and susceptible individuals, which in turn can cause the mean population levels $(\overline{X}, \overline{Y})$ to deviate from the deterministic equilibria.
3. *Increased transients*. Stochastic perturbations away from the equilibrium solution are countered by the generally convergent behavior of the underlying deterministic dynamics. When far from the endemic equilibrium point, these restorative forces are usually strong and dominate, so that the model acts much like the deterministic

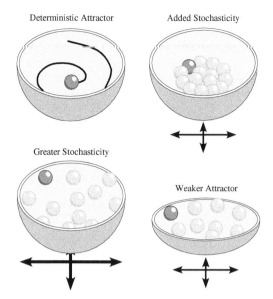

Figure 6.1. Four examples showing how the effects of stochasticity can be conceptualized. Top left: deterministic attractor with no stochasticity. Top right: moderate amounts of stochasticity and a strong underlying attractor. Bottom left: much higher levels of stochasticity lead to greater variation. Bottom right: moderate stochasticity but with a much weaker attractor; this also produces greater variation.

equations. Thus, the dynamics of stochastic models can be conceptualized as resulting from random perturbations away from, and transient-like return toward, the deterministic attractor.

4. *Stochastic resonance*. As most SIR-type disease models approach equilibrium in a series of decaying epidemics, the increased transient-like dynamics of stochastic models leads to oscillations close to the natural frequency (Chapter 2). Thus, stochasticity can *excite* epidemic oscillations around the normal endemic prevalence, leading to sustained cycles.

5. *Extinctions*. Those probabilistic models that are integer-valued may additionally suffer from stochastically driven extinctions. In closed populations, chance fluctuations will always result in eventual disease extinction, irrespective of population size. Long-term persistence requires the import of pathogens from an external source. Similar extinctions may occur during the early stages of invasion.

Figure 6.1 shows an intuitive way to conceptualize the effects of stochasticity. The bowl shape represents the underlying deterministic attractor, which in turn completely defines the dynamics of a ball placed inside the bowl. Stochastic forces can then be visualized as the effects of shaking the bowl. When there are no stochastic forces (top left), the ball follows the deterministic attractor and eventually settles in the bottom of the bowl. When a moderate amount of stochasticity or shaking is added (top right), the position of the ball is, in the long run, less certain although it generally lies close to the base of the bowl. Substantially greater deviation from the deterministically predicted position of the ball occurs when either there is more shaking (greater stochasticity) or a shallower bowl (a weaker underlying attractor), as shown in the two lower panels. Although there are clear phenomenological discrepancies between the gravity-driven dynamics of the ball and the epidemiological dynamics of a disease, this is a useful pictorial analogy that can help us conceptualize the interactions between stochasticity and determinism.

6.1. OBSERVATIONAL NOISE

The simplest form of noise that can be introduced into disease modeling is observational noise. For this approach, the underlying epidemic dynamics remain the standard deterministic differential equations, but there is assumed to be some uncertainty in the recorded data. Information on notifiable diseases, such as measles in England and Wales (Finkenstädt and Grenfell 1998), foot-and-mouth (Keeling et al. 2001b), or classical swine fever (Elbers et al. 1999), tends to be highly detailed and fairly accurate. In contrast, infectious diseases that are largely asymptomatic such as meningocaccol infection (Ranta et al. 1999) or diseases of wildlife (Hudson et al. 2001) may be subject to greater uncertainty in their reporting.

For any specific infectious disease, the stochasticity due to observational noise can most readily be quantified as either incomplete reporting of true cases, or mis-diagnosis (both type I and type II errors). For the measles data from England and Wales, underreporting is well documented, with an estimated 60% of the actual cases being reported (Fine and Clarkson 1982; Finkenstädt and Grenfell 2000). In contrast, during the 2001 foot-and-mouth outbreak in the United Kingdom, misdiagnoses were frequent—some animals infected with other diseases were incorrectly diagnosed as having foot-and-mouth, although it is unlikely that animals actually infected with foot-and-mouth could have escaped detection.

The recorded number of infected individuals, Y_{recorded}, can be represented as the sum of two binomial distributions:

$$Y_{\text{recorded}} = \text{Bin}(p_r, Y_{\text{true}}) + \text{Bin}(p_m, N - Y_{\text{true}}),$$

where p_r is the probability that a case is correctly reported and p_m is the probability that a healthy individual is misdiagnosed; N is the total population size. Although such considerations are obviously important when interpreting case-report data (Finkenstädt and Grenfell 1998; Rohani et al. 2003), they do not interact with or influence the underlying epidemiological dynamics. The true number of cases always tends to the deterministic fixed point and observations are purely determined by the two probabilities, p_r and p_m. Even in recurrent epidemics, it has been shown that observational error affects the amplitude of data, not the periodicity (see, for example, Miller et al. 1992).

Observational noise does not impact epidemiological dynamics, it modifies only the reported data.

6.2. PROCESS NOISE

A more intuitive and fundamentally different way to incorporate noise is to introduce it directly into the deterministic equations (Bjørnstad, Finkenstädt, and Grenfell 2002). As such, the dynamics at each point in time are subject to some random variability and this variability is propagated forward in time by the underlying equations. We are, therefore,

Box 6.2 Inclusion of Noise in Differential Equations

Consider the stochastic differential equation model:

$$\frac{dx}{dt} = \text{Noise}$$

The most simple means of solving such equations is using the Euler method of integration, breaking time into small components, δt.

$$x_{t+\delta t} = x_t + \delta t \frac{dx}{dt} = x_t + \delta t \text{Noise} = x_0 + \delta t \sum_1^{\frac{t}{\delta t}} \text{Noise}$$

Thus, the equations progress as the summation of many small noise terms, with mean zero.

If successive noise terms are independent, then the variance of x at any time decays to zero as the step size is made smaller. Thus, in the limit $\delta t \to 0$, when the updating method is exact, all the noise terms effectively cancel. This is a reflection of the well-known problem that there is no simple mathematical method of expressing the noise term.

The simplest solution to this problem is to scale the noise term with respect to the integration step. Throughout this chapter we shall assume that Noise = RANDN—independent Gaussian with mean 0 and variance 1—and that:

$$\xi = \frac{\text{Noise}}{\sqrt{\delta t}},$$

such that as the time step of integration decreases, the amplitude of the noise used in each step, ξ, increases. If this scaling is used with higher-order integration methods (such a Runge-Kutta; see Press et al. 1988), then the noise should be calculated *before* each integration step, though it is often safer to use forward Euler.

This new definition of ξ has the properties that we require. Such that if

$$\frac{dx}{dt} = f\xi,$$

then, averaged over multiple simulations, the mean of x is zero while the standard deviation grows like $f\sqrt{(t)}$, and therefore the dynamics correspond to a random walk.

concerned with the interplay between deterministic and stochastic forces—how they cancel out or amplify each other.

As a starting point, let us incorporate noise into only the transmission term (Marcus 1991). Let $\xi(t)$ be a time series of random deviates derived from the normal distribution with mean zero and unit variance (see Box 6.2). The basic equations, assuming frequency-dependent (mass-action) transmission, are transformed to:

This is online program 6.1

$$\frac{dX}{dt} = \nu N - [\beta XY/N + f(X, Y)\xi] - \mu X,$$

$$\frac{dY}{dt} = [\beta XY/N + f(X, Y)\xi] - \gamma Y - \mu Y, \qquad (6.1)$$

$$\frac{dZ}{dt} = \gamma Y - \mu Z.$$

For generality, a function $f(X, Y)$ has been included to scale the randomness in response to the current variable sizes.

6.2.1. Constant Noise

Figure 6.2 shows examples of long-term dynamics as the value of f and hence the amount of noise is varied. As expected, when more noise is added, the dynamics deviate further from the deterministic equilibrium. However, rather than these deviations being completely random, there is a distinct oscillatory component to the dynamics. This is caused by the interaction between deterministic and stochastic forces—a common feature in all dynamically stochastic models (Bartlett 1957; Renshaw 1991; Rohani et al. 2002; McKane and Newman 2005). The noise terms allow trajectories to wander away from the equilibrium, but the underlying deterministic clockwork forces them back toward the equilibrium point. This return movement closely matches the deterministic prediction of decaying oscillations with a natural period (see Chapter 2). As pointed out by Grossman (1980), in general, the SIR equilibrium is weakly stable (because the real part of the dominant eigenvalue is small relative to the imaginary part). As a result, the underlying oscillatory dynamics may be "excited" by noise. This process is referred to as stochastic resonance and is a well-known phenomenon in physics. We can therefore see noise (and most forms of dynamic stochasticity) as overcoming deterministic forces when these are weak and close to the equilibrium point, but dominated largely by the deterministic behavior far from the equilibrium point. This tension between deterministic and stochastic forces means that the transient dynamics on the approach to the deterministic equilibrium can dominate the observed stochastic behavior (Rohani et al. 2002; Bauch and Earn 2003). However, if the noise term is very large (greater than shown), it can completely swamp any deterministic component and a pure random-walk is observed.

Noise in the transmission term can cause the number of cases to resonate at (or near) the natural frequency of the system. Hence, the observed stochastic dynamics more closely match the transient behavior of the deterministic equations.

Figure 6.2 also shows how the aggregate properties of the dynamics vary with the noise level. The variance in the number of infectious individuals increases almost linearly with the variance in the noise. For the very highest levels of noise, however, the variance does not increase linearly; this is due to the action of strong nonlinear forces that operate when the amplitude of the resonant epidemics becomes large and the number of infecteds is occasionally forced close to zero. Perhaps more surprisingly, there is a corresponding decrease in the covariance between susceptible and infected individuals. This is because if a particular value of the noise is "good" for the infected population ($\xi \geq 0$), it is "bad" for the susceptibles; hence, increasing noise levels generate increasingly large amplitude fluctuations with a strong negative covariance between X and Y.

Noise in the transmission term causes variance in the number of infecteds and in the number of susceptibles, and negative covariance between them. The magnitude of these values increases almost linearly with the variance of the noise.

This negative covariance has implications for the average dynamics. The rate at which cases arise ($\beta XY/N$) is now reduced, because the product of X and Y is smaller than expected due to the negative covariance. This reduction in transmission (and therefore a reduction in the effective reproductive ratio, $R = R_0 X/N$), is reflected by a slight increase in the mean number of susceptibles. This is a very interesting concept: Background stochasticity does not just cause variations about the deterministically predicted fixed

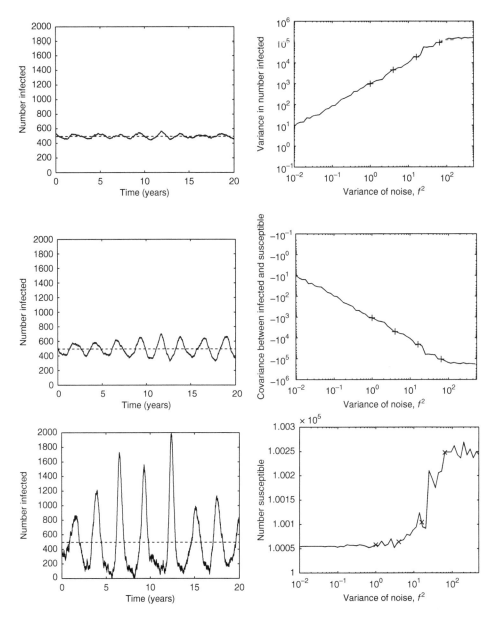

Figure 6.2. The left column gives examples of the dynamics of the SIR model with births and deaths ($\mu = \nu = 0.02$ per year, $R_0 = 10$, $1/\gamma = 10$ days, $N = 10^6$). The amount of noise added to the transmission terms increases from top to bottom ($f = 1, 2$, and 8). The deterministic equilibrium is depicted by the horizontal dashed line. The right column shows how the variance in the number infected, the covariance between susceptible and infecteds, and the average number of susceptibles change with the variance of the noise. From biological considerations, if noise ever forced the number of infected individuals below zero, the number was reset to zero—this only occurs for the largest levels of noise. Dots on the right-hand graphs mark the values of f used to generate the left-hand graphs.

points, but can actually cause significant changes to the mean values (Coulson et al. 2004). The nonlinear nature of the disease equations (in particular the product of X and Y in the transmission term) allows stochasticity to modify the mean values. From the graph it is clear that stochasticity has little effect on the mean values, primarily because the oscillations are relatively small and therefore the localized dynamics are almost linear. However, for small population sizes or diseases that show frequent localized extinctions or strong oscillatory resonance, the impact of the negative covariance on the mean can be far more substantial.

Stochasticity can introduce changes to the mean values as well as causing variation about the means.

6.2.2. Scaled Noise

Although the simple method of including noise outlined above provides variability, the amplitude of such random terms has so far been assumed to remain constant. In practice, this is rarely the case. Generally, the absolute magnitude of the variability increases with increasing population size, although the relative magnitude usually decreases (Keeling and Grenfell 1999; Bjørnstad et al. 2002). If events, such as transmission, occur at random, then in any short time interval the number of events will be Poisson distributed and hence their variance will equal the mean (Kendall 1949). This provides a direct means of determining the magnitude of the noise term, denoted f in the original equation (6.1), such that $f = \sqrt{\text{rate}}$. This concept can be extended still further, such that each event is associated with a noise term:

$$\frac{dX}{dt} = [\nu N + \sqrt{\nu N}\,\xi_1] - [\beta XY/N + \sqrt{\beta XY/N}\,\xi_2] - [\mu X + \sqrt{\mu X}\,\xi_3],$$

$$\frac{dY}{dt} = [\beta XY/N + \sqrt{\beta XY/N}\,\xi_2] - [\gamma Y + \sqrt{\gamma Y}\,\xi_4] - [\mu Y + \sqrt{\mu Y}\,\xi_5], \qquad (6.2)$$

$$\frac{dZ}{dt} = [\gamma Y + \sqrt{\gamma Y}\,\xi_4] - [\mu Z + \sqrt{\mu Z}\,\xi_6].$$

This is online program 6.2

This set of equations now contains six distinct noise terms ($\xi_1, \ldots \xi_6$), one for each event type; the same noise is used for each event even though it may appear in more than one equation. Although at equilibrium the event rates cancel each other (such that $\frac{d}{dt} = 0$), the noise terms add together. Thus, the amount of noise is determined by the absolute magnitude of opposing rates (such as births, which replenishes susceptibles, and death and infection, which decreases susceptibles), rather than by the rates of change themselves.

The effects of noise from events add together, even though the effects of the events themselves may cancel each other.

In Figure 6.3, we show the typical dynamics of models with this more complete form of scaled noise. Large populations (10 million or 100 million) are affected only slightly by such noise terms, with dynamics that are close to those predicted by the deterministic model. In contrast, smaller population sizes (smaller than 100,000) experience proportionally more noise and their behavior lies further from the deterministic ideal (Keeling and Grenfell 1999; Rohani et al. 2002). Once again, the presence of noise can be seen to induce cycles close to the natural epidemic period in the smaller population sizes (focusing on years 15–20 for $N = 10^6$). A log-log plot of variance against the mean number

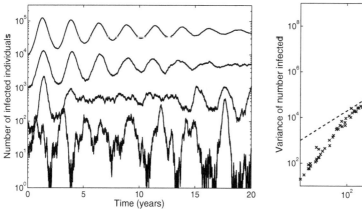
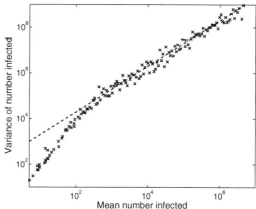

Figure 6.3. The dynamics of SIR epidemics ($\mu = \nu = 5.5 \times 10^{-5}$ per day, $R_0 = 10$, $1/\gamma = 10$ days) with scaled noise. The left-hand graph shows how the expected oscillatory behaviour becomes disrupted by noise in smaller populations, whereas large populations conform close to the deterministic ideal ($N = 10^5$, 10^6, 10^7, 10^8). The right-hand graph show the meanvariance relationship for a range of population sizes from 10,000 to 100 million. The dashed line represents $var = 100 \times mean$, clearly showing how the scaling operates at large population sizes.

infected, Figure 6.3 (right-hand graph), provides a concise method of summarizing the variability across a range of population sizes. For scaled noise, and large population sizes (with more that 200 infected individuals on average), the variance scales linearly with the mean and therefore with the population size as well. This linear behavior is expected for a wide range of stochastic epidemiological (as well as ecological) models, irrespective of the detailed dynamics or the way that stochasticity is introduced (Keeling and Grenfell 1999; Keeling 2000a). Thus, although large populations behave more like the deterministic equations, the absolute level of variation is greater than in small populations. For very small populations, this linear relationship between the mean and variance is destroyed due to the strong nonlinearities that operate when relatively large amplitude epidemics are triggered.

A log-log plot of mean against variance is a useful method of summarizing the variability with population size. For large population sizes, we expect the variance in the number of cases to be proportional to the mean.

6.2.3. Random Parameters

An associated modeling technique comparable to the introduction of noise terms is to consider the governing parameters to be noisy (Rand and Wilson 1991; Roberts and Saha 1999; Mahul and Durand 2000; Dexter 2003). Although mathematically equivalent, this alternative mechanism of including randomness has a very different underlying premise. The previous models assumed that the population levels were subject to random fluctuations, whereas the models in this section consider the basic parameters to vary. This can be for one of two reasons. First, there is variability in the parameters due to some external, unpredictable force. For example, the transmission rate β may be modified by climatic conditions; many viruses survive better in cold, damp conditions with little ultraviolet light (Ansari et al. 1991; Abad et al. 1994; Albina 1997). Fluctuations in temperature or humidity will therefore be translated to fluctuations in β; such variability is independent of disease prevalence or the number of susceptibles. A second source of parameter fluctuations is individual variation, both in terms of the pattern of contacts and the immunological response to infection (resulting in variable transmission rates). In this case, the fluctuations will be determined by which individuals are infected at that time. As such, this form of stochasticity would be expected to decrease when more individuals are infected because the variability between them will be averaged out. We can consider this as a special case of the structured population models discussed in Chapter 3, where mixing is random and all individuals experience the same force (risk) of infection.

Focusing on variability in the transmission rate, the basic equations for the SIR model become:

$$\frac{dX}{dt} = \nu N - \beta(1 + F\xi)XY/N - \mu X,$$
$$\frac{dY}{dt} = \beta(1 + F\xi)XY/N - \gamma Y - \mu Y. \tag{6.3}$$

When the variability is due to an external force, F is generally constant. However, when the noise comes from population heterogeneity in transmission we expect $F \propto Y^{-\frac{1}{2}}$—this is because the variability in the *average* transmission rate becomes smoothed out when many individuals are infected. (If the stochasticity is due solely to heterogeneity in susceptibility, then $F \propto X^{-\frac{1}{2}}$ and in general the effects will be weaker.) The proportionality constant in either case is determined by the magnitude of the external fluctuations or by the variability of the population, respectively.

Parameter noise can either be constant (if due to external factors) or decrease with the population level (if due to differences between individuals).

6.2.4. Summary

A range of models have now been described that can include stochasticity by modifying the basic form of the differential equations. This modification is achieved by inserting appropriately scaled noise, thereby mimicking a variety of random effects. We can now contrast these approaches, and comment on their general suitability, advantages, and disadvantages.

6.2.4.1. Contrasting Types of Noise

Let us contrast the differences between the models (equations (6.2) and (6.3)) and the early versions with simple additive noise (equation (6.1)). We do this by noting that f in the original formulation (equation (6.1)) is equal to $\beta F XY/N$ in the new model (equation (6.2)). Hence, we have four distinct noise scenarios:

1. Plain Additive Noise. f is constant.
2. Scaled Additive Noise. $f = \sqrt{\beta XY/N}$.
3. External Parameter Noise. F is constant, so $f \propto \beta XY/N$.
4. Heterogeneous Parameter Noise. $F \propto Y^{-\frac{1}{2}}$, so $f \propto \beta X\sqrt{Y}/N$.

From this comparison, we may consider how these noise terms (f) scale as the population size, N, changes, noting that in general X and Y are proportional to N. Scenario 1 is independent of the population size, scenarios 2 and 4 scale with the square root of the population size, and scenario 3 is proportional to the population size. Thus, when the population size is large the dominant error term is due to externally driven fluctuations in the parameters (scenario 3). Additionally, when there is considerable variability in an individual's response to infection, the parameter noise due to this factor (scenario 4) can easily exceed the scaled additive noise (scenario 2) due to dynamic variability. However, despite the potential importance of parameter noise, it has received relatively little attention in the epidemiology literature.

Noise can be generated from a variety of sources. The relative magnitude of these noise terms depends on the population size. External parameter noise dominates when the population is large.

6.2.4.2. Advantages and Disadvantages

Before we consider event-driven approaches, it is important to reflect on the relative merits of the methods outlined in this section. Including noise into the standard differential equations has three main benefits. First, the modification to the basic equations is relatively straight forward and therefore (with a little care), the same techniques can be used to iterate the equations and determine the dynamics. Second, the correspondence between the standard deterministic models and the stochastic equations is clear—as the noise terms are reduced to zero, we regain the deterministic dynamics. Finally, the computational overheads associated with this form of stochastic model are small in comparison to the techniques used for demographic stochasticity; in particular, simulation times are independent of population size.

These stochastic equations suffer from one major drawback: They do not incorporate the discrete, individual nature of populations and are therefore not a suitable model when population levels (e.g., the number of infectious individuals) are small. Techniques such as scaled noise (Section 6.2.2) or random parameters due to population heterogeneity (Section 6.2.3) are therefore trapped between two opposing elements. Such models differ substantially from their deterministic counter-part only when population levels are low, but are accurate representations of the dynamics only when population levels are large. Thus, although these noise-based models allow for the extinction of pathogen, their accuracy for such individual-based processes is questionable. For this reason the individual-based nature of demographic stochasticity is the predominant method of capturing stochasticity.

In summary, these noise-based modifications to the standard differential equations are most appropriate when considering the effects on large populations of parameter variation due to external sources. They can also help shed some analytical insights into the effects of stochasticity in general (see Section 6.6.1). However, when dealing with low numbers of individuals, an alternative approach is required.

6.3. EVENT-DRIVEN APPROACHES

Modeling approaches that incorporate demographic stochasticity are becoming increasingly popular. This is in large part due to their highly mechanistic approach to including randomness and the individual nature of their formulation (Bartlett 1956; Tille et al. 1991;

Garner and Lack 1995; Barlow et al. 1997; Smith et al. 2001). Demographic stochasticity is defined as fluctuations in population processes that arise from the random nature of events at the level of the individual. Therefore, even though the baseline probability associated with each event is fixed, individuals experience differing fates due to chance. Additionally, in contrast to the previous methods, the number of infectious, susceptible, and recovered individuals is now required to be an integer—we deal with whole numbers of people, animals, or other organisms.

6.3.1. Basic Methodology

Event-driven methods require explicit consideration of events. For the standard SIR model, we need to consider the six events that can occur, each causing the numbers in the relative classes to increase or decrease by one:

- Births occur at rate μN. Result: $X \to X + 1$.
- Transmission occurs at rate $\beta \frac{XY}{N}$. Result: $Y \to Y + 1$ and $X \to X - 1$.
- Recovery occurs at rate γY. Result: $Z \to Z + 1$ and $Y \to Y - 1$.
- Deaths of X, Y, or Z (three independent events) occur at rate μX, μY, and μZ. Result: $X \to X - 1$, $Y \to Y - 1$ or $Z \to Z - 1$.

There are different ways of implementing this framework, though most practitioners use Gillespie's Direct Method (Gillespie 1977). This scheme first estimates the time until the next event, based on the cumulative rates of all possible events. Then, by converting event rates into probabilities, it randomly selects one of these events. The time and numbers in each class are then updated according to which event is chosen. This process is repeated to iterate the model through time (see Section 6.4.1.1 and Box 6.3). Here, noise affects only the probabilities associated with the fates of individuals and the updating of each consecutive event is independent—there is no assumption concerning environmental stochasticity (e.g., "good" versus "bad" years).

Box 6.3 Event-Driven Approaches: Gillespie's Direct Algorithm

In his seminal 1977 paper, Gillespie outlined two alternative, mathematically equivalent, "stochastic simulation algorithms" (SSAs): the direct method (outlined below) and the first reaction method (Box 6.4). The following pseudo-code provides an efficient and accurate implementation of demographic stochasticity.

1. Label all possible events $E_1, \ldots, E_n$.
2. For each event determine the rate at which it occurs, $R_1, \ldots, R_n$.
3. The rate at which any event occurs is $R_{total} = \sum_{m=1}^{n} R_m$.
4. The time until the next event is $\delta t = \dfrac{-1}{R_{total}} log(RAND_1)$.

5. Generate a new random number, $RAND_2$. Set $P = RAND_2 \times R_{total}$.
6. Event p occurs if

$$\sum_{m=1}^{p-1} R_m < P \le \sum_{m=1}^{p} R_m.$$

7. The time is now updated, $t \to t + \delta t$, and event p is performed.
8. Return to Step 2.

Event-driven approaches require integer-valued variables and a probabilistic fate of individuals.

6.3.1.1. The SIS Model

Before discussing a general algorithm (see Box 6.3), we consider the particular problem of an SIS disease without births or deaths, which is a common model for the dynamics of sexually transmitted diseases (Jacque and Simon 1993; Allen and Burgin 2000; Welte et al. 2000; Chapters 2 and 3). The iteration of this model is simplified because it has only two possible events: transmission $(X \to X - 1, Y \to Y + 1)$ and recovery $(X \to X + 1, Y \to Y - 1)$. Suppose that at time t we have X susceptibles and Y infecteds, which make up the entire population $(N = X + Y)$. The underlying differential equations that inform about the rates at which events occur are:

$$\frac{dX}{dt} = -\beta XY/N + \gamma Y,$$

$$\frac{dY}{dt} = \beta XY/N - \gamma Y.$$

Hence, transmission occurs at rate $\beta XY/N$, recovery occurs at rate γY, and so either event occurs at total rate $(\beta XY/N + \gamma Y)$. Letting $RAND_1$ denote a uniform random number between (but not including) 0 and 1, the time until either event occurs is given by:

$$\delta t = \frac{-\log(RAND_1)}{\beta XY/N + \gamma Y},$$

which is equivalent to the assumption used previously (Section 6.2.2) that in a small time period the number of events that occur is Poisson distributed (for a detailed derivation, see for example Bartlett 1957; Gillespie 1977; Renshaw 1991). All that remains is to determine which event occurs. We choose the event randomly, weighted by the relative event rates. Thus, transmission occurs if

$$RAND_2 < \frac{\beta XY/N}{\beta XY/N + \gamma Y},$$

This is online program 6.3

otherwise an infectious individual recovers. Here, $RAND_2$ is another random number between 0 and 1. Finally, time is updated $(t \to t + \delta t)$ and the appropriate changes are made to X and Y. The process is then iterated for the prescribed period of time we wish to simulate.

The dynamics of the SIS model serve as an interesting example of the differences between deterministic and stochastic systems. Transmission is frequency dependent whereas recovery occurs at a fixed rate and is independent of the number of infecteds. For the logistic growth model commonly used in ecological problems ($\frac{dx}{dt} = rx(1 - x/K)$), the reverse is often considered true, that births are independent of population size whereas deaths are density dependent. In the deterministic setting, the SIS and the logistic growth equations are identical (Chapter 2); therefore, the same dynamics are predicted for both models, asymptoting to a fixed prevalence/population level. However, in a stochastic system, this similarity is lost as the density dependence enters in different ways. The most obvious effect of this is that whereas disease prevalence in the SIS model has an absolute upper bound (the number infectious can never be bigger than the population size), there is no such limit for logistic populations. This is an example of a much more general phenomenon; although there is a single set of deterministic equations underlying each stochastic model, more than one stochastic model may correspond to each deterministic system (Keeling 2000a). In general, it is usually clear from the context what stochastic model is implied by the deterministic equations, although this does not have to be the case.

6.3.2. The General Approach

Box 6.3 generalizes the event-driven approach using the Direct Gillespie Algorithm (Gillespie 1976, 1977; Renshaw 1991), which can easily be adapted to any disease model. In general, the Direct Algorithm has been used in all event-drive simulations given in this chapter, unless explicitly stated otherwise.

Again, this form of stochastic model shows many of the attributes that we have come to expect: the excitation of oscillations at the natural frequency (Bartlett 1957; Renshaw 1991; Rohani et al. 2002; McKane and Newman 2005; Dushoff et al. 2004); variability in the prevalence, which closely mimics the observed patterns and can have a profound impact on the dynamics (Tille et al. 1991; Keeling and Grenfell 1999; Gibson et al. 1999; Smith et al. 2001); and negative covariances between the numbers of susceptible and infectious individuals. In addition, several elements are peculiar to such event-driven models and their method of iteration, which are discussed below.

This is online program 6.4

6.3.2.1. Simulation Time

Using Gillespie's Direct Method (Box 6.3) leads to a scaling of simulation time with population size. Here, the absolute rates of events increase linearly with population size, N. Intuitively, in a large population, individual transmission and recovery events should be much more frequent than in a small population, reflecting the fact that the per capita rates remain constant. This increase in the transition rates leads to a decrease in the time to the next event (δt) and hence an increase in the number of iterations needed to advance the model a specified period of time. As discussed in Box 6.4, an alternative implementation of this model is Gillespie's First Reaction Method, which is perhaps easier to program than the Direct Method, but substantially slower to use (Figure 6.4).

Box 6.4 Event-Driven Approaches: Gillespie's First Reaction Method

The following pseudo-code provides a slower, but often more intuitive, means of modeling demographic stochasticity; this and the Direct Method (Box 6.3) are equivalent and are both exact analogs of the underlying ODE system.

1. Label all possible events $E_1, \ldots, E_n$.
2. For each event determine the rate at which it occurs $R_1, \ldots, R_n$.
3. For each event, m, calculate the time until it next occurs, $\delta t_m = \dfrac{-1}{R_m} log(RAND_m)$.
4. Find the event, p, that happens first (has the smallest δt).
5. The time is now updated, $t \rightarrow t + \delta t_p$, and event p is performed.
6. Return to Step 2.

With either of these popular implementations of stochasticity, the amount of computer time needed to simulate a particular disease scenario increases linearly with the population size (Figure 6.4). Clearly, this has practical implications for the size of population that can be easily studied with this method, especially when multiple simulations may be needed.

Similarly, simulation of a large epidemic with many cases is slower than simulating a disease close to its endemic level, as many more events occur in the same time period. Some approaches to increasing the computational efficiency of such models is discussed in Box 6.5. As shown in Figure 6.4, Gillespie's recently proposed "τ-leap" method is substantially faster than the previous approaches, with simulation time much less affected by population size.

This is online program 6.5

Box 6.5 Event-Driven Approaches: Efficient Algorithms

The overwhelming drawback associated with using the First-Reaction (Box 6.4) and Direct (Box 6.3) Methods of Gillespie is their slow simulation times for large population sizes (see Figure 6.4). Recently, the first reaction method has been modified by Gibson and Bruck (2000). Their scheme (called the Next Reaction method) is substantially more challenging to program but is significantly faster than even the direct method when there are a large number of different event types. The Direct, First Reaction, and Next Reaction methods are all *exact* stochastic analogs of the underlying ODEs. Gillespie (2001) has recently proposed minor sacrifices in simulation accuracy in order to obtain substantial gains in simulation speed. The new algorithm is called the "τ-leap method" and may be explained as follows using the simple SIS model from Section 6.4.1.1:

1. Let the time increment between steps, δt, be "small" and fixed (discussed below).
2. Let $M_T(t)$ and $M_R(t)$ represent the number of transmission and recovery events by time t.
3. Defining $\delta M_i = M_i(t + \delta t) - M_i(t)$ $(i = T, R)$, then

$$P(\delta M_T = 1 \mid X, Y) = \frac{\beta XY}{N} \delta t + o(\delta t),$$

$$P(\delta M_R = 1 \mid Y) = \gamma Y \delta t + o(\delta t)$$

   define the transition probabilities for transmission and recovery events occurring in the time interval δt.
4. For small δt, the increments δM_i are approximately Poisson, such that:

$$\delta M_T \approx \text{Poisson}(\frac{\beta XY}{N} \delta t), \qquad \delta M_R \approx \text{Poisson}(\gamma Y \delta t).$$

5. Now, the variables can be updated:

$$X(t + \delta t) = X(t) - \delta M_T + \delta M_R, \qquad Y(t + \delta t) = Y(t) + \delta M_T - \delta M_R.$$

6. Time is updated, $t = t + \delta t$. Return to Step 4.

The key issue here clearly concerns the size of the fixed integration step, δt. Specifically, how small is "small"? For a range of epidemiological systems we have explored, a "good" value for δt appears to be around $1/10$ day, though this will clearly be inefficient (too small) for very small population sizes and perhaps too large when N is of the order of millions. The leap-size selection question is discussed in some depth by Gillespie and Petzold (2003). Problems arise with this method when multiple events associated with the same individual (infection followed by recovery, or infection followed by transmission to someone else) are likely to occur in the same step; for this reason models with more realistic assumptions about the progression of infection ($SEIR$ or multi-compartmental models, see Section 3.3.1) are more suited to simulation by this technique.

Some event-driven algorithms run much slower as the population size (and in particular the number of infectious individuals) becomes large.

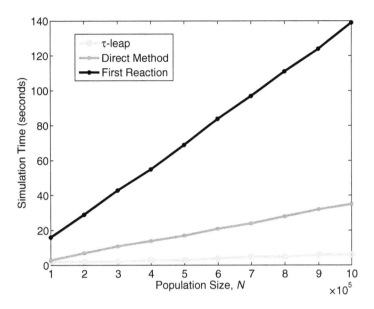

Figure 6.4. The time (in seconds) to simulate 1,000 years of $SEIR$ epidemics ($\nu = \mu = 0.02$ per year, $R_0 = 10$, $1/\sigma = 8$ days, $1/\gamma = 10$ days) on a 3.4 GHz Pentium PC. For Gillespie's Direct (gray line) and First Reaction (black line) methods, as the population size increases so does the simulation time, which, for very large populations, could become prohibitive. In contrast, the τ-leap method (light gray) is fast and largely unaffected by population size.

6.3.3. Stochastic Extinctions and the Critical Community Size

Although extinction of pathogens can happen in noise-based models (Section 6.2), this occurs in a continuous manner and is due to small numbers of infecteds experiencing a run of "bad luck." In contrast, real infectious diseases go extinct because the number of infected individuals drops from one to zero—a stochastic process in which the integer-valued nature of the population is key (Bartlett 1956, 1957; Bolker and Grenfell 1995; Keeling 1997; McKenzie et al. 2001; Dexter 2003; Bozzette et al. 2003).

In deterministic models, when the prevalence of the pathogen becomes low, the infection has a large average growth rate. The same is true for models with demographic stochasticity, except that there is also a chance that the number of infecteds will reach zero. Whether or not a disease is at a high risk of extinction, therefore, depends on several factors. If R_0 is large, then in general there is a large restoring force whenever the number of infectious individuals is low, thus reducing the chance of extinction. Disease extinctions are also much less likely when the average prevalence of the infection is high. Thus, diseases in small populations (Grenfell 1992; Finkenstädt and Grenfell 1998; Bozzette et al. 2003), ones that undergo large amplitude oscillations (see Chapter 5) (Grenfell 1992; Keeling 1997; McKenzie et al. 2001), or diseases with low R_0 (Hagenaars et al. 2001) are the most prone to extinction.

The same reasoning also holds for invading diseases (Bunn et al. 1998). There is a chance that an invading infected individual will recover before passing on the disease to any secondary cases. We can make this argument more mathematically explicit, using what is known as a branching process (Harris 1974; Bartlett 1956; Jacquez and O'Neill

1991; Farrington and Grant 1999; Hagenaars et al. 2001). Suppose that one infectious individual arrives in a large totally susceptible population and let P_{ext} be the probability that the infectious disease goes extinct before it causes a major epidemic (defined as one in which a significant fraction of the population is infected). Initially, one of two events can happen: Either the infectious individual recovers (rate γ) or it causes a secondary case (rate $\beta X/N = \beta$). If the event is recovery, then extinction is guaranteed, otherwise we need to consider the probability of extinction (before a major epidemic) given that two individuals are now infectious. This probability is P_{ext}^2, because it requires the lineages from both transmission events to go extinct independently. The reason the lineages act independently is because neither have yet caused a major epidemic, so the population of susceptibles has not been significantly depleted. Thus:

$$P_{ext} = \frac{\gamma}{\beta + \gamma} \times 1 + \frac{\beta}{\beta + \gamma} \times P_{ext}^2, \tag{6.4}$$

$$\Rightarrow P_{ext} = \frac{\gamma}{\beta} = \frac{1}{R_0}.$$

Therefore, not only is an invading pathogen with a high R_0 difficult to control (Chapters 2 and 8), it also has a low probability of chance extinction.

Similar reasoning and calculations show that if the outbreak starts with n infectious individuals in a population with some immunity, then:

$$P_{ext} = \frac{1}{R^n},$$

where $R = R_0 S = R_0 X/N$ is the effective reproductive ratio. Finally, if the infectious period is a constant interval, rather than the standard assumption that individuals recover at random (Section 3.3), then a similar but more mathematically involved argument reveals that in a totally susceptible population the risk of extinction is:

$$P_{ext} = 1 - R(\infty) < \frac{1}{R_0},$$

where $R(\infty)$ is the final size of the epidemic as predicted by the deterministic equations (Chapter 2). This is an example of a more general phenomenon: Diseases with a more variable infectious period have greater variability in the number of secondary cases, and therefore a greater risk of extinction when infectious numbers are low (Keeling and Grenfell 1997a,b).

From a single infectious individual in a totally susceptible population, the probability of extinction before a major epidemic ensues is equal to $1/R_0$.

A partially resistant population or a highly variable infectious period increases the likelihood of extinction, whereas multiple introductions of pathogen decrease the likelihood.

Although this form of calculation works well for invading diseases when the (initial) growth is exponential, the only way to assess the risk of extinction for endemic diseases is through repeated simulation of the stochastic model. The left-hand graph of Figure 6.5 shows extinction results from demographic (event-driven) stochastic models for an SIR-type infection for various population sizes. As comparison, the right-hand graph gives the corresponding results for measles and whooping cough epidemics in England and Wales

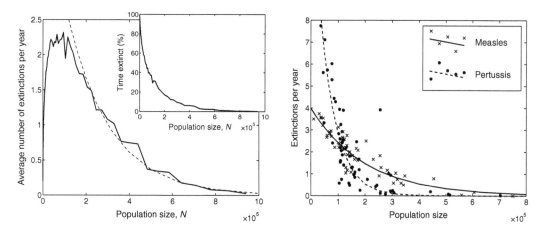

Figure 6.5. The number of extinctions per year from a simple SIR stochastic model (left-hand graph) and from measles and pertussis (whooping cough) data in England and Wales before vaccination (right-hand graph). Exponential functions, $k \sim \exp(-aN)$, are fitted to the point data. The inset graph in the left-hand figure shows how the amount of time the population is disease-free scales with the population size. ($\mu = \nu = 5.5 \times 10^{-5}$ per day, $R_0 = 10$, $\gamma = 1/10$ per day, import rate $\delta = 0.02\sqrt{N}$ per year; see Section 6.3.3.1).

before vaccination. Data and model results all approximate an exponential curve, such that larger populations are less susceptible to stochastic extinctions. The England and Wales data encompasses the effects of age structure (Chapter 3) and seasonality (Chapter 5), which are very strong for these two diseases, whereas the simulated stochastic model ignores such heterogeneities.

Historically, there has been considerable effort in matching the level of persistence predicted by stochastic models to that observed in diseases, with the forced $SEIR$ models suffering much higher levels of extinction (Grenfell et al. 1995; Bolker and Grenfell 1996; Keeling 1997). The principle aim of many stochastic modeling approaches has been to capture the Critical Community Size (see Section 6.3.3.2 for other measures of persistence). The Critical Community Size (CCS) is defined as the smallest population size that does not suffer disease extinction. For measles the CCS has been found to be remarkably consistent, estimated at between 300,000 and 500,000 for England and Wales, the United States, and isolated island communities (Bartlett 1957, 1960; Black 1966). In theory, no population size is ever completely immune from stochastic extinction of the pathogen, but in reality large populations (with substantial imports of infection, see Section 6.3.3.1) tend not to experience disease extinction.

Figure 6.6 shows the approximate CCS (estimated as the population size that experiences one extinction event per year) for the standard SIR model with demographic stochasticity. The infectious period ($1/\gamma$) has the greatest impact on the CCS; diseases with long infectious periods have a much reduced risk of extinction and hence need a much smaller CCS. The basic reproductive ratio, R_0, has a much weaker effect. When the infectious period is short, there is a slight requirement for a larger CCS with increasing R_0; however, when the infectious period is long, smaller R_0 values are associated with a larger CCS. However, to capture the true CCS, the associated stochastic model must also approximate the observed disease dynamics, which for measles or pertussis requires the inclusion of seasonal forcing (Chapter 5) mimicking the opening and closing of schools. This seasonal forcing, and the

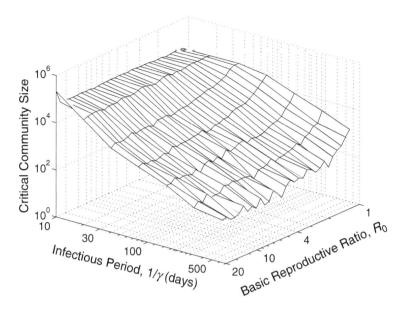

Figure 6.6. The Critical Community Size, approximated as the population size that experiences one extinction event per year, for the stochastic SIR model ($\mu = \nu = 0.02$ per year, import rate $\varepsilon = 0.0213(R_0 - 1)/\sqrt{N}$ per year, this mimics how imports are expected to change with both population size and the average prevalence of infection; see Section 6.3.3.1). Results are from simulations of 100 years after transients.

large-scale epidemics that are generated, leads to frequent extinctions in most stochastic models; the stochastic seasonally forced $SEIR$ model still suffers disease extinctions in populations of 10 million, substantially more than the observed CCS (Bolker and Grenfell 1993). Several modifications have been proposed to increase the persistence of stochastic seasonally forced models, usually involving adding extra levels of heterogeneity such as age structure, spatial structure, or more discrete infectious periods (Bolker and Grenfell 1993, 1995, 1996; Keeling and Grenfell 1997a; Ferguson et al. 1997a; Lloyd 2001).

The question of which ingredients are needed in order to capture the observed measles CCS deserves a little more attention. In Figure 6.7, we plot the extinction frequency as a function of population size for different distributions of latent and infectious periods (denoted by the parameters n and m) in the $SEIR$ model (described in detail in Section 3.3). As shown in the left-hand graph, when identical parameter values are used (notably the amplitude of seasonality, $b_1 = 0.25$), the precise distribution underlying these epidemiological classes does not substantially affect the CCS (cf. Lloyd 2001; Keeling and Grenfell 2002). The figure demonstrates subtle differences when population sizes are small, highlighting the greater extinction frequency of models assuming an exponentially distributed infectious period. As population size increases, however, it is difficult to distinguish between the predictions of models assuming different distributions. Keeling and Grenfell (2002) argued, however, that in order to capture the mechanisms underlying disease persistence, and the CCS in particular, we need to compare "best-fit" models in each case. This means that for any assumed distribution of the latent/infectious period, in order to compare like with like, we need to estimate key parameters such as the amplitude of seasonality by fitting the model to data before examining the predicted CCS. As shown

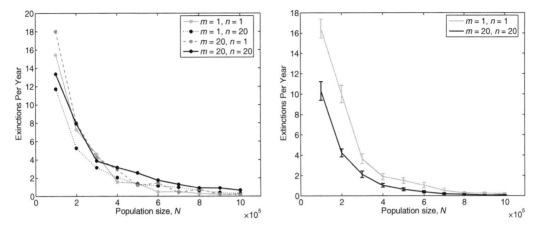

Figure 6.7. The mean number of extinctions per year from 50 stochastic realizations of the $SEIR$ model with different distributions of latent and infectious periods. In the left-hand graph, we plot the mean number of annual extinctions as the number of compartments in the latent and infectious periods are altered. Specifically, we assume both classes conform to a gamma distribution, with different values of the shape parameter, m and n, respectively—simulated using multiple subclasses (Section 3.3). The figure shows that for identical parameter values (specifically the amplitude of seasonality, $b_1 = 0.25$), there is little to distinguish between a purely exponentially distributed model and one containing different distributions. In the right-hand graph, we show that predicted extinction dynamics are significantly different when the best fit model values are used (specifically, $b_1 = 0.29$ for the exponential model and $b_1 = 0.13$ for the gamma distributed model). Error bars demonstrate standard errors. (Other parameter values were $\mu = \nu = 0.02$ per year, $R_0 = 17$, $1/\sigma = 8$ days, $1/\gamma = 5$ days, and the import rate $\delta = 0.02\sqrt{N}$ per year).

in the right-hand graph of Figure 6.7, such an exercise reveals that the exponentially distributed $SEIR$ models are significantly more prone to extinction than models assuming a gamma distribution.

The Critical Community Size (CCS) is the smallest population size observed not to suffer disease extinctions; for measles the CCS is around 400,000. Simple stochastic models that capture the observed seasonally induced measles epidemics fail to show adequate levels of persistence which can be captured only with a greater heterogeneity within the models.

6.3.3.1. The Importance of Imports

Although disease incidence in real populations may frequently undergo stochastically driven fade-outs, imports of pathogen from outside the population can prevent permanent extinction. In human populations, such imports usually take the form of visitors who remain for a limited period but, in doing so, re-introduce the pathogen (Keeling and Rohani 2002). Conceptually, we often think of these visitors as commuters, whose movements between communities prevent the populations from being totally isolated. For wildlife diseases, imports are more likely due to the permanent movement of animals into the group, herd, or area (Clancy 1996), or by exposure to an external source

(Rhodes et al. 1998). Finally, for livestock or crop diseases, imports are due to purchasing infected stock, the mechanical transfer of infected material by humans or vehicles, or by airborne plumes of infection (Cannon and Garner 1999; Ferguson et al. 2001a; Keeling et al. 2001b).

Despite the wide variety of mechanisms, imports are commonly modeled by just two methods. First, by assuming that there is a probability (independent of disease dynamics but dependent on the population size) of an infected individual joining the population:

$$Y \to Y + 1 \qquad \text{Rate } \delta(N).$$

If we also wish to maintain a constant population size, each import can be countered by the loss of either a susceptible or recovered individual. We can equate this mechanism to the permanent immigration of an infected individual.

Second, we may assume there is an external source (again, dependent on the population size) that adds to the force of infection:

This is online program 6.6

$$X \to X - 1, \ Y \to Y + 1 \qquad \text{Rate } \varepsilon(N)X.$$

This formulation corresponds either to susceptibles moving to another location and picking up the pathogen prior to returning to their natal population, or to an infectious transient visitor (commuter) transmitting the pathogen in the community but leaving sufficiently rapidly so as not to affect the population size. For this latter method, the rate of imports is governed by the state of the population. Following an extinction, as the number of susceptibles increases, so does the probabilistic risk of imports triggering a new case. Generally, these two modeling approaches produce similar results when correctly scaled ($\varepsilon X \approx \delta \Rightarrow \varepsilon \approx \delta R_0 / N$) because the fluctuations in the number of susceptibles are usually relatively small. Interestingly, in human populations, it has been estimated that the number of infectious imports generally scales with the square root of the population size (Bartlett 1957), so that small populations have relatively more imports than large ones.

For human populations, imports of pathogen scale with the square root of the population size.

For measles in England and Wales, by looking at the times between extinctions and invasions, it has been estimated that:

$$\text{Rate of imports} \approx 5.5 \times 10^{-5} \sqrt{N},$$

so that a population size of one million experiences about 20 infectious imports per year. If we assume that these imports occur due to the movement of infectious individuals (either permanent movements or commuters), we should expect this rate to scale with both the equilibrium prevalence of infection and the transmission rate. By matching to the known parameters and equilibrium levels for measles in England and Wales, we can therefore estimate the expected level of imports for different diseases, although import rates will to some extent depend upon ease of travel, disease morbidity, and the average age of infection.

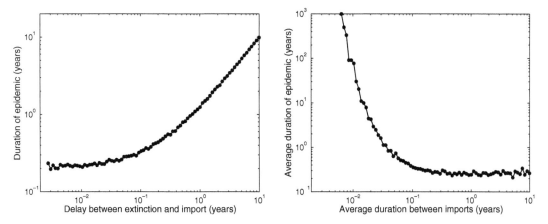

Figure 6.8. Left-hand graph: the average duration of an epidemic caused by an import when there is a fixed delay between extinction and reseeding of infection. Right-hand graph: the average duration of an epidemic caused by an import, when imports are continual and random. ($\nu = \mu = 5.5 \times 10^{-5}$ per day, $R_0 = 10$, $1/\gamma = 10$ days, $N = 10^5$).

$$\text{Rate of imports, measles} = \delta_{\text{measles}},$$

$$\approx 5.5 \times 10^{-5}\sqrt{N} = k\beta I^*_{\text{measles}},$$

$$= k\beta\frac{\mu}{\gamma}\left(1 - \frac{1}{R_0}\right),$$

$$k \approx \frac{5.5 \times 10^{-5}\sqrt{N}}{\mu(R_0 - 1)},$$

$$\approx 0.0625\sqrt{N},$$

$$\delta_{\text{general}} = k\beta I^* = 0.0625\mu(R_0 - 1)\sqrt{N}.$$

and we therefore expect to see an increase in imports with increasing R_0. We can perform a similar operation assuming commuter-type imports:

$$\varepsilon_{\text{general}} = \frac{1.06\mu(R_0 - 1)}{\sqrt{N}},$$

in which case the expected number of imports ($\varepsilon X \approx \varepsilon N/R_0$) shows little variation with R_0.

Extinctions are much more likely to occur if the number of susceptibles has been reduced by recent stochastic fluctuations, such that the effective reproductive ratio, R, drops below one. In such cases, an immediate reseeding of infection is also very likely to fail. Thus, the *frequency* of imports has a profound effect on the success of the subsequent epidemics. Figure 6.8 (left-hand graph) shows how the average duration of the ensuing epidemic increases with the delay between extinction and reseeding, because this allows the level of susceptibles to rise. The increase in epidemic duration with delay is slightly less than linear, and even with a very rapid reintroduction of infection sustained epidemics can still arise.

Although imports are noticed only once the disease is extinct, they are in fact a continual process happening even when the disease is present. These continual imports can substantially lessen the risk of extinction by increasing the average level of infection and rescuing the disease from times of low prevalence. Continual imports can also interfere with the disease dynamics, damping any fluctuations (Ferguson et al. 1996a) or even supporting infection when it would normally die out (Rhodes et al. 1998). The right-hand graph of Figure 6.8 illustrates this point. If a population has many imports (e.g., when a small town is tightly coupled to a large city (Finkenstädt and Grenfell 1998)), extinctions are rare. When imports are less frequent, the risk of extinction during an epidemic quickly reduces to an asymptotic level. Thus, although infrequent imports may allow the susceptible population to recover between epidemics, they do little to support the persistence of infection.

Imports of infection into a community can significantly modify the extinction process in a complex manner. Frequent imports may prevent extinctions, whereas long delays between extinctions and imports improve the chance that an import successfully triggers a large epidemic.

In many structured models (see Chapters 4 and 7), the early dynamics of infection in a new population or subpopulation are defined by two attributes; the rate that infection arrives in the population, and the subsequent growth within that population. In deterministic (continuous population) models, these imports occur continually at very low levels. Therefore, minute fractional imports can be rapidly increased by the within-population dynamics, leading to a large epidemic despite the fact that less than one infectious individual has arrived in the population. The stochastic framework overcomes this problem, because an entire individual needs to be imported before an epidemic can occur. Thus, when the coupling between populations is low, the stochastic formulation can substantially increase the delay between an epidemic in one population and the epidemics that are triggered in other coupled populations (see Chapter 7 and in particular Figure 7.8).

When dealing with very low levels of imports between populations, it is essential that this is modeled stochastically.

6.3.3.2. Measures of Persistence

Measuring persistence, or the lack of it, in models is far from straightforward. There are three main measures, each of which has its advantages and pitfalls:

1. *Extinctions with Imports*. The most biologically plausible measure of persistence is to simulate the stochastic dynamics of the population, including the random import of infection from external sources, and count the average number of extinctions during a given period (Bartlett 1957; Grenfell 1992; Keeling 1997). Figure 6.5 uses this approach. The main advantage is that it corresponds to biological reality, and therefore the results can be compared to observations. However, as shown above, the persistence of diseases is strongly influenced by the pattern of imports and in general these are very difficult to measure or parameterize.
2. *Time to Extinction*. A more mathematically pleasing approach is to start the stochastic simulations at, or near to, the deterministic equilibrium and measure the average time until the population goes extinct—called the "first passage time." Alternatively,

one could measure the proportion of simulations that have gone extinct after a given time (Farrington and Grant 1999). The appeal of this method is that imports are not required—populations act in isolation. However, this measure does not conform to any observed biological process and the results are sensitive to the initial conditions.

3. *Conditional Extinctions.* An alternative measure is to look at the asymptotic rate of extinctions, conditional on the disease currently being present in populations without imports. In practice, this is calculated by first simulating populations for many generations and then discarding those in which the pathogen has already died out. The remaining simulations, which still possess infection, can then be simulated further to ascertain the rate of extinction. The initial run-in time followed by discarding means that the calculations start from a natural population distribution due to the stochastic dynamics, rather than the somewhat artificial starting condition of Method 2. This measure is theoretically appealing and is often used whenever analytical results are feasible (Nasell 1999; Andersson and Britton 2000).

6.3.3.3. Vaccination in a Stochastic Environment

One of the primary aims of epidemiological modeling is to inform public health intervention measures. The relationship

$$V_C = 1 - \frac{1}{R_0}$$

between the critical level of vaccination needed to eradicate an infectious disease, V_C, and the basic reproductive ratio, R_0, is a fundamental tenet of modern epidemiology (Chapter 8). For stochastic populations this threshold still applies; with vaccination coverage at or above V_C, the disease cannot invade or cause an epidemic. However, stochastic dynamics can work against the pathogen by causing extinctions even though the vaccination level is below the required threshold (Bolker and Grenfell 1996; Smith et al. 2001).

Figure 6.9 considers the effects of vaccinating a proportion of children at birth, such that the underlying differential equations would be:

$$\frac{dX}{dt} = \nu N(1 - V) - \beta XY/N - \mu X,$$

$$\frac{dY}{dt} = \beta XY/N - \gamma Y - \mu Y,$$

$$\frac{dZ}{dt} = \nu NV + \gamma Y - \mu R.$$

Vaccination (below the eradication threshold) has two main effects. As predicted by the deterministic model, vaccination reduces the average level of infection within the population, which in itself has clear benefits. In addition, associated with the lower prevalence is an increased risk of extinction. Thus, as vaccination level is increased, the disease is likely to be eliminated (as least within the local population) by stochastic extinctions before the deterministic threshold is reached. Hence, stochasticity actually benefits control programs, allowing diseases to be eradicated at lower vaccination levels (Smith et al. 2001); however, see Bolker and Grenfell (1996) for a plausible counterexample. Note, however, that the threshold level of vaccination must be achieved if reinvasion of infection is to be prevented.

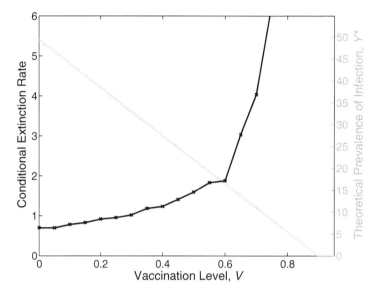

Figure 6.9. The dual effects of vaccination. The conditional extinction rate per year (black line, see Section 6.3.3.2) shows a sharp increase as the deterministically predicted level of infection (gray line) decreases. ($v = \mu = 5.5 \times 10^{-5}$ per day, $R_0 = 10$, $1/\gamma = 10$ days, $N = 10^5$.)

In many situations, stochasticity may be beneficial, leading to extinction of a disease well before the vaccination threshold is reached.

6.3.4. Application: Porcine Reproductive and Respiratory Syndrome

One application where population sizes are often relatively small and therefore stochasticity can play a major role is for epidemics within a farm (Stark et al. 2000; MacKenzie and Bishop 2001). Although some farms may have tens of thousands of livestock, even at these numbers stochasticity and chance extinctions can be important. More often, the number of animals is far lower and therefore stochasticity all the more important. Livestock epidemics are becoming an increasingly active area of research, motivated in part by the 2001 foot-and-mouth epidemic in the United Kingdom, the 1997 classical swine fever outbreak in the Netherlands, and the risk of agro-terrorism. For diseases, such as foot-and-mouth, that spread rapidly within the farm it may be plausible to ignore the within-farm dynamics treating the farm as a single unit (Keeling et al. 2001b). However, for diseases with a more prolonged infection and slower spread, the dynamics within a farm may play a critical role in the wider transmission of pathogen.

The example discussed here is Porcine Reproductive and Respiratory Syndrome (PRRS), a viral disease of pigs (Albina 1997; Wills et al. 1997). PRRS was first observed in the United States in 1986, but was only identified following an outbreak in the Netherlands in 1991 when it spread rapidly through the pig-producing areas of the country (Terpstra et al. 1991). Since then, PRRS has been endemic within both the United States (prevalence 60–80%) and the Netherlands. PRRS has economic and welfare implications for the pig industry because it may reduce the reproductive success of sows and cause respiratory problems in piglets (Terpstra et al. 1991). Clinical signs may include anorexia, fever, or lethargy together with cyanotic (blue) ears, vulva, tails, abdomens, or snouts—giving rise to the disease's alternative name, blue-eared pig disease. Reproductive failure is characterized by late-term abortions, increased numbers of stillborn fetuses, and/or premature weak pigs.

The initial PRRS epidemic spreads slowly within a farm, after which the infection may persist at low levels for many years. This slow spread and low prevalence means that within-farm modeling and a stochastic approach are vital. Nodelijk et al. (2000) modeled this disease based upon the observed dynamics on a closed breeding-to-finish herd of around 115 sows belonging to the University of Utrecht. Blood samples were taken from all sows just before the 1991 outbreak, again during the initial epidemic, and then twice yearly after that; all sera were tested for antibodies to the PRRS virus. It was shown that the observed pattern of seropositives matched SIR-type dynamics with waning immunity, given that imports and exports of livestock were included.

$$
\frac{dX}{dt} = \mu N + wZ - \mu X - \beta XY,
$$

$$
\frac{dY}{dt} = \beta XY - \gamma Y - \mu Y, \tag{6.5}
$$

$$
\frac{dZ}{dt} = \gamma Y - wZ - \mu Z,
$$

where, as usual, β is the contact parameter and γ is the recovery rate. (Note that although we have chosen to model transmission as density dependent, the fact that the number of pigs on a farm remains constant makes this assumption irrelevant.) Other parameters are μ, the replacement rate of livestock and w, the rate at which immunity wanes and the pigs revert to being susceptible. The longitudinal data collected were sufficient to parameterize this model; most notably R_0 was estimated to be 3.0 (95% confidence interval 1.5 to 6.0), such that in a deterministic setting the disease would always take off and readily persist.

Breeding-to-finish pig farms, such as the one in the study by Nodelijk et al. (2000), possess additional structure in that sows and rearing pigs are often physically separated. Nodelijk et al. (2000) incorporated this extra detail into their model; however, for clarity we take the simpler approach that the farm consists of one freely mixing herd. For other farm types that specialize in a single age range of pigs, this simplifying assumption is more likely to be true. Figure 6.10 (top row of graphs) shows the distribution of the number of cases (left-hand graph) and duration (right-hand graph) of a series of simulated outbreaks on a farm with 115 sows ($N = 115$). The number of cases (including multiple reinfections of a given pig) is bimodal in nature because some seed infections fail to trigger a significant outbreak. As stated earlier, for three initial infections we would expect a fraction $R_0^{-3} \approx 0.037$ to fail—this is lower than the value from simulations (≈ 0.057), which is attributable to the small population size. There is also a significant probability that the epidemic will be both very large and protracted, because the distributions of both the size and duration have long tails. It is important to recognize that no single measure can encompass the risks described by these distributions, thus it is vitally important that, at the very least, both means and some measure of the variation are always stated.

This model can easily be extended to consider the impact of herd size on the stochastic epidemic dynamics. As expected from the theoretical results shown above and the work on extinctions of measles in human communities (Bartlett 1956, 1957), the average duration and hence the size of epidemics increases dramatically with the number of pigs. For farms

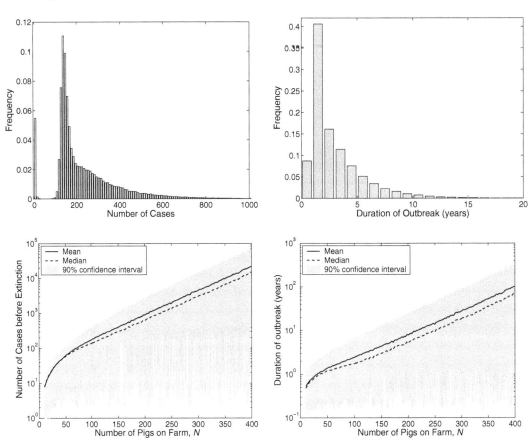

Figure 6.10. The simulated behaviour of PRRS epidemics with event-driven (demographic) stochasticity. The top left-hand graph shows the distribution of the total number of cases, including reinfections, from an introduction of 3 infected animals in a farm with 115 sows. The top right-hand graph shows the duration for a similar scenario. Results are from 100,000 simulations. The lower graphs consider how these quantities change with the number of animals on the farm; both show an exponential increase with the number of sows. ($R_0 = 3$, $\gamma = 6.517$ per year, $\mu = 0.6257$ per year, and $w = 0.2607$ per year). Shaded regions give the 90% confidence intervals associated with a single epidemic.

with more than 50 pigs, the increase in both cases and duration is exponential, suggesting that very large farms can act as persistent reservoirs of infection. In contrast, very small farms cannot sustain the epidemic and the size and duration of the outbreak is even smaller than would be expected by extrapolating from larger farms. Again, the distribution of epidemic sizes and hence the confidence intervals on the model predictions provide useful additional information. It appears that the mean number of cases and duration of the outbreak are dominated by the long tail of both distributions, whereas the lower 5% bound is very noisy but shows little variation with the number of livestock.

These results have clear implications for the farming community. Even very large pig farms may often have very short, limited epidemics that last for less than a year and cause few cases. Such epidemics are extinguished rapidly by stochastic effects due to the small number of infectious animals that seed the epidemic. However, once the epidemic becomes established it can persist for many years in the larger farms. Thus, these farms should be the target of control measures, because smaller farms are much less likely to sustain the virus and consequently less likely to spread PRRS. Such considerations may also be appropriate for diseases such as bovine tuberculosis, which spreads slowly between cattle with a low R_0, where farm-level dynamics and the persistence in large premises may be key to the national pattern of cases.

6.3.5. Individual-Based Models

Although all the models of demographic (event-driven) stochasticity consider the population to be composed of individuals, they are not strictly individual-based models. This type of model is dealt with in some detail in Section 7.5, though here we focus on the differences between stochastic and individual-based models as well as highlighting a few "tricks" to speed the computation of such models.

Models with demographic stochasticity force the population to be made up of individuals, but all individuals of the same type are compartmentalized and measured by a single parameter—there is no distinction between individuals in the same class. In contrast, individual-based models monitor the state of each individual in the population. For example, in models with demographic stochasticity we know how many individuals are in the infectious class but not *who* they are, whereas in individual-based models we know the state of each individual and hence *which* individuals are infectious. Individual-based models are therefore often more computationally intensive (especially in terms of computer memory) because each individual in the population must have its status recorded.

Although individual-based models can be readily iterated with any of the standard algorithms (see Boxes 6.3, 6.4, and 6.5), there can be computational difficulties due to the vast number of different events—we need a different event for each individual concerned, especially for the fast τ-leap method where a Poisson distribution needs to be realized for each event. A variety of methods can be used to overcome this difficulty; here we outline two that are readily usable.

One solution is to first use either the Gillespie direct method (Box 6.3) or the τ-leap method (Box 6.5) with the total rate of each event (i.e., combining the rates of all infection events), and then determining which individuals were involved in the event. For the SIS model outlined in Section 6.3.1.1, the individual-based equivalent would be:

$$\delta t = \frac{-\log(RAND_1)}{\text{sum of all infection and recovery rates}},$$
$$= \frac{-\log(RAND_1)}{\sum_{i=S} T_i + \sum_{i=I} G_i},$$

which gives the time to the next event. Here T_i and G_i are the transmission and recovery rates of individuals at that moment, and the two sums select only those individuals who are susceptible and infectious, respectively. As before, we determine that the next event would be transmission if:

$$RAND_2 < \frac{\sum_{i=S} T_i}{\sum_{i=S} T_i + \sum_{i=I} G_i}.$$

Suppose the next event is recovery; we then let individual p recover if:

$$\sum_{i=I}^{i \leq p-1} \bar{G}_i < RAND_3 \times \left(\sum_{i=I}^{i \leq p} \bar{G}_i\right) \leq \sum_{i=I}^{i \leq p} \bar{G}_i. \qquad (6.6)$$

This latter step can be greatly simplified if all the recovery (or transmission) events occur at the same rate. Although this modification is formulated for Gillespie's direct method, it can be equally applied to the τ-leap method, with the individual involved in each event determined by equation (6.6). One potential difficulty with this method is that for each event, an additional random number needs to be calculated to determine which individual is involved.

As an alternative—which works particularly well when the time since infection is important (Section 3.3)—we can combine the Gillespie direct and first reaction methods (Boxes 6.3 and 6.4). We again use the example of the SIS model, but suppose that the times to recovery, δG, for each individual are determined as soon as they are infected. The time to the next recovery event is therefore $\min(\delta G_i) = \delta G_{\min_G}$ (where $\min_G$ refers to the individual that will recover first), whereas the time to the next infection is:

$$\delta t = \frac{-\log(RAND_1)}{\sum_{i=S} T_i}.$$

The time to the next event is then the minimum of these two values. If recovery happens first, the individual with the shortest time $(\min_G)$ recovers, all the recovery times are reduced by $\delta G_{\min_G}$, and a new minimum recovery time is calculated. If infection happens first, then the individual involved can be calculated as above, all the recovery times are reduced by δt, and a new minimum recovery time is calculated *if* the newly infected individual is predicted to recover before any others.

This concept of storing a recovery time for each individual is very powerful, and can be readily extended to a death time, both of which allow for more realistic assumptions than the standard exponential distribution that arises from constant rates. However, we can also utilize such a method to predict transmission. For each individual we can set a cumulative level of pathogen, C_i, that they need to encounter before they are infected. In a model with frequency-dependent transmission (and equal levels of susceptibility and transmissibility per individual), the cumulative pathogen levels would be chosen from an exponential distribution:

$$C_i = -\log(RAND)$$

and at each time step, say δt, these values get reduced by $\delta t \beta Y / N$. Transmission then occurs when an individual's C_i value drops below zero. Using this approach, the fate of an individual can be determined by the selection of random numbers (from appropriate distributions) at their birth—which may limit the number of times a random number generation routine is needed. However, there is an additional benefit. For normal methods of updating, even when starting with the same random number seed, a small deviation in the epidemic patten (due to slightly different parameters) can initiate major changes because the same stream of random numbers will inevitably be applied to different processes. In contrast, when the fate of an individual is set at birth, the use of the random numbers is independent of the epidemic dynamics. This provides us with a fascinating opportunity to test the effects of parameters or controls on the *same* stochastic epidemic process. Rather than having to phrase all comparisons in terms of averages, we can now test controls on replicated epidemics. Hence, we may provide definitive answers to questions such as, "If this exact epidemic occurred again, would we be better to vaccinate"?

6.4. PARAMETERIZATION OF STOCHASTIC MODELS

As with all epidemiological problems, accurate parameterization is the key to useful and applied modeling. For stochastic systems, the parameterization approach can often be made much more statistically rigorous. In deterministic settings, parameters are generally chosen that minimize the deviation between the observed and simulated epidemics. However, this minimization requires many arbitrary choices (e.g., the weighting between errors in susceptibles and errors in infecteds, or whether the absolute or proportional errors are used), which can influence the choice of parameters. The simplest fitting procedure chooses parameters so that the deterministic equilibrium predictions match the average observed dynamics. In general, this may often give a reasonable approximation of the true parameter values; however, as we have shown in this chapter already (Section 6.2.1.1), stochasticity can have a significant effect on the mean and thus bias even this simple form of estimation. For example, the negative correlations that develop between susceptible and infected individuals would lead to an underestimation of the basic reproductive ratio, R_0, if the standard deterministic parameterization was used. This problem becomes significantly worse when localized extinctions are frequent.

By including stochasticity in our models, an alternative methodology can be used. Parameters can be found that maximize the *likelihood* (or probability) of the model generating the observed data (Finkenstädt and Grenfell 2000). Such maximum likelihood estimates are a well-defined and well-understood statistical tool. In general, the likelihood is calculated sequentially. Thus the probability, P_t, of observing the data recorded at time t is calculated given that the model agrees with the observations recorded at time $t - 1$. The total likelihood is then found from multiplying together all the probabilities. Often these likelihoods will be very small, because the probability of seeing exactly the same number of cases at exactly the same times is highly unlikely. These sort of likelihood calculations lend themselves naturally to an MCMC (Monte Carlo Markov Chain) approach (Ranta et al. 1999; O'Neill et al. 2000; Neal and Roberts 2004), which considers all the plausible stochastic pathways between two observations.

6.5. INTERACTION OF NOISE WITH HETEROGENEITIES

In the above examples, the differences between the deterministic and stochastic models were due either to the noise itself, or the interaction of noise with the nonlinear transmission dynamics. This section considers the interaction of noise with other components of disease models, such as temporal forcing, age, or risk structure and spatial heterogeneities.

6.5.1. Temporal Forcing

As discussed more fully in Chapter 5, temporal forcing can occur for a variety of reasons—generally due to climatic variations or changes in social mixing (such as the differences between school terms and holidays). The main effect of such regular forcing is to induce periodic cycles in disease prevalence. Even when the forcing is annual, multi-annual cycles are common (Keeling et al. 2001a).

However, as discussed in this chapter, the inclusion of stochasticity can often cause the disease to resonate at its natural frequency due to the greater role of transient dynamics. In many cases, a clash can occur between the deterministic period due to forcing and the natural resonant period due to stochasticity (Rohani et al. 2002). Both measles and whooping cough experience strong seasonal forcing due to the school year. The deterministic attractor for measles is biennial, which is very close to the natural resonant period; hence, there is no conflict between the two frequencies. In contrast, whooping cough possesses an annual deterministic attractor whereas the resonant period is closer to three years. This tension between the two periods is observed in the number of reported cases; large populations (which experience relatively little demographic stochasticity) display predominantly annual epidemics, whereas small populations are more likely to experience epidemics every 2–3 years in keeping with the stochastic concept of increased transient-like dynamics (Rohani et al. 2002). This phenomenon is discussed in more detail in Chapter 5 on temporal forcing.

6.5.2. Risk Structure

Risk- and age-structured models subdivide the population into groups of individuals, such that all individuals within a group experience a similar risk of infection (see Chapter 3). The classic example is of sexually transmitted diseases where some individuals, due to their lifestyle, are at a much greater risk of catching and transmitting infection. Such core groups will also be subject to different levels of stochasticity. Although all the models and methods developed in this section transfer directly to such structured populations (Sleeman and Mode 1997; MacKenzie and Bishop 2001), the subtle differences compared to unstructured models needs to be addressed.

We have already seen that in a stochastic environment new infections can fail to invade and endemic infections can be driven extinct. How does structuring the population into risk-groups affect this? The high-risk population is smaller, suggesting that it experiences more stochastic effects, but the reproductive ratio within this group is larger suggesting reduced chance of extinction. Simulation of structured and unstructured SIS-type models with event-driven stochasticity is shown in Figure 6.11. When the models share the same R_0, the structured model shows greater variability relative to the disease prevalence. However, if the two models were parameterized to have the same equilibrium prevalence of infection, then the reverse is true and the unstructured model is more variable. This illustrates a crucial difficulty when comparing models of different forms; the comparison between the models is sensitive to which characteristics are matched. In practical situations, the available data should always constrain our choice of model and parameters (Keeling and Grenfell 2002). If data exists on both the early growth rate (hence R_0) and the equilibrium prevalence, it may be impossible to fit an unstructured model, in which case no fair comparison can be made.

In some cases, the presence of extra structure will always increases the level of variability. Consider a population that has a few very rare super-spreaders or super-shedders. On average, these super-infectious individuals are not likely to be infected, but in those rare simulations where one of them does become infected, the prevalence of the disease can increase dramatically. Such rare events with large consequences will clearly increase the variability between model simulations and reduce our ability to predict the outcome and hence control the disease in real situations.

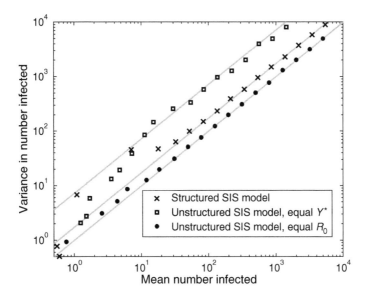

Figure 6.11. A log-log plot of the variance against the mean number of infected individuals for a *SIS* disease in various population sizes. Randomness was included as individual-based event-driven demographic stochasticity. The structured population has 20% of its individuals in the high-risk group, and assumes the following transmission matrix which was used throughout Chapter 3:
$\beta = \begin{pmatrix} 10 & 0.1 \\ 0.1 & 1 \end{pmatrix}$, $\gamma = 1$, $\delta = 10^{-3} \times N$. The unstructured population has the same basic reproductive ratio (circles $\beta \approx 2.0$), or the same equilibrium infection level (squares $\beta \approx 1.158$), as the structured model.

6.5.3. Spatial Structure

Spatial structure (Chapter 7), the subdivision of the population due to geographical location, is another situation where stochasticity plays a vital and dominant role. Without underlying heterogeneities in the parameters at different locations, most deterministic models asymptote to a uniform solution at all locations, thereby negating the necessity of a spatial model. However, in a stochastic setting, different spatial locations experience different random effects, and spatial heterogeneity may therefore be maintained.

The dynamics of spatially structured models is discussed more fully in Chapter 7. However, one facet of the interaction between spacial partitioning and stochasticity is worth emphasizing here. For the same total population size, does one large homogeneously mixed population suffer more or fewer stochastic effects than many smaller weakly interacting populations? The answer is that it very much depends on the nature of the disease and the host population (Keeling 2000b). However, in many situations a spatially structured population suffers less stochastic effects (and extinctions) than a large well-mixed population, despite the fact that each spatial subunit experiences a high level of stochasticity due to its small size (Keeling et al. 2004a; Section 7.2.3).

6.6. ANALYTICAL METHODS

Although the methods given above provide a robust means to simulate the stochastic behavior of a disease, they lack the analytical tractability that provides an intuitive understanding of the behavior. The following three formulations provide this analytical link, but require a certain level of mathematical sophistication. Hence, some readers may wish to skip the details of this section and read only the highlighted conclusions. In addition, all three methods provide a deterministic description of stochastic processes by considering the distribution or average of many stochastic realizations. In all the following examples, we shall focus primarily on the SIS model (without births or deaths) because this model is only one-dimensional (as $X + Y = N$), which makes the mathematics simpler and more transparent.

6.6.1. Fokker-Plank Equations

These equations are closely linked to the addition of process noise to the standard equations introduced in Section 6.2 of this chapter. Rather than simulate multiple realizations of the same set of noisy equations, it is possible to derive equations for the probability distribution of values (Bartlett 1956; van Herwaarden 1997; Andersson and Britton 2000). Thus $P(X, Y, t)$ is the probability of a stochastic simulation having X susceptible individuals and Y infectious individuals at time t. In deterministic models, the probability distribution is zero everywhere except a single point, and the dynamics of this point obey the standard differential equations. The key question is how noise should be introduced. Intuitively, noise should lead to the probability distribution spreading out, such that the more noise that is experienced, the greater the spread of values around the deterministic solution that are likely to be encountered.

Standard mathematical theory allows us to make this relationship more exact (Øksendal 1998). For a very general differential equation together with additive noise:

$$\frac{dx}{dt} = F(x) + f(x)\xi,$$

the probability distribution $P(x, t)$ is given by the partial differential equation:

$$\frac{\partial P(x, t)}{\partial t} = -\frac{\partial}{\partial x}(F(x)P(x, t)) + \frac{1}{2}\frac{\partial^2}{\partial x^2}(f(x)^2 P(x, t)).$$

Hence, noise in the stochastic simulations is replaced by diffusion in the partial differential equations.

The use of such partial differential equations has three advantages over noisy simulations: (1) it needs to be solved only once to capture the entire range of possible dynamics, (2) by setting $\frac{\partial P}{\partial t} = 0$ we can readily find the long-term stationary distribution of values, and (3) the model is deterministic and therefore does not rely upon a time-consuming random number generation routine. However, these three advantages must be offset against the additional difficulty of solving partial differential equations, which in itself is a computationally demanding task.

The Fokker-Plank equation provides an *exact* solution to multiple simulations of a model with noise. Noise in the equations translates to diffusion of the probability distribution.

We now extend this concept to a more epidemiologically relevant situation, the dynamics of an SIS model of the sort commonly used to describe the behavior of sexually transmitted diseases; a similar form of model was used by Roberts and Saha (1999) to study bovine tuberculosis in the possum population of New Zealand. The differential equation for the number of infected individuals, with scaled additive noise, becomes:

$$\frac{dY}{dt} = [\beta XY/N + \sqrt{\beta XY/N}\xi_1] - [\gamma Y + \sqrt{\gamma Y}\xi_2]$$

$$= \beta XY/N - \gamma Y + \sqrt{\beta XY/N + \gamma Y}\xi.$$

Here the two noise terms have been combined by noting that the sum of two normal distributions with variances v_1 and v_2 is another normal distribution with variance $v_1 + v_2$. An equation for X is unnecessary because $X = N - Y$. The probability, P, of having exactly Y infectious individuals is given by:

$$\frac{\partial P(Y)}{\partial t} = -\frac{\partial}{\partial Y}([\beta XY/N - \gamma Y]P(Y)) + \frac{1}{2}\frac{\partial^2}{\partial Y^2}([\beta XY/N + \gamma Y]P(Y)).$$

This expression provides an exact solution to the distribution obtained from infinitely many simulations of the SIS model with scaled additive noise (Figure 6.12).

If we are primarily concerned with the long-term behavior of the pathogen, rather than initial dynamics, then an analytical expression for the equilibrium distribution $P^*(Y)$ can be found by setting $\frac{\partial P}{\partial t} = 0$:

$$\Rightarrow [\beta XY/N - \gamma Y]P = \frac{1}{2}\frac{\partial}{\partial Y}([\beta XY/N + \gamma Y]P),$$

$$= \frac{1}{2}\frac{\partial}{\partial Y}([\beta(N-Y)Y/N + \gamma Y]P),$$

$$\Rightarrow \quad \frac{\partial P}{\partial Y} = \left[\frac{2\beta XY - 2\gamma YN - \beta(N-2Y) + \gamma N}{\beta XY + \gamma YN}\right]P.$$

This is a linear equation in P, which although complex can be solved numerically with considerable ease, Figure 6.12 (lower graphs).

Although this method of understanding the role of noise can be extended to the traditional SIR-type models (Clancy and French 2001), it necessitates a two-dimensional probability distribution (specifying both X and Y), which requires much more sophisticated tools to integrate. Therefore, we will leave the formulation and solution of such models as an exercise for the keen and mathematically able. However, as computational power continues to advance, it may be preferable to solve the exact PDE equations—which capture even the rarest events—rather than perform multiple simulations of the differential equations with noise.

6.6.2. Master Equations

Master equations (also known as Ensemble or Kolmogorov-forward equations) are the integer-valued and event-driven equivalent of the Fokker-Plank equations described above. Essentially, they require the formulation of a separate differential equation for the probability of finding the population in every possible state (Stollenwerk and Briggs 2000; Stollenwerk and Jansen 2003; Viet and Medley 2006). For example, $P_{X,Y}(t)$ is the probability of having X susceptibles and Y infected, where X and Y are both integers,

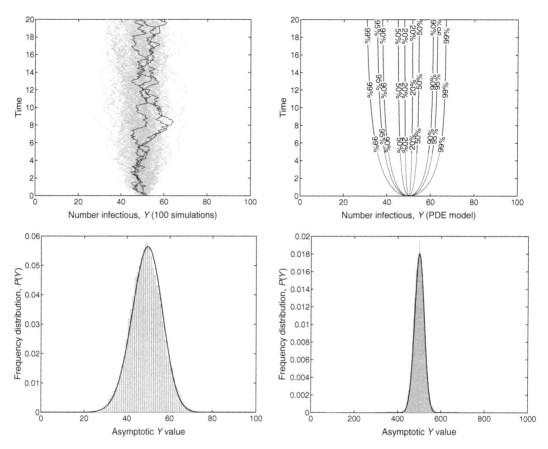

Figure 6.12. Comparison of 100 stochastic simulations and the Fokker-Plank PDE model, for the SIS model starting at the equilibrium point ($Y^* = N/2$). The top-left graph shows the time series of 100 simulations, with three random trajectories highlighted for greater clarity—note that time is plotted on the y-axis. The top-right graph gives the probability contours for the associated PDE model, corresponding to the probability of Y within a given range. The bottom graphs compare the long-term asymptotic results from both simulations and the PDE model for population sizes $N = 100$ and $N = 1000$; the results from the simulations (gray bars) being the average of 100 replicates between times 80 and 100 and the results from the PDE model (solid line) being the solution to $\frac{\partial P}{\partial t} = 0$. (All results are for $R_0 = 2$, $\gamma = 0.1$, $\mu = \nu = 0$).

at time t. Events, such as birth, infection, or recovery, move populations between various states and this is reflected in the equations for the probabilities.

The master equation provides an exact solution to multiple simulations of a model with demographic (event-driven) stochasticity.

Again, because of its simplicity and the fact that only two events can occur, we focus on the SIS equation. We let $P_Y(t)$ be the probability that Y individuals are infectious, noting again that there is no need to explicitly keep track of the susceptibles. It is often conceptually easier to think of a large (infinite) number of simulations, and to consider P_Y to be the proportion of simulations that have Y infecteds. Four processes can occur that modify the proportion of simulations in state Y:

1. A simulation in state Y can have an infected individual recover at rate γY, such that there are now only $Y - 1$ infecteds.
2. A simulation in state Y can have a susceptible individual become infected at rate $\beta XY/N = \beta(N - Y)Y/N$, such that there are now $Y + 1$ infected.
3. A simulation in state $Y + 1$ can have an infected individual recover at rate $\gamma(Y + 1)$, such that Y infected remain.
4. A simulation in state $Y - 1$ can have a susceptible individual become infected at rate $\beta(X + 1)(Y - 1)/N = \beta(N - Y + 1)(Y - 1)/N$, such that there are now Y infecteds.

Processes 1 and 2 cause the loss of a simulation in state Y, whereas processes 3 and 4 are associated with the gain of a state Y. Formulating an explicit equation for these processes, we obtain:

$$\frac{dP_Y}{dt} = -P_Y[\gamma Y] - P_Y[\beta(N - Y)Y/N] + P_{Y+1}[\gamma(Y + 1)]$$
$$+ P_{Y-1}[\beta(N - Y + 1)(Y - 1)/N]. \qquad (6.7)$$

This generates $N + 1$ differential equations ($Y = 0 \ldots N$), each of which is coupled to the two nearest probabilities. We insist that P_{-1} and P_{N+1} are zero, for obvious biological reasons.

As well as being able to iterate the probabilities forward from any particular initial conditions, one of the great benefits of these models is in understanding how stochasticity affects the final distribution of disease prevalence and the expected variation. However, a significant difficulty exists with this model, which also occurs for many models that use event-driven stochasticity—the fact that extinction events are permanent. Let us consider the dynamics of P_0, noting that P_{-1} is zero by definition:

$$\frac{dP_0}{dt} = -P_0[\gamma \times 0] - P_0[\beta N \times 0/N] + P_1[\gamma \times 1] = \gamma P_1.$$

Hence, P_0 keeps increasing, there is no escape from extinction, and eventually every simulation falls into this absorbing state, although this may take a very long time. So, the modeler is often trapped between the mathematical certainty of extinction and the epidemiological observation that extinctions in large populations are rare. Two solutions exist: either modify the equations such that a low level of infectious imports arrive in the population, or ignore those simulations that have gone extinct and find the distribution of the disease conditional on pathogen persistence (see Section 6.3.3.2). This latter method is often preferable because it retains the simplicity of the standard equations.

If we are interested in the final distribution of cases, then rather than having to iterate all the equations forward it is often far simpler to solve for when $\frac{dP_Y}{dt} = 0$. For this one-dimensional SIS model, the equilibrium solution can be found by equating the proportion of simulations moving from Y to $Y + 1$ with the proportion moving from $Y + 1$ to Y. If the model is at equilibrium, then these two "movements" must be equal or else some of the P_Y would be changing. This observation leads to iterative equations for the equilibrium

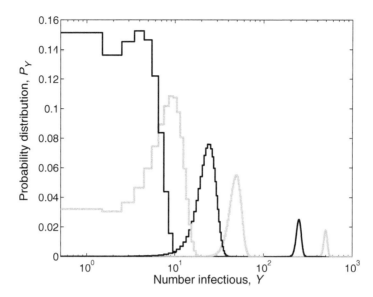

Figure 6.13. Final equilibrium probability P_Y^*, from the master equations for the SIS model, conditional on the infection being present. $R_0 = 2$, $N = 10, 20, 50, 100, 500$, and 1000. Note that the number of infectious individuals is plotted on a log scale. The integrals beneath all the curves are equal to one, even though the log scale makes the distributions for large populations appear to occupy a smaller area.

distribution:

$$P_{Y+1}^*[\gamma(Y+1)] \;=\; P_Y^*[\beta(N-Y)Y/N] \qquad \text{such that } \sum_{Y=1}^{N} P_Y^* = 1,$$

$$\Rightarrow \qquad P_{Y+1}^* \;=\; P_1^* \prod_{J=1}^{Y} \frac{\beta(N-J)J}{N\gamma(J+1)},$$

$$P_Y^* \;=\; P_1^* \frac{(N-1)!}{(N-Y)!Y} \left(\frac{\beta}{\gamma N}\right)^{Y-1}. \tag{6.8}$$

Hence, from either the initial differential equations (6.7) or the final explicit form (6.8), the equilibrium distribution of infection (conditional on nonextinction) can be calculated (cf Andersson and Britton 2000). Such conditional distributions are shown in Figure 6.13 from which several key factors emerge. For large population sizes (N greater than 500), the conditional equilibrium distribution is approximately Gaussian. This is because over the range of likely values the deterministic dynamics are approximately linear and the integer-valued nature of the population is largely irrelevant—in such cases the Fokker-Plank approximations may be very effective. (The smaller values of P_Y^* for these large population sizes occurs because the distribution is spread over a wider range of Y values). For smaller population sizes the distributions are much more skewed because the nonlinear behavior further from the fixed point plays a more dominant role and individuals are more important. For the very smallest population sizes, the distribution becomes bimodal with a peak at $Y = 1$ as well as near the expected mean. These results are equivalent to the earlier observation (Section 6.2.4.2) that for large population sizes the addition of noise to the differential equations may be an adequate means of capturing stochasticity, whereas for small population sizes an individual-based event-driven approach is required.

This mechanism can be extended to examine the stochastic dynamics of the SIR equation; this is conceptually straightforward, but algebraically awkward. The probability is now two-dimensional, depending on both the number of infected and susceptible individuals. In a population of N individuals, the master equations for the SIR model has $\frac{1}{2}(N+1)(N+2)$ equations, whereas for the SIS model there were only $N+1$ equations, therefore iterating the SIR master equations is far more computationally intensive. However, when trying to understand the stochastic disease dynamics within farms (or small communities) with only a few hundred individuals, the calculation may be feasible. The equations are also slightly more complex to formulate because there are now six possible events (infection, recovery, birth, death of susceptible, death of infected, and death of recovered) that need to be considered. Results from this equation with a population size of $N = 10,000$ are shown in Figure 6.14. There are clearly many differences between the average dynamics of the master equation (black) and the behavior of the standard deterministic equations (gray) (top-left graph). The master equation has a higher level of susceptibles at equilibrium as expected, due to the negative correlations that develop between the numbers of susceptibles and infecteds. Interestingly, due to the higher dimensionality of the master equations, the trajectories of the average quantities can "cross"—this is not observed for the standard model. We also notice that in the master equations the mean values of X and Y asymptote far quicker to their equilibrium values compared to the deterministic model—thus, in some sense stochasticity is acting to increase convergence. Looking at the distributions in more detail, the first snapshot (top-right) shows the initial development of negative correlations such that high prevalence is associated with low levels of susceptibles. At later times (lower graphs), there is a much wider distribution of values and the effects of disease extinction can be observed leading to high levels of susceptibles with few or no infecteds.

Such master equations are clearly a powerful but computationally intensive tool for understanding the dynamics of stochastic disease models. In many respects they should act as the template against which other modeling strategies are judged. With the ever-increasing power of computers, the simulation of such large sets of differential equations will become increasingly feasible. The more mathematically able reader may like to reflect on the fact that equation (6.7) and master equations in general are linear in terms of the probability distributions and therefore can be written in matrix form ($\frac{d\underline{P}}{dt} = \boldsymbol{M}\underline{P}$), which substantially simplifies their analysis and computation.

6.6.3. Moment Equations

A convenient way to calculate the effect of noise or stochasticity on the dynamics of a disease is to develop moment equations (Nasell 1991,1996; Keeling 2000b). This approach has also been highly successful at studying the worm burden associated with a variety of macro-parasitic infections (Michael et al. 1998; Herbert and Isham 2000). The process starts by considering the effects of second-order moments (variances and covariances) on first-order moments (mean values). We introduce the convenient notation that $\langle.\rangle$ defines the average value of a quantity across infinitely many stochastic simulations. Looking at

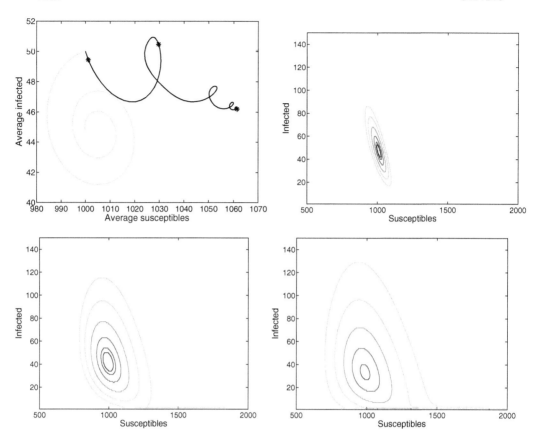

Figure 6.14. Dynamics of the SIR master equation ($R_0 = 10$, $\gamma = 0.1$ per day, $\nu = \mu = 5 \times 10^{-4}$ per day, $N = 10^4$, imports of infection occur at the rate of two per year, $\varepsilon \approx 0.055$ per day). Note that the values of ν and μ lead to a much faster demographic turnover than we have previously assumed; the average life expectancy is around 5.5 years. This has been done to increase the prevalence and hence improve disease persistence. The top-left graph shows the dynamics of the average number susceptible and infected predicted by the master equations (black) and predicted by standard differential equations (gray). The remaining graphs show the predicted distribution of infected and susceptibles at three times (indicated with stars on the top-left graph). The contours correspond to 95%, 90%, 75%, 50%, 25%, and 10% confidence intervals. The master equation was initialized with $P_{1000,50} = 1$ and all other terms zero. The equations were simulated within the rectangle $500 \leq X \leq 3000$, $0 \leq Y \leq 500$, and hence contain 1.25 million individual equations.

the equation for the number of infected individuals:

$$
\begin{aligned}
\frac{d\langle Y \rangle}{dt} &= \langle \beta XY/N - \gamma Y \rangle, \\
&= \beta \langle XY \rangle /N - \gamma \langle Y \rangle, \\
&= \beta \langle X \rangle \langle Y \rangle /N + \beta\, Cov_{XY}/N - \gamma \langle Y \rangle.
\end{aligned}
$$

Hence, the covariance between X and Y enters the equations and modifies the average transmission rate, which in turn modifies the mean values. Occasionally there are clear biological reasons to assume that the covariance has a given form (Keeling et al. 2000),

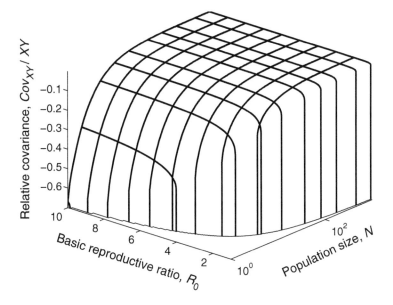

Figure 6.15. The relative magnitude of the covariance compared to the product XY at equilibrium for the SIS model as estimated from the moment closure equations. For small population sizes and small R_0 the disease cannot persist due to the strong negative correlation effects.

or we may wish to leave the covariance terms as a free parameter. However, frequently the next step is to use the disease dynamics to formulate a differential equation that describes the behavior of the covariance (Keeling 2000b). For the SIS model, after a fair amount of algebra, we find that the covariance rapidly converges to a value that depends on X and Y:

$$Cov_{XY} = -\frac{\beta XY + \gamma YN}{2\gamma N + 2\beta Y + \beta - 2\beta X},\qquad (6.9)$$

where closure has been achieved by assuming a normal distribution (Keeling 2000b). Therefore, we find that the covariance is large relative to the standard product (XY) whenever R_0 or the population size N is small (Figure 6.15). In fact, for very small populations or very small R_0, the effect of the covariance is so strong that the disease is always forced extinct—this is partly due to the normality assumption that was used to construct the covariance equation.

   Although the amount of algebra involved in determining the moment equations and equilibria is somewhat daunting, there are some significant benefits. In particular, we now have an explicit relationship between the variance, the covariance, and the disease parameters which can often provide a more intuitive understanding than simulations alone. In the example provided by Figure 6.15, it is clear where the covariance is relatively large and hence where stochasticity is likely to affect both the effective transmission of infection and also the mean values. These insights can often provide novel interpretations of the results of stochastic simulations which would have been difficult to extract from the simulation results alone.

6.7. FUTURE DIRECTIONS

Almost all natural populations experience some degree of stochasticity. This is being increasingly recognized by epidemiologists and modelers who seek to account for the random nature of transmission and recovery. In recent years there has been a huge increase in the use of stochastic models in the literature; event-driven stochasticity is proving by far the most influential method due to its mechanistic underpinning and the ability to incorporate complex rules. This trend is likely to continue, as making large individual-based stochastic models become ever more computationally feasible (Levin et al. 1997).

Future work may also consider whether the stochastic nature of disease transmission can be turned to a public health advantage. We have clearly seen that stochasticity can lead to the rapid, and premature, extinction of infection. Harnessing and even amplifying this effect would be highly beneficial and could improve the efficacy of vaccination programs (see Chapter 8).

Stochasticity offers a new form of uncertainty that should be quantified by researchers. It is as important to consider the sensitivity of model results to stochasticity as well as the more common sensitivity to parameters. Associated with this is the difficulty of quantifying and explaining stochastic results. Mean results can often be biased by rare (but still stochasticity feasible) events; thus, at the very least, confidence intervals or distributions should always be specified when describing the outcome of stochastic models.

The analytical models addressed in Section 6.6 offer a bridge between the purely deterministic approaches and computer simulations. Thus, although mathematically daunting, these methods offer an analytical, and therefore often more intuitive, insight into the role of stochasticity. It is very likely that the development of such models, by skilled mathematicians, will play a pivotal role in the way epidemiologists think about stochasticity and stochastic modeling in the future.

6.8. SUMMARY

In this chapter we have shown how the random nature of transmission events and the individual nature of populations can affect disease dynamics. Although there are several methods of simulating noisy dynamics, we have concentrated on event-driven demographic stochasticity as a robust means of capturing random events and the discrete nature of individuals.

➤ Observational noise does not impact the epidemiological dynamics; it modifies only the reported data.

➤ Stochasticity in the transmission term can cause the number of cases to resonate at (or near) the natural frequency of the system. Hence, the observed stochastic dynamics more closely match the transient behavior of the deterministic equations.

➤ Stochasticity in the transmission term causes variance in the number of infecteds and in the number of susceptibles, and the negative covariance between them. The magnitude of these values increases almost linearly with the variance of the noise. Stochasticity can also introduce changes to the mean values as well as cause variation about the means.

➤ A log-log plot of mean against variance is a useful method of summarizing the variability with population size. For large population sizes, we expect the variance in the number of cases to be proportional to the mean.

➤ For large population sizes, where the individual nature of the population is unimportant, noise can be added to the differential equations to mimic the impact of stochasticity.

➤ Noise can be generated from a variety of sources. The relative magnitude of these noise terms depends on the population size. External parameter noise dominates when the population is large.

➤ Demographic stochasticity requires integer-valued variables and a probabilistic fate of individuals.

➤ Models using demographic stochasticity run much slower as the population size (and in particular the number of infectious individuals) becomes large; the τ-leap method provides a possible solution to this difficulty.

➤ From a single infected individual in a totally susceptible population, the probability of extinction before a major epidemic ensues is equal to $\frac{1}{R_0}$. A partially resistant population or a highly variable infectious period increases the likelihood of extinction, whereas multiple introductions of infection decrease the likelihood.

➤ The Critical Community Size (CCS) is the smallest population size observed not to suffer disease extinctions; for measles the CCS is around 400,000. Simple stochastic models that capture the observed seasonally induced measles epidemics fail to show adequate levels of persistence which can be captured only with a greater number of classes within the models and appropriate parameterization.

➤ Imports of infection into a community can significantly modify the extinction process in a complex manner. Frequent imports may prevent extinctions, whereas long delays between extinctions and imports improve the chance that an import successfully triggers an epidemic. When dealing with very low levels of imports between populations it is essential that this is modeled stochastically.

➤ In many situations, stochasticity may be beneficial, leading to the extinction of a disease well before the vaccination threshold is reached. Optimal vaccination strategies should take advantage of such extinctions.

➤ The Fokker-Plank equation provides an *exact* solution to multiple simulations of a model with noise. Noise in the equations translates to diffusion of the probability distribution.

➤ The master equation provides an exact solution to multiple simulations of a model with demographic (event-driven) stochasticity.

Chapter Seven

Spatial Models

It is intuitive that in most circumstances disease transmission is predominantly a localized process. For directly transmitted diseases, for example, transmission is most likely between individuals with the most intense interaction, which generally implies those in the same location. Additionally, movement of individuals between population centers facilitates the geographical spread of infectious diseases. This chapter is concerned with capturing these host population characteristics, enabling us to address issues such as: determining the rate of spatial spread of a pathogen, calculating the influence of large populations on smaller ones, and finding optimally targeted control measures that take into account the local nature of spatial transmission. Generally, models of this sort operate by partitioning the population according to the spatial position of hosts, such that nearby hosts are grouped together and interact more strongly. A wide variety of model formats have been developed to accomplish this, with the primary differences being the scale at which hosts are aggregated. Section 7.7 outlines when each of the model types can be used, although no definitive rules exist. Rigorous analytical results for spatial epidemiological models remain rare. Since the late 1980s, however, the increasing ease of access to computational power has permitted the detailed simulation of such models (Levin et al. 1997). Frequently, these models incorporate stochasticity, so readers may wish to familiarize themselves with Chapter 6 before continuing.

We will introduce the main types of spatial models in some detail, outline the general methodology, comment on their strengths and weaknesses, and consider how such models have been used to explore the spatial spread of real-world epidemics. The following three motivating examples will illustrate how the behavior of the infectious disease and the host, and the amount of knowledge available, dictates the form of model.

The 2001 foot-and-mouth epidemic in the United Kingdom was a clear example of how space can play a significant role in disease dynamics. Cases were predominantly restricted to three regions: Cumbria, Devon, and the Welsh borders, with most transmission occuring within 3 kilometers from the source farm. Given that the location of all farms in the United Kingdom was known, and that accurate predictions (rather than generic insights) were required, two of the models used during the epidemic (Keeling et al. 2001b; Morris et al. 2001) were individual-based and explicitly spatial, using the known location of farms and the estimated rate of transmission as a function of distance to simulate the epidemic.

If we are interested in the spread of human disease, such as pandemic influenza (Grais et al. 2003) or SARS (Riley et al. 2003), then models need to focus on the dynamics at a large geographical scale. Given our limited knowledge about human movement patterns and interactions, it is typically impractical to simulate each individual within a population, although this has recently been attempted for both small (Halloran et al.

2002; Longini et al. 2005) and large population sizes (Ferguson et al. 2005, Ferguson et al. 2006). One practical limitation of these models is the difficulty with which we can assess the sensitivity of their predictions to perturbations in the social network structure. Instead, it is often plausible to assume random mixing within a localized community and reduced mixing between the communities—a so-called metapopulation model. We may, for example, wish to subdivide the population by town/city, by county, or by state, depending upon our level of knowledge and the detail of results required (see, for example, Viboud et al. 2006).

Finally, for diseases of wildlife or plants, such as tuberculosis (TB) in badgers (Shirley et al. 2003), rabies in foxes (Murray et al. 1986) or raccoons (Smith et al. 2002), Dutch elm disease (Swinton and Gilligan 1996), or Sudden Oak Death (Kelly and Meentemeyer 2002), there may not be a natural partitioning of the host population. Instead, it is frequently assumed that individuals are either uniformly or randomly distributed, with their density reflecting landscape and environmental factors. In such cases, continuous-space models, phrased as partial differential equations (PDEs) or integro differential equations (IDEs), can be used (Kot 2001). Again, it would be unfeasible to model every badger sett within the United Kingdom or every raccoon in the United States, and naive to assume that either creature respects county or state boundaries. However, due to limitations arising from the spatial resolution of empirical data, which may be aggregated at the county or state level, it may be necessary to model the host population at a similarly coarse scale (Smith et al. 2002).

The type of model used is directly dependent on the host organism, our degree of knowledge about its behavior, and the scale we wish to consider.

7.1. CONCEPTS

A variety of models can be used to study the spatial spread of pathogens, and although each has its own specific aspects, a range of concepts are shared. We first discuss these elements, so that the similarities and differences between the models will be more apparent, and to introduce the language of spatial processes.

7.1.1. Heterogeneity

Spatial heterogeneity refers to differences between populations or individuals at different geographical locations. Such heterogeneities can arise from two sources. Underlying (environmental) heterogeneities describe spatial differences in the fundamental forces governing the population dynamics. For example, wildlife populations in different locations may experience differing habitat conditions that may affect demographic rates, or different human populations may have different social structures leading to variation in disease transmission rates (Finkenstädt and Grenfell 1998; Grenfell and Bolker 1998; van Buskirk and Ostfeld 1998; Auvert 2000). Such underlying heterogeneities are common in the real world, but are frequently ignored in models due to the extra complexity they introduce and due to a lack of available data. If quantitatively precise predictions are required from models, however, it is often vitally important that such underlying heterogeneities are considered (Keeling et al. 2001b; Smith et al. 2002). The second form of heterogeneity is emergent and describes observed differences in population structures arising from dynamical processes, such as stochasticity, or differences in movement between populations (Hassell et al. 1991; Rhodes and Anderson 1996; Green and Sadedln 2005). In general, this second form of heterogeneity is greatest between populations that experience large amounts of stochasticity, have very different underlying parameters, and have little transfer of infection between them.

Heterogeneity can describe either the underlying differences between two populations, or the emerging dynamic differences in the population levels (such as the proportion of the population that are infectious).

A convenient measure of observed heterogeneity is provided by estimating *correlations*—they quantify the degree to which the dynamics in two (or more) populations behave in the same manner. Simply put, correlations help to establish whether epidemics in different populations are synchronized or out of phase (Grenfell and Bolker 1998; Rohani et al. 1999; Grenfell et al. 2001). If we let I_i denote the time series documenting the prevalence of an infection in population i, then the correlation between epidemics in two populations is calculated as:

$$C_{12} = \frac{(I_1(t) - \overline{I_1})(I_2(t) - \overline{I_2})}{\sqrt{\text{var}(I_1)\text{var}(I_2)}}. \tag{7.1}$$

Here, $\overline{I_i}$ refers to the mean infection prevalence (averaged over time) in population i. If the fluctuations in prevalence over time in the two populations are either identical or directly proportional ($I_1 \propto I_2$), then the correlation attains its maximum value of 1. If epidemics in the two populations are independent, then the correlation is zero. If the outbreaks are out of phase, then C_{12} is negative. Given time-series data on the number of cases in two populations, we are predominately interested in the average correlation over a given period, rather than the instantaneous value that is subject to short-term stochastic fluctuations. The correlation cannot be defined for deterministic populations at their equilibrium values (because both population levels are constant and the variance is zero), and therefore, in general, correlations are usually associated with stochastic or seasonally forced systems.

Correlations provide a quantitative measure of the differences between populations: A positive/negative correlation indicates that epidemics are spatially synchronous/asynchronous.

Although the standard correlation (equation (7.1)) measures the heterogeneity generally derived from the stochastic nature of the epidemic process, heterogeneities can also arise due to *traveling-waves*. Consider the spread of West Nile virus across the United States from New York in 1999 to the West Coast in 2003 (see Chapter 4 for a more detailed description of West Nile virus). The observed heterogeneities in incidence on the East and West Coast are unlikely to be a result of either inherent habitat differences, or due to stochasticity. They simply reflect the fact that the disease emerged in the east and travelled west. To quantify this traveling-wave type of heterogeneity we need to use *lagged correlations*:

$$C_{12}^{\tau} = \frac{(I_1(t+\tau) - \overline{I_1})(I_2(t) - \overline{I_2})}{\sqrt{\mathrm{var}(I_1)\mathrm{var}(I_2)}}. \tag{7.2}$$

If a traveling wave is observed, then the value of τ that maximizes the lagged correlation, C_{12}^{τ}, should increase with the separation, d, between two populations. If the traveling wave moves with constant velocity, c, then $\tau_{max} = d/c$, which can be derived from the fact that the time taken for the wave to travel a distance, d, is the distance divided by the velocity. More advanced versions of this approach have been successfully used by Grenfell et al. (2001) to identify traveling waves of measles infection in England and Wales spreading from large population centers like London to the surrounding smaller communities.

7.1.2. Interaction

Consider the behavior of an infectious disease within several human populations. If there is no interaction (movement) between the populations, then their dynamics will be independent and hence the correlation between them will be zero (assuming no other synchronizing mechanisms, such as seasonal forcing or climatic factors). However, movement of hosts between populations, with the associated risk of disease transmission, can couple dynamics. The way in which we choose to model this interaction should reflect the behavior of the host and the scale at which our model operates.

One of the simplest means of modeling the interaction between (for example) two populations is for susceptible individuals at one location to experience an additional force of infection due to infectious individuals at the other. This would represent a phenomenological approach to spatial modeling and we frequently refer to the strength of such interactions as the level of *coupling* between the populations. The greater the coupling, the more each population is impacted by the transmission dynamics of the other and the higher the level of correlation and synchrony.

Interaction or coupling between different spatial locations allows infection to spread and acts to synchronize the epidemic dynamics at the two locations.

It is intuitive that in many situations the interaction between two populations should decrease with the distance, d, between them. This type of behavior can be captured by introducing a *transmission kernel*, K, which modifies the coupling term and is a function of the distance between two populations. Common examples of transmission kernels include exponential ($K \propto \exp(-Ad)$), Gaussian ($K \propto \exp(-Ad^2)$), or power-law ($K \propto d^{-A}$) (Erlander and Stewart 1990; Gibson 1997a,b; Keeling et al. 2004b; Xia et al. 2004), with the precise form chosen determined by the observed dynamics. Estimating the kernel form and parameters is a very difficult but important problem; although the more common, short-distance transmission events determine the basic reproductive ratio, it is the long-distance tail of the kernel that determines the eventual speed of a traveling wave of invading pathogen (Diekmann 1978; van den Bosch et al. 1990; Mollison 1991; Shaw 1995; Lewis 2000; Xia et al. 2004). However, at long distances the kernel is usually very small, so there will be only limited amounts of data for the estimation processes. These rare jumps to new areas cannot be ignored, however, because they are often vitally important to the invasion process.

The reduction in transmission risk with distance is captured by a transmission kernel, which is frequently assumed to be either exponential, Gaussian, or power-law.

7.1.3. Isolation

Isolation is another factor that is common in a wide range of models and real scenarios. It simply refers to the situation when a group of hosts is protected from the risk of transmission due to their spatial separation from an infectious source. For example, we might consider communities that have few contacts with the outside world as being isolated from the general pool of infectious individuals. Alternatively, animal populations that are separated by large distances can be epidemiologically isolated from other populations. The existence of isolated populations can have a profound impact on parameterization; if isolated populations are included in an estimate of disease parameters, their rarity of transmission will bias the results.

7.1.4. Localized Extinction

In any stochastic population model, there is always the risk that the disease will, by chance, become extinct and this risk increases as the host population size gets smaller (see Chapter 6). This is where the spatial resolution of study becomes important. For example, if we consider the aggregate epidemics of an infectious disease like measles in the prevaccine era in the whole of England and Wales, then the probability of witnessing a fade out is almost zero. However, as we examine case reports in increasingly smaller cities and towns, the frequency of local extinctions increases, with the smallest population centers experiencing fade outs in between epidemics sparked by the introduction of infection from cities where it is endemic. The likelihood of such "recolonization events" is influenced by the synchrony of measles epidemics and the coupling between subpopulations (Bolker and Grenfell 1996; Earn et al. 1998; Rohani et al. 1999; Keeling 2000b; Hagenaars et al. 2004). As a result, overall population persistence is determined by a key relationship between the subpopulation size, the degree of interaction, and the strength of asynchrony. The precise details of this relationship remain largely unclear, but some of the complexities are discussed below when metapopulations are described (Section 7.2.3).

Due to the smaller subpopulation sizes often involved in spatial models, localized extinctions are common. Large-scale eradication is prevented by coupling between subpopulations leading to the reintroduction of infection into disease-free areas.

7.1.5. Scale

Two forms of scale are important for spatial models: (1) the scale of interaction, and the (2) the scale of simulation.

The majority of spatial models make some assumption about the spatial scale of interaction and the scale at which the population can be subdivided. Although there is rarely a "correct" scale, it is clear that using too fine a scale and hence creating many subpopulations can be computationally prohibitive, whereas aggregating at too large a scale can eliminate the spatial effects that are of primary interest. For individual-based models (which operate at the level of the individual) and metapopulation models (which consider communities as the aggregate unit), the scale is generally fixed. However, for both lattice-based and continuous-space models (PDEs and IDEs), the question of scale is much more subtle. In such models, it is frequently assumed that each population interacts only with a limited number of other populations (usually within a prescribed distance), and this set of populations is referred to as the (interaction) *neighborhood*. In general, the use of a finite neighborhood is an approximation to the true dynamics, but may greatly increase the speed of any spatial simulation.

When dealing with a particular "real-world" problem, such as foot-and-mouth in the United Kingdom or SARS in North America, the spatial extent of our model is fixed by the problem. However, in more abstract or generic situations, the scale at which the spatial model operates is less constrained. Intuitively, we wish to model at a sufficiently large scale so that the full range of dynamics is observed and the scale of the model has limited impact on our results. However, models that are too "big" may be slow to compute, thereby limiting their usefulness. Finding the correct scale usually involves assessing whether the salient behavior of large-scale models is still captured by models of a smaller, more computationally manageable scale. Therefore, at least initially, large-scale simulations are required to determine the unconstrained dynamics. Although a number of dynamical systems techniques exist to find the optimal, most informative scale (Mead 1974; Keeling et al. 1997a; Pascual et al. 2001), the answer is often context dependant.

Assessing the fine scale at which individuals are aggregated and the larger scale at which simulations are performed should either be based on sound epidemiological knowledge or achieved by comparing simulated results across a range of scales.

7.2. METAPOPULATIONS

Metapopulations are one of the simplest spatial models, but are also one of the most applicable to modeling many human diseases. The metapopulation concept is to subdivide the entire population into distinct "subpopulations", each of which has independent dynamics, together with limited interaction between the subpopulations. This approach has been used to great effect within the ecological literature; a comprehensive guide to the broader applications of this approach is given by Hanski and Gilpin (1991, 1997) and Hanski and Goggiotti (2004). For disease-based metapopulation models, a suitable modified version of the *SIR* equation (Chapter 2) would be:

$$\frac{dX_i}{dt} = \nu_i N_i - \lambda_i X_i - \mu_i X_i,$$
$$\frac{dY_i}{dt} = \lambda_i X_i - \gamma_i Y_i - \mu_i Y_i,$$
(7.3)

where the subscript i defines parameters and variables that are particular to subpopulation i. The force of infection, λ_i incorporates transmission from both the number of infecteds within subpopulation i and the coupling to other subpopulations. In this general formulation, the demographic and epidemiological parameters may vary between subpopulations, reflecting differences in the local environments (Finkenstädt and Grenfell 1998; Grenfell and Bolker 1998; Langlois et al. 2001; Broadfoot et al. 2001)

Metapopulations provide a powerful framework for modeling disease dynamics for hosts that can be naturally partitioned into spatial sub-units.

The precise relationship between the force of infection for population i and the number of infectious individuals in population j depends on the assumed mechanism of transmission and the strength of interaction between the two populations. In general terms, the force of infection can be written as a sum:

$$\lambda_i = \beta_i \sum_j \rho_{ij} \frac{X_j}{N_i},$$

where the coefficients, ρ, are a measure of the strength of interaction between populations. Specifically, ρ_{ij} measures the relative strength of transmission *to* subpopulation i *from* subpopulation j. An important aspect of this formulation concerns the precise scaling with population size in the expression of λ_i. The equation above contains N_i in the denominator, which reflects the implicit assumption that transmission takes place in population i, presumably resulting from the movement of an infectious individual from population j. Alternatively, the assumption that transmission is due to a susceptible individual from population i picking up the infection during a temporary visit to population j would be incorporated by placing N_j in the denominator. The assumptions implicit in the coupling interaction are discussed more fully in Section 7.2.1.

The force of infection within a subpopulation can be expressed as a weighted sum of the prevalence in all populations.

We now explore the differences between deterministic and stochastic versions of this metapopulation model. Consider two large, fully susceptible populations ($S_1 = S_2 = 1$), with $\rho_{ii} = 1$ and ρ_{ij} much less than 1. (We will assume that the two populations are the same size, which simplifies the form of the coupling interaction, and ignore the effects of demography.) We start with an infectious disease solely in population 1, which exhibits a standard epidemic curve (Chapter 2), because the coupling between populations is assumed to be small. In the deterministic framework, ignoring births and deaths, the early dynamics of population 2 (before the proportion of susceptibles drops significantly, $S_2 \approx 1$) will be given by:

$$\frac{dI_2}{dt} = \beta_2 \rho_{21} I_1 + \beta_2 I_2 - \gamma_2 I_2.$$

This equation can be solved using the "integrating factor" (see, for example, Strang 1986) to obtain an expression for I_2 through time

$$I_2(t) = \int_0^t \beta_2 \rho_{21} I_1(s) \exp([\beta_2 - \gamma_2]s) ds. \tag{7.4}$$

This expression represents the exponential growth of prevalence in population 2 from time 0 to time t, due to the presence of $I_1(s)$ infecteds in population 1 at time s. The deterministic equation (7.4) has two main implications: the disease is "present" in population 2 from the start of the epidemic (in population 1), and the early infinitesimal infections that arrive in population 2 trigger an exponential growth at rate $\beta_2 - \gamma_2$.

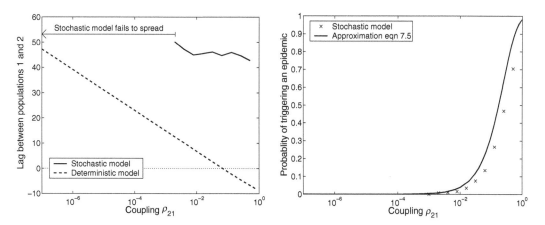

Figure 7.1. For a metapopulation with just two populations we examine the effect of coupling ρ_{21} on the dynamics of the epidemic in population 2. Population 1 is modeled deterministically and is initialized with $I(0) = 10^{-5}$, $S(0) = 1 - I(0)$. The left-hand graph shows the delay between the peak of the epidemic in population 1 and the peak in population 2, where population 2 is modeled either stochastically (solid line) or deterministically (dashed line) and is initially disease-free. The right-hand graph is the probability that a major epidemic is triggered in the stochastic population (crosses) compared to the analytical approximation, equation (7.5). ($\mu = \nu = 0$, $1/\gamma = 14$ days, $\beta = 0.3571$ per day $\Rightarrow R_0 = 5$, $\rho_{ii} = 1$, $\rho_{12} = 0$. Stochastic results are the average of 1,000 realizations, with population 1 treated deterministically, $N_2 = 10^5$).

In the analogous stochastic formulation, model behavior is significantly different. The probability that a major epidemic is triggered in population 2 is given by:

$$\mathbb{P}(\text{epidemic}) = \sum_{n=1}^{N} \mathbb{P}\,(\text{cases in 1 cause } n \text{ cases in subpopulation 2}) \times$$

$$\mathbb{P}\,(n \text{ initial cases lead to a major epidemic}),$$

$$= \sum_{n=1}^{N} \exp\left(-\beta_2 \rho_{21} \int_0^\infty I_1(s)ds\right) \frac{\left(\beta_2 \rho_{21} \int_0^\infty I_1(s)ds\right)^n}{n!} \times \left[1 - \left(\frac{\gamma}{\beta_2}\right)^n\right],$$

$$= 1 - \exp\left(-\beta_2 \rho_{21}\left[1 - \frac{\gamma}{\beta_2}\right]\int_0^\infty I_1(s)ds\right) \qquad (7.5)$$

$$< 1 - \exp\left(-\beta_2 \rho_{21}/\gamma\right).$$

So, in the stochastic formulation, if the coupling ρ_{21} between populations is small enough, there is a good chance that the epidemic will fail to spread (Park et al. 2002). When ρ_{21} is larger, although the pathogen may eventually spread, there may still be a significant delay before other populations are exposed. Hence, in stochastic metapopulation models, the spread of infectious disease is slower than for the deterministic counterpart.

This principle is illustrated in Figure 7.1. The pathogen is introduced in population 1 only, with $\rho_{ii} = 1$, and we measure the lag between the peak of the epidemic in population 1 and the subsequent peak in population 2. In both models, the delay between the peaks decreases as the coupling ρ_{21} increases. For these parameters, however, it is not until the coupling exceeds approximately 0.01 that an epidemic has a significant chance of being triggered in the stochastic model. Two interesting aspects should be noted. The first is that for high levels of coupling, the lag in the deterministic model is negative (population 2 peaks first); this is because the force of infection in population 2 is boosted by the coupling from population 1, giving it a faster rate of epidemic growth (note that we have assumed that $\rho_{12} = 0$, so population 1 cannot receive infection from population 2). Second, the expected average lag predicted by stochastic simulations is far longer than the lag from the deterministic equations due to the chance nature of the initial transmission between populations.

With stochastic metapopulations, the spread of pathogen between subpopulations is reduced compared to the equivalent deterministic model.

7.2.1. Types of Interaction

We now turn our attention to the types of spatial transmission that are associated with different hosts and different types of movement. Although for convenience we label these as plant, animal, and human-commuter interactions, these descriptions are not rigid but rather illustrate the underlying host dynamics that generate the transmission terms. Hence, animal subpopulations that actually interact via occasional encounters would be better modeled as commuters, whereas animal subpopulations where the interaction is due to the wind-borne pathogen spread are more realistically modeled using the plant formulation.

In some circumstances, the interaction between subpopulations can be parameterized from direct observation, as is the case for commuter movements between communities (see Section 7.2.1.3) or from comparison to the recorded epidemic patterns (Wallace and Wallace 1993; Smith et al. 2002; Xia et al. 2004). Alternatively, the interaction terms between subpopulations are often assumed to obey a simple distance relationship (Erlander and Stewart 1990; Finkenstädt and Grenfell 1998).

7.2.1.1. Plants

The most obvious defining feature of plants (from an epidemiological perspective) that separates them from other hosts is that they do not move. This means that any spatial transmission must be wind- or vector-borne. We therefore retain the formulation

$$\lambda_i = \beta_i \sum_j \rho_{ij} I_j \tag{7.6}$$

and consider the coupling, ρ, as a function that decreases with the distance between the subpopulations, and for simplicity set $\rho_{ii} = 1$ (Park et al. 2001, 2002; Thrall et al. 2003). We can now calculate R_0 for infectious individuals in population i, as the expected number of secondary cases generated in all subpopulations:

$$R_0^i = \sum_j \frac{\beta_j \rho_{ji}}{\gamma_i}. \tag{7.7}$$

Note that the coupling term is now ρ_{ji} because we are concerned with transmission to j from i. It is important to realize that with this model, the addition of extra subpopulations (with extra hosts) increases R_0. Intuitively, this is because more external populations can "capture" wind-borne pathogen particles that otherwise would not have contributed to the transmission process. However, dividing one subpopulation into two should not have the same effect—if population j is divided forming two new populations k and l, then $\rho_{ji} = \rho_{ki} + \rho_{li}$ so that R_0 remains constant. It is only the inclusion of additional host populations that can raise R_0.

The concept of a metapopulation is one that has been readily used by the plant research community, such that a variety of data exist on the distribution of diseases in plants (Burdon et al. 1995; Ericson et al. 1999) with a strong focus on the genetic specialization of the pathogen to local populations (Burdon and Thrall 1999; Bergelson et al. 2001).

Such coupling mechanisms are not exclusively for plant hosts, but can be applied to any sessile population with wind- or vector-borne pathogens. Hence, this form of metapopulation model is ideal for describing the spatial dynamics of livestock diseases, where each farm is a subpopulation and transmission between farms can either be wind-borne or due to the movement of people, vehicles, or animals (Keeling et al. 2001b; Ferguson et al. 2001a; Section 7.5.2).

For plants and other sessile hosts, coupling generally decreases with distance, mimicking the effects of wind- or vector-dispersal. Adding an extra subpopulation generally increases R_0 because more pathogens can be intercepted by the additional hosts.

7.2.1.2. Animals

For many animal populations, it is plausible to assume that the spread of disease is due to the migration or permanent movement of individuals. The simplest means of modeling this is to allow animals to randomly move between subpopulations (Foley et al. 1999; Broadfoot et al. 2001; Fulford et al. 2002), although other assumptions based on known dispersal behavior of specific species leading to different spatio-temporal dynamics may be more appropriate (Gudelj et al. 2004). The metapopulation SIR-type model is then:

$$\frac{dX_i}{dt} = \nu_i - \beta_i X_i Y_i - \mu_i X_i + \sum_j m_{ij} X_j - \sum_j m_{ji} X_i,$$
$$\frac{dY_i}{dt} = \beta_i X_i Y_i - \gamma_i Y_i - \mu_i Y_i + \sum_j m_{ij} Y_j - \sum_j m_{ji} Y_i. \tag{7.8}$$

This is online program 7.1

Here, coupling is governed by the parameter m_{ij}, which measures the rate at which hosts migrate to subpopulation i from j—and therefore captures both emigration and immigration. We have assumed density-dependent transmission in equation (7.8), reflecting the common assumption about wildlife diseases—although frequency-dependent transmission could easily be accommodated. It is frequently assumed that the movement rates balance, $m_{ij} = m_{ji}$, so that the subpopulation sizes are maintained—there is no reason why this has to be the case and some subpopulations could act as sources of animals that colonize less favorable habitats. In this coupling framework (and assuming that β and γ are population independent and population sizes are equal or transmission is frequency dependent), the basic reproductive ratio is $\frac{\beta N}{\gamma + \mu}$ and is independent of the coupling strength. Intuitively, this is because each infectious animal transmits at a constant rate irrespective of which population it is in, and so always generates the same average number of secondary cases.

Compared to the plant models above, the movement of animals has two very different components: (1), it is the direct movement of infected animals that spreads the pathogen, and (2) the movement of susceptibles can help prevent stochastic extinctions (see Section 6.3.3) in heavily infected subpopulations. In this way, the coupling implied by animal movements is more variable than that due to wind or vector transmission assumed for plant infections. For the plant model (7.6), there is a continual force of infection from subpopulation 1 to subpopulation 2 (and vice versa); however, for the animal-based model (7.8) transmission between subpopulations occurs only following the movement of an animal. Therefore, in a stochastic framework either no infected animals have moved recently, in which case there is no transmission between populations, or an infected animal has moved, in which case the transmission is reasonably strong.

Models of animal diseases usually capture the transmission of infection by the permanent immigration and emigration of hosts. In these models R_0 is generally independent of the coupling because each host transmits infection at a constant rate.

7.2.1.3. Humans

For human populations, permanent relocation from one population to another is sufficiently rare that it may be ignored as an epidemiologically significant force. Instead, it is more natural to think about commuters spreading the disease (Keeling and Rohani 2002). Commuters live in one subpopulation but travel occasionally to another subpopulation. We therefore label X_{ij}, Y_{ij}, and N_{ij} as the number of susceptibles, infecteds, and total hosts currently in population i that live in population j. When there are multiple communities within the metapopulation, and when the populations are of different sizes or the strengths of interaction differ, it is more informative to return to first principles to calculate the dynamics. From the standard SIR models (Chapter 2) we consider the number of individuals of each type (S, I, and R) in each spatial class:

$$\frac{dX_{ii}}{dt} = \nu_{ii} - \beta_i X_{ii} \frac{\sum_j Y_{ij}}{\sum_j N_{ij}} - \sum_j l_{ji} X_{ii} + \sum_j r_{ji} X_{ji} - \mu_{ii} X_{ii},$$

$$\frac{dX_{ij}}{dt} = \nu_{ij} - \beta_i X_{ij} \frac{\sum_j Y_{ij}}{\sum_j N_{ij}} + l_{ij} X_{jj} - r_{ij} X_{ij} - \mu_{ij} X_{ij},$$

$$\frac{dX_{ii}}{dt} = \beta_i X_{ii} \frac{\sum_j Y_{ij}}{\sum_j N_{ij}} - \gamma Y_{ii} - \sum_j l_{ji} Y_{ii} + \sum_j r_{ji} Y_{ji} - \mu_{ii} Y_{ii},$$

$$\frac{dY_{ij}}{dt} = \beta_i X_{ij} \frac{\sum_j Y_{ij}}{\sum_j N_{ij}} - \gamma Y_{ij} + l_{ij} Y_{jj} - r_{ij} Y_{ij} - \mu_{ij} Y_{ij}, \qquad (7.9)$$

$$\frac{dN_{ii}}{dt} = \nu_{ii} - \sum_j l_{ji} N_{ii} + \sum_j r_{ji} N_{ji} - \mu_{ii} N_{ii},$$

$$\frac{dN_{ij}}{dt} = \nu_{ij} + l_{ij} N_{jj} - r_{ij} N_{ij} - \mu_{ij} N_{ij},$$

This is online program 7.2

where l_{ij} measures the rate that individuals leave their home population j and commute to population i, and r_{ij} measures the rate of return. Frequently, in many human disease scenarios the parameters l and r can be found from commuter movement data or travel statistics (Grais et al. 2003; Cliff and Haggett 2004). Other parameters are allowed to depend on both the home and current location—hence v_{ij} refers to individuals who live in location j but are born in location i. In these equations we have assumed frequency dependent transmission—as is the normal for human diseases—with $\sum_j N_{ij}$ giving the number of individuals currently in population i.

Using the formulation of equation (7.9), Figure 7.2 provides an example of an infectious disease spreading through the 67 counties of Great Britain. The simulations are initialized with the entire population of each county (as recorded in the 1991 census) being susceptible, and 10 infectious cases are placed in Inner London. The rates at which individuals commute l_{ij} are also taken from the 1991 census, while individuals are assumed to return home relatively quickly, at a rate of $r_{ij} = 2$ per day. Although this model is parameterized at a national level in terms of regular commuter movements, a similar modeling framework could be used to deal with the more irregular long-distance travel that can spread infection around the globe (Wilson 2003; Grais et al. 2003; Cliff and Haggett 2004).

Many factors emerge from Figure 7.2 that could be intuitively obvious without resorting to a spatial model: The first county to suffer a major epidemic is the source of initial infection, Inner London. Outer London peaks next due to its tight coupling to Inner London, and Outer London experiences the largest epidemic due to it having the largest population size. However, some elements may be more surprising: Greater Manchester and the West Midlands (along with West Yorkshire and Strathclyde) are among the last counties to have a sizable number of cases—despite their frequent interactions with London. This is attributable to the size of the population in these counties, such that although the infection may arrive relatively early it takes many weeks before the epidemic reaches its peak. Despite the clear delays between epidemics in different counties, within the first 50 days the infection has reached most areas (middle graph), after which time commuting plays a very minor role in the dynamics. This highlights the importance of rapidly imposing movement restrictions if we wish to curtail the spatial spread of infection. Additionally from this graph, it can be seen that the natural time to extinction is more than 300 days despite the fact that prevalence has dropped to very low levels within half this time. Finally, and as expected from simple models (such as shown in Figure 7.1), a deterministic version of this model predicts a faster speed of disease spread.

7.2.1.4. Commuter Approximations

Although equation (7.9) provides a full mechanistic description of the disease behavior, a large number of equations are frequently involved ($3n^2$ equations for n populations), and so it is informative to relate this model to the simpler ones defined earlier. In fact, Keeling and Rohani (2002) showed that for two populations of equal size and equal epidemiological characteristics, equation (7.9) can be simplified by assuming that all commuter movements are very rapid. In this case, the force of infection experienced by population i can be written as:

$$\lambda_i = \beta_i\big((1-\rho)I_i + \rho I_j\big) \qquad j \neq i. \tag{7.10}$$

The coupling parameter, ρ, can be defined in terms of the mechanistic movement of individuals by:

$$\rho = 2q(1-q), \tag{7.11}$$

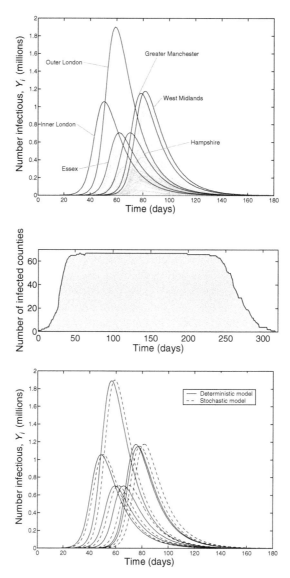

Figure 7.2. Deterministic and stochastic results for an infection spread through the 67 counties of Great Britain. The epidemic is initialized with 10 cases in Inner London, and is spread by commuter movements. The population size and rate of commuting is taken from the 1991 census database, and all trips are considered to be of short duration, $1/r = 0.5$ days. The top figure shows the county-level epidemics from a single stochastic iteration, with six counties highlighted. The middle graph shows the number of counties with infection from the same stochastic model. The bottom graph compares the deterministic solution (solid line) of equations (7.9) with the stochastic model (dashed line) for the six counties highlighted in the top graph. ($\mu = \nu = 0$, $1/\gamma = 14$ days, $\beta = 0.3571$ per day $\Rightarrow R_0 = 5$).

where q is the proportion of the time that individuals spend away in the other population,

$$q = l_{ij}/(r_{ji} + l_{ij}) = l_{ji}/(r_{ij} + l_{ji}).$$

Equation (7.11) is derived from the fact that when *either* the susceptibles of one population or infecteds of the other (but not *both*) move, there is a transfer of pathogen. We therefore find that ρ and hence the transfer of infection is maximized when individuals spend equal amounts of time in both home and away populations, $q = \frac{1}{2}$.

From this commuter approximation (equation (7.10)) it is clear that although increased coupling, ρ, leads to greater transmission between subpopulations, it weakens the transmission within subpopulations, therefore making R_0 independent of the coupling strength. Again, this type of disease transmission does not apply only to human commuters; Swinton et al. (1998) used a similar model to describe the spread of phocine distemper virus through harbour seals in the North Sea. In this context, haul-out beaches, where seals leave the water and congregate, act as natural subpopulations and infection was spread spatially by the occasional visits of seals to nearby beaches—thus the spread of infection is much closer to commuter-type movements than permanent migration usually associated with animals.

The spread of human diseases is best captured by the rapid commuter movements of individuals from their home subpopulation to another subpopulation and back again—requiring us to model both the current location and home location of individuals. When commuter movements are of short duration, this can be approximated by simple coupling. In these models, R_0 is independent of the coupling.

7.2.2. Coupling and Synchrony

Although coupling and the interaction between populations is key for the spatial invasion and spread of a disease, it also affects the endemic dynamics. In particular, the correlation between the disease dynamics in two subpopulations is generally a sigmoidal function of the interaction between them. Figure 7.3 shows the correlation against interaction strength for the four coupling mechanisms discussed above: equations (7.6), (7.8), (7.9), and (7.10). These results echo a general finding within metapopulation models, that the "interesting" spatial dynamics occur when the interaction is between 10^{-3} and 0.1 (Bolker and Grenfell 1995). When the interaction between subpopulations is too small the dynamics are effectively independent and spatial structure is unimportant, whereas when the interaction is too large the dynamics are synchronized, therefore, the subpopulations act like one large well-mixed population and again spatial structure is epidemiologically unimportant. The most notable factor in Figure 7.3 is the similarity between the results for different coupling mechanisms; once correctly scaled the correlation is a function of the interaction strength, with both the form of interaction and population size playing a relatively minor role. The scalings between interaction strengths are approximately:

$$\rho_{\text{plants}} \sim \frac{m}{\gamma + m} \sim 2\frac{l}{r+l}\left(1 - \frac{l}{r+l}\right) \sim \rho_{\text{commuters}}.$$

Hence, whereas commuter movements get scaled by their effective length of stay, permanent migration is multiplied by the average infectious period within the new subpopulation. In addition, although movement of susceptible commuters can lead to the

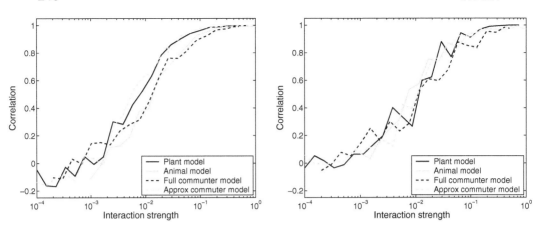

Figure 7.3. The correlation between the levels of an endemic disease in two stochastic populations for a range of interaction strengths and a variety of forms of interaction. It is clear that the correlation increases with the strength of interaction, and eventually asymptotes to one—when the population dynamics are synchronized. For the plant model (equation (7.6), solid black line) and the approximate commuter model (equation (7.10), dashed gray line), the x-axis is the coupling strength ρ. For the animal model (equation (7.8), solid gray line) the x-axis is $m/(\gamma + m)$ whereas, for the full commuter model (equation (7.9) dashed black line, $r_{ij} = r_{ji} = 1$), the x-axis is $2(p)(1 - p)$, where $p = l/(l + r)$, which using equation (7.11) is equal to ρ for the approximate commuter model. ($\gamma = 0.1$, $\beta = 1$, stochastic imports at a rate $\varepsilon = 10^{-2}$, $\mu = 10^{-4}$, left-hand graph $\nu = 1 \Rightarrow N_i = 10^4$, right-hand graph $\nu = 10 \Rightarrow N_i = 10^5$).

transfer of infection between populations—because they can get infected while away and bring the disease home—leading to an additional factor of 2, the same is not true for the permanent migration of animals. Finally, although R_0 increases with coupling in the plant model (equation (7.7)) this effect is not strong enough to affect the correlation between populations.

The correlation between disease prevalence in two subpopulations increases sigmoidally with the strength of interaction between the populations. In general, the change from largely independent dynamics to synchrony occurs for interaction strengths from 10^{-3} to 0.1.

7.2.3. Extinction and Rescue Effects

Fundamental to the dynamics of any spatially segregated population is the pattern of localized extinction and subsequent colonization by rescue events (Foley et al. 1999; Boots and Sasaki 2002; Onstad and Kornkven 1992). As discussed in Chapter 6, smaller populations suffer a greater risk of extinction—intuitively, smaller populations will have fewer infected individuals and therefore be more prone to stochastic individual-level effects. In particular, the extinction risk has been seen to decrease exponentially with population size (Chapter 6; Bartlett 1957, 1960). This has clear implications for spatially segregated populations (and spatial models) where one large population has been subdivided into many smaller ones, with each small subpopulation facing a far greater risk of extinction. Therefore, in a metapopulation where there is no interaction or coupling, the disease in each of the small subpopulations will be rapidly driven extinct, leading to much swifter global eradication than if coupling is large and the metapopulation was effectively randomly mixed. If we assume that the rate of extinction is proportional to $\exp(-\epsilon N)$, where N is the total population size, then the average time to extinction for the completely mixed (large coupling) metapopulation is:

$$\text{Time to extinction} = \frac{k}{e^{-\epsilon N}} = ke^{\epsilon N},$$

where k is the proportionality constant. In contrast, if the population is divided into n independent, noninteracting subpopulations (zero coupling), then the time to global extinction (in all subpopulations) becomes:

$$\text{Time to extinction} = \frac{k}{ne^{-\epsilon N/n}} + \frac{k}{(n-1)e^{-\epsilon N/n}} + \cdots + \frac{k}{e^{-\epsilon N/n}},$$

$$< k(1 + \log(n))e^{\epsilon N/n},$$

which for persistent diseases ($e^{\epsilon N}$ much greater than 1) is far shorter as n becomes large. The above formula comes from calculating the average time to the first extinction when n subpopulations are infected, followed by the average time to the next extinction given that now only $n-1$ subpopulations are infected, proceeding iteratively until all populations are disease free.

Without interaction between the subpopulations, a spatially segregated metapopulation generally suffers a faster rate of stochastic extinction than its randomly mixed counterpart.

When the subpopulations interact, the behavior is far more complex. The transmission of infection to a disease-free subpopulation—termed a rescue event in the metapopulation literature (Hanski and Gilpin 1991, 1997; Hanski and Goggiotti 2004)—can allow the subpopulation to recover from extinction. In this way, long-term persistence can be achieved because infection is constantly reinvading subpopulations that have gone extinct. However, two antagonistic forces are in operation. Rescue events are clearly maximized when the rate that infection enters a subpopulation is large. Intuitively this requires that the interaction or coupling between the subpopulations is large. However, it also requires the populations to be asynchronous, such that when the disease is extinct in one population some of the others have ample infectious individuals; this occurs only when the interaction is fairly weak. As a result, tension exists between high levels of interaction but maintaining asynchrony, and therefore persistence can be maximized at an intermediate level of coupling. A comprehensive understanding of this problem has still not been reached. Keeling (2000b) and Hagenaars et al. (2004) have studied the level of stochastic extinction in two different models of disease metapopulations. The basic structure of the models is very similar, with only minor differences in the mechanism of spatial coupling; however, the two models are focused on diseases with very different properties and time scales. Keeling (2000b) focuses on acute infections comparable to measles, whereas Hagenaars et al. (2004) consider more persistent infections with longer infectious periods and lower R_0. In Figure 7.4, we show examples of the similarities and differences in model behavior reported in these two papers. Both models predict that, at the local scale, higher levels of coupling lead to the disease being present in the subpopulations for longer (less time disease-free, as shown in the top-left graph). However, at the global

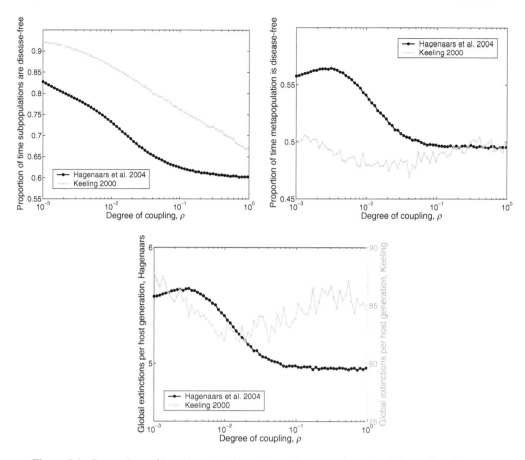

Figure 7.4. Comparison of how the extinction patterns change with the level of coupling from two models: Keeling (2000b) in gray and Hagenaars et al. (2004) in black. The top-left graph shows the proportion of the time that a subpopulation is disease free, whereas the top-right graph shows the proportion of the time that the entire metapopulation is disease free. The bottom graph shows the number of discrete global extinction events (the number of times the entire metapopulation becomes disease free) per host generation; note that the results from the two models are plotted at very different scales. The model of Hagenaars et al. (2004) deals with a small total population size ($N = 10,000$), with a relatively protracted infection ($1/\gamma = 125$ days, $R_0 = 5$) with infrequent imports ($\delta = 5 \times 10^{-5}$ day^{-1}; see Section 6.3.3.1), whereas the model of Keeling (2000b) the population size is slightly larger ($N = 30,000$), and the disease parameters are measles-like ($1/\gamma = 13$ days, $R_0 = 17$, $\delta = 9.3 \times 10^{-3}$ day^{-1}). In both models the birth and death rates are ($\nu = \mu N$, $\mu = 5 \times 10^{-5}$ day^{-1}), where the other parameters of the Hagenaars model have been scaled accordingly. In both models the population is split into 10 subpopulations.

scale, the two model predictions diverge. The model of Keeling (2000b) shows very little variation in the proportion of time the entire metapopulation is disease free and has a shallow minimum point at $\rho \approx 0.02$. In contrast, the model of Hagenaars et al. (2004) shows more variation and a clear minimum when the coupling is large, $\rho = 1$. Finally, when we focus on the number of global extinction events the Keeling model has again an interior minimum, whereas for the Hagenaars model extinctions are minimized at $\rho = 1$. In addition, extinctions are far more likely in the Keeling model. It is still a major challenge to incorporate such results into a general framework, understanding how disease characteristics and spatial coupling interact to determine persistence.

When interaction between the subpopulations is included, the level of local (subpopulation scale) and global (metapopulation wide) extinctions is an emergent property of the dynamics and cannot be easily predicted from the disease parameters.

Rather than being an abstract concept, this problem has strong public health implications. The last century has seen far more national and international travel, increasing the coupling between populations (Grais et al. 2003; Cliff and Haggett 2004)—the implications of this for disease persistence and eradication needs to be studied using the form of metapopulation model developed here. Much more work is still needed to determine how population structure, disease dynamics, population size, infectious import rate, and coupling interact to determine disease persistence.

Two other applied problems are strongly related to the issues of coupling, synchrony and extinction. It has long been realized in conservation biology that increasing the mixing between populations, by introducing linking corridors of suitable habitat, is an effective means of conserving a species (Hanski 1999; Earn et al. 2000). However, these corridors may also act as a conduit for the spread of infection, and when facing a highly virulent pathogen, the presence of corridors may exacerbate the extinction of the host (Hess 1996). More recent, detailed work has called into question the strength of this result (Gog et al. 2002; McCallum and Dobson 2002), but the concept is still one that conservationists should consider.

A second applied problem concerns the dynamics of vaccinated metapopulations. Vaccination reduces the prevalence of infection within a population (Chapter 8), which increases its risk of extinction, and also reduces the effective interaction strength because less transmission of infection can occur. This reduction in the effective coupling in turn leads to less synchrony in the disease dynamics and therefore more effective rescue events (Earn et al. 1998; Rohani et al. 1999). This behavior is seen to some degree in the measles data set for England and Wales, where despite vaccinating at around 60% from 1970 to the mid-1980s the rates of stochastic extinction remained unchanged (Keeling 1997). Therefore, the effects of lower infection levels but greater asynchrony of epidemics effectively cancel. Pulsed vaccination (see Chapter 8) has been suggested as a mechanism of overcoming this problem where national vaccination campaigns for a few weeks each year will help to synchronize the dynamics and hence weaken the effect of rescue events (Agur et al. 1993; Nokes and Swinton 1997; Shulgin et al. 1998; Earn et al. 1998). Figure 7.5 shows the correlation (which measures the degree of synchrony) between two coupled populations under standard and pulse vaccination; clearly pulsed vaccination maintains the synchrony of epidemic behavior over of wide range of vaccination coverage.

Pulsed vaccination campaigns act to synchronize epidemics in coupled populations, and may lead to an increase in global extinction rates.

Extending these theoretical vaccination results into a practical public health tool is a challenge for the future. Pulsed vaccination clearly has the ability to synchronize epidemic behavior, which in turn limits the potential of rescue effects between coupled populations. It is hoped that such synchronization may increase the level of global extinctions at the metapopulation level and therefore promote the eradication of infection. Pulsed vaccination can also have significant logistical benefits when individual health care is limited, because

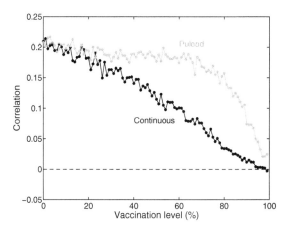

Figure 7.5. The correlation between two coupled populations with SIR-type infection as the level of vaccination is varied. For continuous vaccination (black), individuals are vaccinated at birth, whereas for the pulsed vaccination (gray) a comparable number of individuals are vaccinated every four years. Vaccination is assumed to offer lifelong protection. ($N = 100,000$, $\rho = 0.01$ with approximate commuter coupling, $1/\gamma = 10$ days^{-1}, $R_0 = 10$, stochastic imports $\delta = 5$ per year; see Section 6.3.3.1).

dedicated vaccination teams can be deployed. However, in the delay between the pulses there is the potential for many individuals to become infected. Clearly there is a trade-off: Frequent pulses (or continuous vaccination) limit the buildup of susceptibles, whereas infrequent, and therefore large, pulses have a greater synchronizing action. Determining the appropriate timing of vaccination pulses (or even a mixture of continuous and pulsed vaccination) requires the use of well-parameterized stochastic spatial models that also include age-structure—although the construction of such models can be pieced together from this chapter as well as Chapters 6 and 3, the parameterization and analysis of the possible control permutations is beyond the scope of this book.

7.2.4. Levins-Type Metapopulations

In the metapopulation models considered so far, the population levels within each subpopulation have been modeled explicitly—this may be computationally intensive. An alternative formulation was proposed by Levins (1969) where each subpopulation is simply defined as being either empty (disease-free) or occupied (having infection). There are clear parallels between this new classification of subpopulations, which ignores the precise prevalence and the traditional SIR classification of hosts, which ignores the level of pathogen. The intuitive way to conceptualize Levins-type metapopulations is to assume that localized extinctions and successful recolonization events are extremely rare, so that each subpopulation spends the vast majority of its time either disease free or close to the endemic equilibrium.

If we have a large number of subpopulations, and the coupling is global (so that each subpopulation has an equal probability of reinfecting any other subpopulations), then the probability that a subpopulation is infected, P, is given by:

$$\frac{dP}{dt} = \rho(1 - P)P - eP, \tag{7.12}$$

where ρ measures the reinfection (coupling) rate from an infected subpopulation to an uninfected one, and e is the rate of local extinction. Equation (7.12) is structurally identical to the SIS equation (Chapter 2), reflecting the fact that after a localized extinction the subpopulation is once again susceptible to infection.

When the subpopulations are of different sizes and the interactions are unequal, the above formulation can be refined such that P_i now refers to the probability that subpopulation i is infected.

$$\frac{dP_i}{dt} = \sum_j \rho_{ij}(1 - P_i)P_j - e_i P_i, \qquad (7.13)$$

where different couplings can occur between different populations and the extinction rate (e_i) can also vary, often reflecting the population size. These rates can easily be used to translate this differential equation model into a stochastic one, with P_i being either zero or one.

Levins metapopulation models ignore the internal dynamics within each subpopulation, and instead classify each subpopulation as either infected or disease free.

Although the Levins formulation is intuitively appealing, the accuracy of the results is highly dependent on the assumption that extinction and successful recolonization events are rare compared to the standard epidemiological dynamics (Keeling 2000b). When this assumption breaks down, two confounding factors associated with the internal subpopulation dynamics become important. First, the conditions that lead to an extinction are unlikely to allow an immediate successful reinfection of the subpopulation. Second, following the extinction of infection the level of susceptibles increases; when there is a high proportion of susceptible individuals any subsequent infected is likely to be large and short-lived, rapidly returning to the disease-free state. Therefore, the timing between extinction and recolonization is vital (see Chapter 6, Figure 6.8). However, these discrepancies between the Levins approximation and the full stochastic metapopulation behavior are often during the early colonization dynamics, making the Levins models an ideal tool to study the spatio-temporal invasion of infection.

Despite differences between the equilibrium-level results of Levins and full metapopulation models, the Levins model still remains a useful and simple tool for studying invasion dynamics.

7.2.5. Application to the Spread of Wildlife Infections

We contrast the Levins metapopulation model with the full metapopulation model by considering two examples of invading wildlife diseases that have been tackled by the two differing approaches. The full metapopulation model of Swinton et al. (1998) was used to investigate the spread of phocine distemper virus around the North Sea coastline. Similar formulations have been used to describe the spread of bovine tuberculosis in badgers (White and Harris 1995), parapoxvirus in squirrels using a grid of stochastic subpopulations (Rushton et al. 2000), and rabies in foxes again using a grid of stochastic subpopulations (Tischendorf et al. 1998). The Levins metapopulation model of Smith et al. (2002) was used to explain the spread of rabies in the raccoon population of Connecticut.

7.2.5.1. Phocine Distemper Virus

In 1988, an epidemic of phocine distemper devastated the harbour seal (*Phoca vitulina*) populations in the North Sea (Dietz et al. 1989). Starting in Anholt, Denmark, in early April, the disease spread in a wave-like manner around the North Sea coastline, triggering epidemics from Norway to Ireland. The UK populations were the last to lose the disease in August of 1989, although the bulk of the cases occurred before the end of 1988.

Swinton et al. (1998) modeled this epidemic as a set of 25 subpopulations, mimicking the 25 locations (seal colonies) where infection was recorded. The dynamics of phocine distemper in this metapopulation were modeled as:

$$
\frac{dX_i}{dt} = -\beta X_i \left[(1-\rho)\frac{Y_i}{N_i} + \rho \frac{\sum_{j=i-1,i,i+1} Y_j}{\sum_{j=i-1,i,i+1} N_j} \right] - \mu X_i,
$$

$$
\frac{dW_i}{dt} = \beta X_i \left[(1-\rho)\frac{Y_i}{N_i} + \rho \frac{\sum_{j=i-1,i,i+1} Y_j}{\sum_{j=i-1,i,i+1} N_j} \right] - \sigma W_i - \mu W_i, \tag{7.14}
$$

$$
\frac{dY_i}{dt} = \rho W_i - \gamma Y_i - m Y_i - \mu Y_i,
$$

$$
\frac{dZ_i}{dt} = \gamma Y_i - \mu Z_i,
$$

where $\frac{m}{m+\gamma} \approx 0.2$ gives the probability of mortality from the infection. Swinton et al. simulated this model stochastically and were able to estimate the majority of the parameters needed from good observational data; however, the coupling parameter, ρ, could be found only by matching the model to the observed wavespeed, leading to $\rho \approx 0.1$. The precise form of coupling used in these equations differs from the standard assumptions and therefore requires some explanation. The first term reflects that a fraction $(1-\rho)$ of infectious seals remain at their haul-out site and can therefore infect susceptible seals— the transmission is assumed to be frequency dependent, due to the types of interaction that occur at haul-out sites. The second transmission term is due to encounters with seals at sea, away from the main haul-out sites; again, frequency-dependent transmission is assumed. Therefore, although the formulation of the transmission terms is unusual it reflects the form of interaction between seals. Figure 7.6 shows the wavelike progress of the invading infection; because this is a stochastic model, no two simulations will be identical but the general wave-speed is largely invariant.

Two difficulties exist with this model formulation. First, all subpopulations (seal colonies) are given identical parameters and population sizes. Therefore, heterogeneities that may account for epidemiologically important factors, such as the prolonged epidemic observed in Tayside, are ignored. Second, the model is essentially one-dimensional with nearest-neighbor coupling (see Section 7.3.1); therefore, the true spatial structure has been neglected and the different distances between seal colonies ignored. However, despite these limitations, the model provides a simple means of assessing the spatio-temporal dynamics of phocine distemper epidemics.

7.2.5.2. Rabies in Raccoons

The spread of rabies has been extensively studied using spatial models building upon the pioneering work of Murray et al. (1986), which formulated PDE models (see Section 7.4)

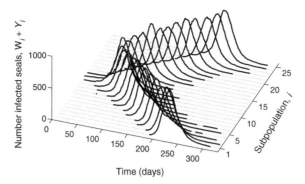

Figure 7.6. Results of a stochastic simulation of the phocine distemper metapopulation model, equation (7.14), as proposed by Swinton et al. 1998. The number of infected seals in each subpopulation is shown, with black lines used when at least one seal is infected. In this model births and deaths have been ignored for simplicity. ($\beta = 0.4$, $\rho = 0.1$, $\sigma = 0.1428$, $\gamma = 0.1143$, $m = 0.0286$, $\mu = 0$, $N_i(0) = 2000$).

to understand the spread of rabies in the fox population of Europe. At a smaller scale, Smith et al. (2002) used a Levins-type metapopulation model to study the spread through Connecticut of rabies in racoons from 1991–1996, concentrating on the underlying spatial heterogeneity of the habitat. This spread is part of a larger wave of infection that began along the Virginia/West Virginia border in the mid-1970s.

Using a similar format as equation (7.13), Smith et al. model the state of the 169 townships in Connecticut, and for township i define the stochastic rate of infection as

$$\varepsilon_i(1 - P_i) + \sum_j \rho_{ij}(1 - P_i)P_j, \tag{7.15}$$

where P_i is one if the racoons in the township are infected and zero otherwise. Unlike the previous Levins metapopulation model (7.13), the localized extinction of infection has been ignored. Here ε_i measures the random long-distance dispersal of rabies due to racoon translocation and ρ_{ij} measures local transmission between adjacent townships (see Figure 7.7). Here $\rho_{ij} = A$ if townships i and j share a land-boundary, $\rho_{ij} = B$ if the townships are separated by a river, otherwise $\rho_{ij} = 0$. This stochastic model (which is an *SI* model because there is no recovery or local extinction of infection) was compared to the observed first recorded case of rabies in each township, in order to determine suitable coupling and long-range transmission parameters. The simplest model that provides a reasonable fit to the observed data had $\varepsilon_i = 2 \times 10^{-4}$, $A = 0.66$, and $B = 0.09$ (all rates in months), showing that rivers reduce transmission by 87% compared to land boundaries and that local transmission accounts for the vast majority of spatial spread. Other, more complex model formulations included the size of the human population as a proxy for the density of racoons within a township; such models suggest that population density plays a small but positive role in transmission.

Figure 7.8 shows the impact of these three different coupling parameters. Rivers clearly have a significant impact on the spatial spread of infection, especially when long-range translocations are impossible (gray line and middle map). When the impact of rivers is ignored (dashed line and right map), the speed of spatial spread is far more rapid, infecting the entire state in two-thirds of the time. Interesting, although the value of ε is very small, its effect is significant because it breaks the assumption of strict local transmission.

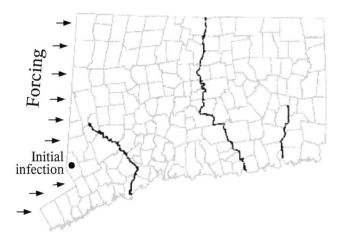

Figure 7.7. The 169 townships within Connecticut showing the initial site of the first recorded cases in Ridgefield Township, the forcing on the western townships due to infection from neighbouring New York, and the position of rivers (black) which act as a natural barrier to the movement of rabid racoons.

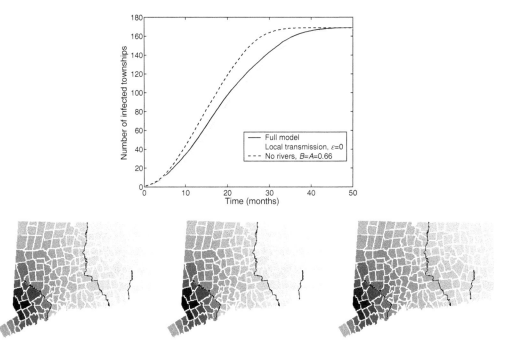

Figure 7.8. The effect of the various coupling terms on the spatial spread of rabies in Connecticut, based on the model and parameterization of Smith et al. (2002). To simplify the dynamics, forcing from New York state is set to zero. The top graph shows the average number of infected townships as a function of time from the initial seeding of infection. Three different assumptions about coupling are shown: the full model ($A = 0.66$, $B = 0.09$, $\varepsilon = 2 \times 10^{-4}$), a model with local transmission only ($A = 0.66$, $B = 0.09$, $\varepsilon = 0$), and a model where rivers had no impact on transmission ($A = 0.66$, $B = 0.66$, $\varepsilon = 2 \times 10^{-4}$). The spatial pattern for these three models is illustrated on the three maps, with darker colors representing earlier average infection times.

By the nature of the Levins-type metapopulation used here, all epidemiological dynamics within a subpopulation are ignored. This means that as soon as a township is infected it can transmit infection as strongly as when endemic equilibrium has been reached. In an ideal situation, the within-subpopulation dynamics should also be included and modeled stochastically. However, the Levins approximation allows for very rapid computation and therefore far richer parameterization; it also circumvents the difficulty of estimating the racoon population levels within each township. This is a good example of where limitations of the available data determine the appropriate model choice, and where simple models can be used to derive the main conclusion that rivers act as significant but permeable barriers to the spread of rabies in racoons.

7.3. LATTICE-BASED MODELS

There are many occasions when the spatial location of the hosts is seen to be important, but there is no natural means of partitioning the entire population into discrete subpopulations. In such situations, lattice- or grid-based models are often used, where individuals within a grid site are grouped together into a subpopulation. Traditionally, these grid-based models take two forms: coupled lattice models that can be considered as a grid of subpopulations with coupling between adjacent grid sites (neighbors), and cellular automata where at most one host can occupy each grid site and again interactions occur only locally with the neighboring sites. In essence, both of these models are special forms of the metapopulation model, with a tightly constrained set of interactions. In general, these models are conceptual tools, used to inform about the effects of spatial separation and nonrandom mixing; they are seldom used as accurate predictive tools, because with few exception (such as orchards) most hosts do not exist on regular lattices. However, the insights provided by this type of model have proved to be invaluable in understanding the spatial spread of infection (White and Harris 1995; Tischendorf et al. 1998; Keeling and Gilligan 2000; Rushton et al. 2000; Kao 2003).

Two main types of lattice exist, which determine how the entire population is divided into subpopulations. Probably the most intuitive is the two-dimension square lattice, so that an individual's position (on the surface of the earth) is translated into x and y coordinates which in turn determine the grid site (White and Harris 1995). Modifications to this standard grid arise by changing the number of dimensions; many theoretical problems can be studied more precisely if the populations are arranged along a one-dimensional line, with interactions between nearest (or nearest and next-nearest) neighbors (Harris 1974; Watts and Strogatz 1998). Alternatively, higher dimensional lattices can also be used to replicate the more complex higher-dimensional social structure of humans (Rhodes et al. 1997). Square lattices impose a definite direction on the space (mathematically we lose the property of isotropy) such that the directions "north," "south," "east," and "west" play a more major role than any others. To overcome this problem, some researchers have used hexagonal grids, where each site has six neighbors; this often generates more natural spatial patterns (van Baalen and Rand 1998; Kao 2003).

7.3.1. Coupled Lattice Models

Although coupled lattice models are simply a special case of the metapopulation formulation, it will be worth considering some of the particular differences in more detail. For a disease with SIR-type dynamics and "commuter-like" interaction terms (see Section 7.2.1), the governing equations become:

$$\frac{dX_i}{dt} = \mu - \beta X_i \frac{(1 - \sum_j \rho_{ji})Y_i + \sum_j \rho_{ij}Y_j}{(1 - \sum_j \rho_{ji})N_i + \sum_j \rho_{ij}N_j} - \nu X_i,$$

This is online program 7.3

$$\frac{dY_i}{dt} = \beta X_i \frac{(1 - \sum_j \rho_{ji})Y_i + \sum_j \rho_{ij}Y_j}{(1 - \sum_j \rho_{ji})N_i + \sum_j \rho_{ij}N_j} - \gamma Y_i - \nu Y_i, \tag{7.16}$$

$$\rho_{ij} = \rho_{ji} = \begin{cases} \rho & \text{if } i \text{ and } j \text{ are neighbors} \\ 0 & \text{otherwise,} \end{cases}$$

where i refers to a grid cell within the lattice. (Given that lattices are usually two-dimensional, some researchers prefer to specify a subpopulation by its coordinates; therefore, X_{ij} is the number of susceptibles at location (i, j). However, this notation can be very cumbersome when we wish to specify the level of interaction between two locations (e.g., $\rho_{(i,j)(k,l)}$).

Equation (7.16) assumes that all populations have identical demographic and epidemiological parameters, and therefore we have moved further away from data-driven models such as those illustrated in Section 7.2.1.3. However, this assumption of homogeneity of parameters allows for a more detailed understanding of the spatio-temporal dynamics. In addition, it is frequently assumed that all populations are of equal size ($N_i = N$), which simplifies the denominator in equation (7.16) to N.

Coupled lattice models are specialized metapopulation models, where subpopulations are arranged on a grid and coupling is generally to the nearest neighbors only.

The most notable feature of this form of lattice model is the clear wave-like spread of invading infections. From a point source of infection, the disease must spread to the neighboring sites before it can spread to the rest of population. Figure 7.9 shows a pictorial example of the wave-like spread for a square lattice, together with a graph of how coupling (ρ) and the basic reproductive ratio ($R_0 = \frac{\beta}{\gamma+\nu}$) both determine the speed of the invading wave.

As with the metapopulation models, a stochastic version of the lattice model has a slower wave speed than the deterministic model, and when the deterministic wave speed is very low the stochastically spreading disease may even fail to colonize the entire lattice. This reduction in speed is because in an integer-based stochastic model transmission of infection into a new subpopulation is a stochastic process and therefore may, by chance, be considerably delayed; by contrast, in a deterministic model very low prevalence will always spread to the new subpopulations. As expected, increasing the level of interaction between neighboring subpopulations, or increasing the within-subpopulation growth rate, allows the infection to spread more rapidly.

The wave speed of an invading epidemic in a coupled-lattice model increases almost linearly with the initial growth rate of the infection, $\beta - \gamma - \nu$; increases nonlinearly with the level of coupling, ρ; and is slightly more rapid in deterministic compared to stochastic models.

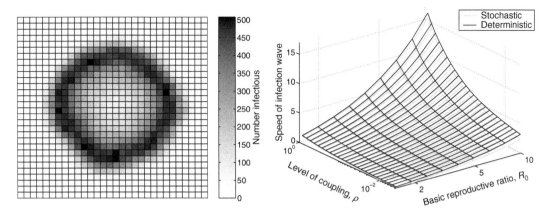

Figure 7.9. Results from a coupled lattice model using the approximate commuter-type coupling (equation (7.16)). The left-hand figure shows a snapshot of the lattice at time $t = 5$, with darker shades illustrating a higher level of infectious individuals ($R_0 = 5$, $\rho = 0.1$). The circular wavelike spread of infection is clear. The right-hand graph shows the speed of the invading wave of infection as the basic reproductive ratio and the level of coupling vary. The model simulates the spread of infection on a 100×100 lattice; the local subpopulations are all identical and contain 1,000 individuals. ($\gamma = 1$; changing this would scale the speed of the invading wave).

In contrast to metapopulations where space is often rather abstract, coupled lattice models provide a definite method of including the spatial location of individuals. This allows us to predict the expected wavelike spread of infection across a homogeneous spatial landscape. As such, coupled lattice models play a vital role in improving our understanding of the spatial spread of infection; they can be readily derived from the standard nonspatial models, and due to the simplistic assumptions about spatial interaction require very little extra information to parameterize. However, such coupled lattice models break down if we attempt to match them too rigorously to the underlying individual-level spatial dynamics. This difficulty arises because of the assumption that mixing within a grid site is completely random (and full strength), whereas mixing between adjacent cells is far weaker. Figure 7.10 shows an example of how the coupled-lattice assumption can fail to capture the expected individual-level behavior. Because individuals A and B are within the same grid cell, the coupled lattice assumption means that the interaction between them is strong, whereas the interaction between B and C (and A and D) is far weaker or zero because they lie in separate cells despite the fact that the separations involved are smaller. This highlights a fundamental flaw with coupled lattice models: The act of artificially aggregating populations into grids can lead to some artificial results and hence the grid size must be chosen with extreme care.

7.3.2. Cellular Automata

Cellular automata also use a lattice-based arrangement of sites. However, in contrast to the lattice-based models discussed above, cellular automata have only a finite, and usually small, number of population states. Most frequently we consider each lattice site to represent a single host (a population size of one) and so each site is generally either empty,

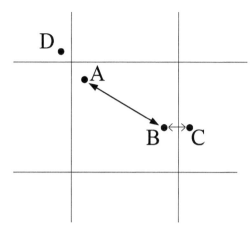

Figure 7.10. A representation of when the coupled-lattice framework may fail to describe the expected interaction between individuals. Although A and D and (B and C) are close, the interaction between them is weak because they occupy different grid locations.

or occupied by a susceptible, infectious, or recovered individual. Due to this finite nature almost all cellular automata disease models are stochastic.

Cellular automata are clearly an abstraction of reality. Other than in agricultural settings (Maddison et al. 1996; Klecakowski et al. 1997; Gibson 1997b), individuals do not often exist in fixed lattice arrangements. In addition, the small number of interaction neighbors that are usually assumed (just the nearest 4 or 8 lattice sites) are unrepresentative of the complex and heterogeneous contacts through which human and animal infections pass. However, cellular automata are superb tools to understand how the individual and spatial nature of populations causes epidemic dynamics to deviate from their deterministic random-mixing ideal.

Cellular automata operate on a lattice of sites, with each site generally assumed to hold a single host. Interaction is usually stochastic and with the neighboring (4 or 8) lattice sites.

Two basic forms of cellular automata dominate the early literature in this subject area, and have clear parallels to disease models. These are the contact process, which is equivalent to the SIS model, and the forest-fire model, which is equivalent to an SIR-type infection.

7.3.2.1. The Contact Process

The contact process, which dates back to 1974, is traditionally formulated in one dimension so that individuals are positioned in a row with contact between adjacent individuals (Harris 1974). The stochastic rules that govern the behavior of this model, when translated into a disease metaphor are biologically intuitive. Infectious individuals transmit infection at a rate τ to any neighboring susceptible and infectious individuals recover at a rate γ, becoming susceptible to the disease once more—this is clearly the SIS model (Chapter 2) in a spatial context. By considering a central individual and both neighbors, we can explicitly define the possible rates of change:

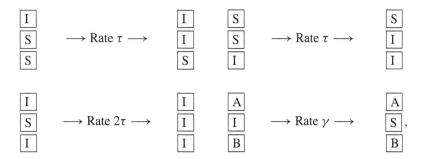

where A and B can be either S or I.

For this contact process we want to calculate R_0, thereby linking the individual level parameters τ and γ to population-level dynamics. If we consider R_0 as the rate that secondary cases are initially produced, multiplied by the infectious period (Chapter 2), then it is clear that $R_0 = 2\frac{\tau}{\gamma}$, where 2 comes from the fact that every individual has 2 contacts. This would suggest that the disease can invade and expand throughout the population if $\tau > \frac{1}{2}\gamma$. However, such arguments do not take into account the spatial structure that develops during the early epidemic process. Consider the initial seed infection, where by chance it infects a neighboring contact; however, this neighbor now has only one susceptible neighbor (unless the seed infection recovers) and therefore has a much reduced potential for spreading the infection. We therefore observe a common phenomenon in many spatial disease models: The local pool of susceptibles is rapidly depleted, which can dramatically reduce the early growth rate.

In many spatial models, the depletion of the locally available susceptible population can reduce the early growth rate of the epidemic and the speed of the invading wave front.

Even for this most simple of individual-based spatial models, the precise threshold value of τ that allows an epidemic to spread can be calculated only by repeated large-scale numerical simulation. The latest estimates suggest that $\tau > 1.64896\gamma$ for the infection to have a nonzero probability of long-term spread and persistence (de Mendonça 1999).

7.3.2.2. The Forest-Fire Model

A second common form of cellular automata is the forest-fire model, which is closely associated with spread of $SIRS$-type infection and is usually simulated on a two-dimensional lattice. In the original notation, lattice sites can be empty, occupied by a healthy tree, or occupied by a burning tree. Burning trees die to leave empty spaces, fire can spread between neighboring trees, trees can colonize empty spaces, and occasional random lightning strikes can cause spontaneous fires. In epidemiological notation, healthy trees are susceptibles, burning trees are infectious, empty sites are recovered (and immune), colonization by trees mimics either the birth of new susceptibles or waning immunity, and lightning represents the import of infection. Again we describe the dynamics in terms of the rates of change of lattice sites:

 $\boxed{S} \longrightarrow \boxed{I}$ Rate $= \tau n + \varepsilon$ where n is the number of infectious neighbors.

$\boxed{I} \longrightarrow \boxed{R}$ Rate $= \gamma$.

This is
online
program
7.4

$\boxed{R} \longrightarrow \boxed{S}$ Rate $= \nu$.

This model was developed by Per Bak and coworkers (Bak et al. 1990), and for statistical physicists displays a range of interesting power-law scaling and self-organized critical behavior. In general, distributions such as the size of patches of susceptibles (trees) or the size of individual epidemics are observed to follow a power-law relationship (frequency $\propto$ size$^{-\alpha}$), with the power-law exponent, α, being largely independent of the precise parameter values. This behavior occurs whenever certain rates of change are much bigger than others; in particular, transmission is much faster than recovery, which is much faster than births, which is much faster than random imports of infection ($\tau \gg \gamma \gg \nu \gg \varepsilon$). Fortunately, this natural ordering holds in most epidemiological examples, so it is hoped that the same power-law scaling and ideas of self-organized criticality will hold also. Much more information on self-organized criticality can be found in the following publications: Tang and Bak (1988); Sole et al. (1999); Allen et al. (2001); Pascual and Guichard (2005).

The forest-fire model typifies many stochastic spatial models. The fact that transmission is faster than recovery, which is faster than births, which is faster than imports of infection, leads to power-law relationships between epidemic size and frequency.

7.3.2.3. Application: Power-Laws in Childhood Epidemic Data

The work of Rhodes and coworkers (Rhodes and Anderson 1997; Rhodes et al. 1997) provides a good example of how cellular automata models can be used to develop deeper insights into the roles of spatial structure and individual-based populations in the dynamics of infectious diseases. They were interested in the dynamics of childhood diseases, and the distribution of outbreak sizes in small isolated populations. The Fareo Islands are isolated islands in the North Atlantic between Scotland and Iceland, and have a population of around 25,000. These island have extremely good historical records of infectious disease outbreaks stretching back for around 100 years (Cliff et al. 1993), and so are an ideal source of data for epidemiological study. In a number of papers, Rhodes and coworkers showed that the outbreaks of childhood diseases in the Faroes follow a power-law relationship, and that a similar scaling can be obtained from a biologically simple cellular automaton model.

In Figure 7.11, we show how the frequency of epidemics greater than or equal to a given size decreases, following a power-law like relationship. For epidemics of fewer than 500 cases, the probability that the epidemic is greater than or equal to size s is given by:

$$\mathbb{P}(\text{epidemic} \geq s) = s^{-\alpha},$$

with $\alpha \approx 0.265$ for measles, $\alpha \approx 0.255$ for whooping cough, and $\alpha \approx 0.447$ for mumps. This is reminiscent of the power-law relationships that are seen in the traditional

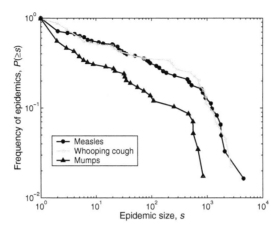

Figure 7.11. The power-law scalings observed in the epidemic outbreak data from the Faroe Islands, 1870–1970. The frequency of epidemics greater than or equal to a given size is shown for three major childhood infections: measles, whooping cough (pertussis), and mumps. It is postulated that a power-law scaling holds for epidemics up to 500–1,000 cases. The graph is plotted on a log-log scale, so that straight lines equate to power-laws. (From Rhodes et al. 1997.)

forest-fire model; therefore, Rhodes and Anderson modified the standard model to overcome some of the limitations and incorporate more realistic human behavior. In particular, the susceptible and infectious hosts are no longer fixed but are able to wander across the lattice by moving into neighbored unoccupied cells. This movement of individuals means that the behavior of adjacent sites is related (because the movement from one cell must be balanced by the movement *to* another), breaking the independence assumptions of formal cellular automata. However, the same basic techniques and results still translate between true cellular automata and this more complex model.

Figure 7.12 shows examples of the Rhodes and Anderson model in two and three dimensions. In two dimensions, the lattice is a 158×158 square, and each site has four nearest neighbors involved in the transmission of infection and movement; in three-dimensions, the lattice is a $29 \times 29 \times 29$ cube, and each site has six neighbors. Although the power-law distribution of (relatively) small epidemic sizes is largely independent of the precise parameter values, the behavior does vary with the dimension of the system. In three dimensions, the distribution of epidemic sizes is a much closer fit to a power-law of $\alpha \approx 0.2$ than when the model is two-dimensional, and therefore seems a plausible representation of measles or whooping cough. When the lattice is made five-dimensional, the power-law changes to $\alpha \approx 0.4$ (Rhodes et al. 1997), and therefore is comparable with the scaling exponent observed for mumps.

It is important to ask what these theoretical results mean at a practical level. Rather that telling us that models in three and five dimensions are required to capture the behavior of measles, whooping cough, and mumps, these results indicate that simple nearest-neighbor transmission in two dimensions is insufficient to describe the true nature of human social contacts. Three and five dimensions are simply methods of introducing the added elements of complexity, clustering, and interaction that exist in human social networks. This reinforces our original contention that most lattice models are abstract tools that can be used to improve our understanding of transmission in a spatial environment, rather than detailed predictive models.

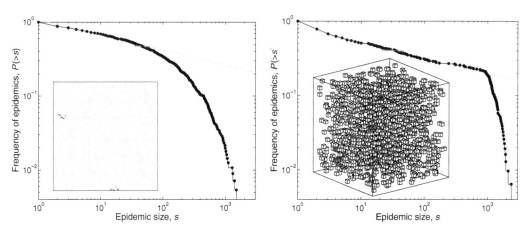

Figure 7.12. Distribution of epidemic sizes and examples of the lattice configuration for 2D and 3D versions of the Rhodes and Anderson model. In the lattice example, gray sites are susceptible, black are infectious, and empty sites are recovered. For both models, the best power-law fit to the distribution of epidemic sizes ($\mathbb{P}(\text{epidemic} \geq s) = s^{-\alpha}$) is calculated to be $\alpha \approx 0.2$. ($\tau = 5$, $\gamma = 0.1$, $\nu = 4 \times 10^{-5}$, imports $\varepsilon = 3.3 \times 10^{-6}$, movement rate $m = 0.1$. $N \approx 25000$.)

7.4. CONTINUOUS-SPACE CONTINUOUS-POPULATION MODELS

One major disadvantage of the lattice-based models is the discretization of space that is introduced by the lattice structure. In consequence, the resolution at which we know an individual's position is limited by the scale of each grid cell. An alternative formulation is to treat both space and the population as continuous, therefore specifying a density of individuals at all locations. We can think of this as the limit of a lattice model as the grid size becomes infinitely fine scale.

The natural way to describe the dynamics of continuous populations in continuous space is using partial differential equations (PDE), or integro-differential equations (IDE). The mathematics behind these formalisms is complex and often highly technical; the details of this approach are excellently described by Murray (2003), so here we review the salient epidemiological implications. Although some notable examples of continuous space models are being used for applied modeling purposes (Noble 1974; Murray et al. 1986; Caraco et al. 2002), such models are generally used to provide theoretical predictions and a generic understanding of the spatial spread of infection (Lopez et al. 1999; Beardmore and Beardmore 2003; Reluga 2004). The main theoretical advantage is the deterministic and tractable nature of the continuous-space models, whereas the assumption of continuous-population levels tends to be a disadvantage in many applied situations. We now focus on the two forms of models, PDEs and IDEs, to illustrate how they are derived from simple nonspatial models (Chapter 2).

7.4.1. Reaction-Diffusion Equations

The standard PDE models are derived from the assumption that infectious individuals transmit the disease only to susceptibles at their current location, and that all individuals are free to move at random (or diffuse) through the landscape. We first show an example of a PDE model, before explaining the individual terms and how such a model is formulated.

A typical PDE model for a disease with SIR-type dynamics would be:

$$\frac{\partial X}{\partial t} = \nu - \beta XY/N - \mu X + D_X \nabla^2 X,$$

$$\frac{\partial Y}{\partial t} = \beta XY/N - \gamma Y - \mu Y + D_Y \nabla^2 Y, \qquad (7.17)$$

$$\frac{\partial Z}{\partial t} = \gamma Y - \mu Z + D_Z \nabla^2 Z,$$

where X, Y, and Z are functions of both space and time, and represent the local density of susceptible, infectious, and recovered individuals, and as always $N = X + Y + Z$. Hence, if we're dealing with a two-dimensional landscape, $X(x, y, t)$ is the density of susceptibles at location (x, y) at time t. We now have to specify the rates of change with partial derivatives (e.g. $\frac{\partial}{\partial t}$), because our variables are now multi-dimensional, being functions of both space and time—this, however, is a technicality and does not change our understanding of what these derivatives mean.

The term ∇^2 is introduced to model the local diffusion of individuals through space. ∇ is shorthand for the rate of change of the quantity across space, so ∇^2 is the change in the rate of change. In two dimensions, the diffusion term for susceptibles becomes:

$$\nabla^2 X = \frac{\partial^2 X}{\partial x^2} + \frac{\partial^2 X}{\partial y^2}.$$

The inclusion of these spatial derivatives mimics the diffusion of individuals across the environment. For greater generality, susceptible, infectious, and recovered individuals are assumed to diffuse at different rates (D_X, D_Y, and D_Z), reflecting the fact that sick individuals may be less likely to move.

We can understand the role of diffusion in such PDE models, by considering the diffusion of a group of susceptibles initially piled at a single point $(0,0)$. Ignoring demography, our equation will be:

$$\frac{\partial X}{\partial t} = D_X \nabla^2 X.$$

This has the solution:

$$X(x, y, t) \propto \frac{1}{2\pi D_X t} \exp\left(-\frac{(x^2 + y^2)}{2 D_X t}\right),$$

which is an ever-expanding bell-shaped (Gaussian) distribution, with the total density of individuals remaining constant. The diffusion parameter D_X governs the speed at which the variance of the Gaussian grows. We can therefore see that diffusion away from a point source leads to Gaussian-like distributions of individuals.

Reaction-diffusion models, which use a PDE formulism, assume local transmission of infection and rely on spatial diffusion of hosts to spread the infection.

Very few PDE models have an exact analytical solution and therefore numerical methods are required for their simulation. Although PDEs are formulated as continuous space models, determining their solution by computer necessitates discretizing space, usually into a regular grid (see Box 7.1). Therefore, for the vast majority of situations, PDE models are approximated by coupled lattice models with a very fine resolution lattice.

Box 7.1 Solving Diffusion PDEs

Although PDE models are formulated in continuous space, numerical solution of the equations often requires us to discretize space in some manner. The most common method of achieving this is to subdivide space into a regular grid, therefore we are again dealing with a lattice-based model. Here we explain how to translate the differential terms into a lattice formulation.

We impose a lattice structure onto our space, where adjacent lattice points are separated by a distance d; we therefore want the lattice solution $X_{i,j}(t)$ to approximate the PDE solution $X(i \times d, j \times d, t)$. We are familiar with ways to treat temporal derivatives (e.g., $\frac{d}{dt}$), integrating forward in time using methods such as forward Euler or Runge-Kutta. However, the PDE model also contains spatial derivatives and these need to be expressed in terms of the lattice structure. Considering the x spatial second derivative for the number of susceptibles:

$$\frac{\partial^2 X_{i,j}}{\partial x^2} \approx \frac{\frac{\partial X_{i+\frac{1}{2},j}}{\partial x} - \frac{\partial X_{i-\frac{1}{2},j}}{\partial x}}{d}.$$

In words, this means the second derivative can be approximated as the change in the first derivative of X between $(i + \frac{1}{2}, j)$ and $(i - \frac{1}{2}, j)$ divided by the distance, d, between $(i + \frac{1}{2}, j)$ and $(i - \frac{1}{2}, j)$. We now perform a similar approximation for the first derivatives in this term:

$$\frac{\partial^2 X_{i,j}}{\partial x^2} \approx \frac{\left(\frac{X_{i+1,j} - X_{i,j}}{d}\right) - \left(\frac{X_{i,j} - X_{i-1,j}}{d}\right)}{d},$$

$$\approx \frac{X_{i+1,j} - 2 X_{i,j} + X_{i-1,j}}{d^2}.$$

Therefore, if we consider the full diffusion term:

$$D_X \nabla^2 X_{i,j} = D_X \frac{\partial^2 X_{i,j}}{\partial x^2} + D_X \frac{\partial^2 X_{i,j}}{\partial y^2},$$

$$\approx \frac{D_X}{d^2} \left(X_{i+1,j} + X_{i-1,j} + X_{i,j+1} + X_{i,j-1} - 4 X_{i,j} \right).$$

This spatial approximation can now be substituted in the PDE equations to give an ODE equation for each lattice point, which we can solve in the usual manner. Diffusion therefore acts like the movement of individuals between the four nearest-neighbor lattice sites. The rate of this movement (or coupling) is $\frac{D_X}{d^2}$, such that it is proportional to the diffusion coefficient but increases quadratically with the number of lattice sites that represent one unit length. It is therefore clear that PDEs can be approximated by very fine scale lattices, with very high levels of movement between neighboring sites.

Figure 7.13 gives an example of the types of spatio-temporal dynamics that can be observed using a PDE model for the spread of an SIR-type infection. Starting at a point source, infection spreads as an expanding epidemic wave, leaving secondary oscillations around the endemic equilibrium in its wake. The left-hand graph of Figure 7.13 shows a snapshot of this circular wave front. However, the right-hand graph provides a more intuitive understanding of the spatial pattern, plotting disease prevalence as a function of the distance from the initial source (black solid line). This is compared to the solution of the standard (nonspatial) SIR model (gray dashed line); clearly there is good agreement between these two, although due to the movement of susceptibles into infected regions, the PDE shows a slightly slower decay of the epidemic because diffusion of susceptibles is playing a comparable role to births, allowing infection to persist locally.

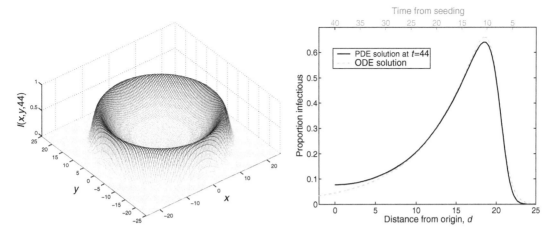

Figure 7.13. Results from the SIR-type PDE model equation (7.18). The left-hand figure is a snapshot, at time $t = 44$, of the density of infecteds in the PDE model. The right-hand figure compares the distribution of infection (at time $t = 44$) as a function of distance from the initial source, with the results from a standard (nonspatial) SIR model with the same basic parameters. For the nonspatial model the x-axis represents the time from the start of the simulation, whereas for the PDE model the x-axis represents the distance from the initial point of infection. The values on the x-axis have been scaled by the wave speed so that the two curves coincide. (The PDE was simulated using a 501×501 lattice, $N = 1$, $\nu = \mu = 10^{-3}$, $\gamma = 0.1$, $\beta = 1$, $D_X = D_Y = D_Z = 0.1$.)

The comparison between the diffusion-based PDE and the nonspatial SIR model hints at a deeper relationship. Involved mathematical calculation shows that, once transient dynamics have died away, the PDE leads to a traveling wave with constant velocity, c (Box 7.2). All such traveling waves (initiated at the origin) can be written as:

$$Y(x, y, t) = \widehat{Y}(r - ct), \qquad \text{where } r = \sqrt{x^2 + y^2}.$$

Therefore, if we stand in one place and record the wave moving past us, we observe the same profile as looking at the spatial wave form at a given time.

For PDE models (in two dimensions), infection spreads as a growing circular wave of near constant velocity.

7.4.2. Integro-Differential Equations

Integro-differential equations (IDEs) share the continuous-space and continuous-population assumptions of the PDE models, but provide far greater flexibility in the way in which infection is transmitted. The PDE model was concerned with the very localized spread of infection and the movement (diffusion) of individuals; in contrast, the IDE models focus on longer-range transmission from static individuals—although this latter constraint is not always true. Although a wide variety of model forms exist, the following

---

Box 7.2 Speed of Infection Wave

Calculating the invading epidemic wave speed analytically for a given set of diffusion-based PDEs is a complex procedure, which is covered comprehensively and in great detail by Murray (2003). Here we give a brief outline of the concept for the simple epidemic (with no births or deaths) in a two-dimensional population. We start with the basic equations together with host diffusion:

$$\frac{\partial X}{\partial t} = -\beta XY/N + D\nabla^2 X,$$

$$\frac{\partial Y}{\partial t} = \beta XY/N - \gamma Y + D\nabla^2 Y.$$

We now assume that the number of susceptibles and infecteds are circularly distributed $(X(x, y, t) = \hat{X}(r, t), Y(x, y, t) = \hat{Y}(r, t)$ where $r^2 = x^2 + y^2)$; this changes the equations to:

$$\frac{\partial \hat{X}}{\partial t} = -\beta \hat{X}\hat{Y}/N + D\frac{1}{r}\frac{\partial}{\partial r}\left(r\frac{\partial \hat{X}}{\partial r}\right),$$

$$\frac{\partial \hat{Y}}{\partial t} = \beta \hat{X}\hat{Y}/N - \gamma \hat{Y} + D\frac{1}{r}\frac{\partial}{\partial r}\left(r\frac{\partial \hat{Y}}{\partial r}\right).$$

Next, we assume that the dynamics can be written as a traveling wave solution with velocity c $(\hat{X}(r, t) = \tilde{X}(z), \hat{Y}(r, t) = \tilde{Y}(z)$ where $z = r - ct)$:

$$-c\frac{d\tilde{X}}{dz} = -\beta \tilde{X}\tilde{Y}/N + \frac{D}{r}\frac{d\tilde{X}}{dz} + D\frac{d^2\tilde{X}}{dz^2},$$

$$-c\frac{d\tilde{Y}}{dz} = \beta \tilde{X}\tilde{Y}/N - \gamma \tilde{Y} + \frac{D}{r}\frac{d\tilde{Y}}{dz} + D\frac{d^2\tilde{Y}}{dz^2}.$$

Looking at the $\tilde{Y}$ equation, and assuming that r is large so that we are looking for the long-term large-radius dynamics:

$$D\frac{d^2\tilde{Y}}{dz^2} + c\frac{d\tilde{Y}}{dz} + \frac{\beta \tilde{X}\tilde{Y}}{N} - \gamma \tilde{Y} = 0$$

At invasion, when $X = N$, this differential equation only has wavelike solutions when $Ds^2 + cs + (\beta - \gamma) = 0$ has real solutions for s. This provides a lower bound for c; this lower bound is the observed wave speed:

$$c = 2\sqrt{D(\beta - \gamma)} = 2\sqrt{D(R_0 - 1)\gamma}.$$

Therefore, the wave speed is proportional to the square root of the diffusion coefficients, the square root of $(R_0 - 1)$, and the square root of the rate of recovery. This is an *asymptotic* wave speed, which only holds at large radii and once an invading wave front has fully developed.

---

relatively simple example illustrates the salient points of integro-differential equations:

$$\frac{dX(x, t)}{dt} = \nu(x) - \lambda(x, t)X(x, t) - \mu(x)X(x, t),$$

$$\frac{dY(x, t)}{dt} = \lambda(x, t)X(x, t) - \gamma Y(x, t) - \mu(x)Y(x, t), \qquad (7.18)$$

$$\text{where } \lambda(x, t) = \beta \int Y(y, t)K(x - y)dy,$$

where the dependence on location, x (which could be in one, two, or more dimensions), and time, t, have been explicitly stated, and the demographic parameters are allowed to vary between locations. The equation for the number of susceptible and infectious individuals is the same as in Chapter 2; it is only through the force of infection, λ, that spatial interactions enter the dynamics.

The force of infection, $\lambda(x, t)$, models the transmission of infection from all points in space (labeled y in the integral) to the point x that we are considering. The transmission rate is assumed to vary with the distance between the susceptible and infectious individual $(x - y)$, and is described by a transmission kernel, K. In simple terms, K defines how infectivity decreases with distance. Clearly, this gives far greater flexibility than achieved by the PDE model; rather than transmission being a local event, it can now occur over a variety of scales. Equation (7.18) models transmission as a density-dependent process; making transmission frequency dependent is more complex because it depends on how we expect the transmission to operate. Two possible alternatives are:

$$\lambda_1(x, t) = \beta \int \frac{Y(y, t)}{N(y, t)} K(x - y) dy, \qquad \lambda_2(x, t) = \beta \frac{\int Y(y, t) K(x - y) dy}{\int N(y, t) K(x - y) dy}.$$

In the first formulation, it is the local proportion infectious at each point that is important, which most closely mimics the situation where interactions at each point y take place sequentially, so that transmission is based on the point frequency. In contrast, the second formulation corresponds to simultaneous interaction with individuals from a range of points, such that it is the averaged proportion infectious that is important.

With integro-differential equations, the spatial spread of infection is via a transmission kernel that defines how transmission risk decays with distance.

Various properties can now be defined for this type of model. Here we will assume that demographic and epidemiological parameters are invariant across space and that space is infinite and two-dimensional, but this does not necessarily have to be the case. To simplify the calculations we rescale the parameters such that the population density at each point is one, $N = 1$. First, we consider R_0, which again will predict the likely success of an epidemic.

$$R_0 = \beta \int_{\mathbb{R}^2} K(y) dy = 2\pi\beta \int_0^\infty r K(r) dr.$$

Therefore, for the basic reproductive ratio to be finite we require that the kernel, $K(r)$, eventually decreases faster than r^{-2}. If the kernel decays more slowly (has a "fat tail"), then the integral is infinite; it is difficult to envisage situations where this is a reasonable assumption. In a similar manner, the average dispersal distance, D, is given by:

$$D = \frac{\beta \int_{\mathbb{R}^2} \|y\| K(y) dy}{\beta \int_{\mathbb{R}^2} K(y) dy} = \frac{2\pi\beta}{R_0} \int_0^\infty r^2 K(r) dr,$$

and for this to be finite requires that the tail of $K(r)$ decays faster than r^{-3}. Finally, when the variance of the dispersal distance $var(D) (= \frac{2\pi\beta}{R_0} \int_0^\infty r^3 K(r) dr - D^2)$ is also finite ($K(r)$ decreases faster than r^{-4}), we observe a wave of infection that moves with a constant speed. However, if the variance is infinite but the average is finite ($K(r)$ decays slower than r^{-4} but faster than r^{-3}), the wave front accelerates indefinitely (Diekmann 1978; van den Bosch et al. 1990; Mollison 1991; Shaw 1995). Therefore, Gaussian and exponential kernel distributions always lead to a traveling infectious wave front moving at a constant speed, whereas power-law kernels can give rise to a wide variety of behavior.

The shape of the tail of the transmission kernel (which defines the long-range transmission) determines the eventual spatial pattern of invasion, from wavelike spread, to scattered local foci, to highly probable extremely long-range jumps.

These results have worrying implications. It is the long-range transmission of infection, defined by the tail of the kernel distribution, that determines the ultimate dynamics of the spatial spread of infection. However, (assuming R_0 is finite) very few secondary cases actually occur at long range and therefore the precise shape of the kernel is difficult to estimate from observational data. Instead, the long-range behavior of the kernel is often estimated only from the population-level properties of the invading wave.

As with the diffusion-based PDE models, very few of these integro-differential equation models have exact analytical solutions, so numerical simulations are required. Again this computational exercise will require the discretization of space into a fine-resolution grid of subpopulations, breaking the continuous space assumption.

7.5. INDIVIDUAL-BASED MODELS

Individual-based models can encompass a wide range of model forms, and can be designed to include a variety of complex and detailed host behavior that could not be readily expressed within the other model types (Nielen et al. 1999; Mangen 2002; Bates et al. 2003; Noordegraaf et al. 2000; Stacey et al. 2004; see also Section 6.3.5). As the name suggests, these models consider the dynamics of individuals that occupy a spatial landscape. In general, these individual-based models have properties in common with both continuous-space models and stochastic metapopulation models; a transmission kernel is generally used to capture the spatial spread of infection but this is tempered by the stochastic, individual-based nature of the population processes leading to a slower rate of spatial spread (Lewis 2000). Here, as an example of this methodology, we will formulate a general stochastic individual-based model where each host is capable of localized movement and transmission is distance dependent; this gives rise to five distinct probabilistic events: transmission, recovery, birth, death, and movement.

Transmission. Transmission is captured using a technique similar to the integro-differential equations models. The rate of transmission (or force of infection) to a susceptible individual, i, is given by:

$$\lambda_i = \beta \sum_{j \in \text{infectious}} K_T(d_{ij}),$$

where d_{ij} is the distance between the susceptible individual i and an infectious individual j; K_T is the transmission kernel that measures how transmission decreases with distance.

Recovery. As is common in almost all the models of this book, the recovery of an infectious host is independent of its environment. Therefore, the recovery rate of infectious individual j is constant: $G_j = \gamma$.

Birth. For many species, the birth rate is a function of two local density components. First, there must be sufficiently many others in the local environment so that the individual can find a mate. Second, fecundity is generally a function of an individual's fitness and hence decreases with the amount of competition it suffers—a further two kernels may therefore be required.

Death. The natural demographic processes will depend on the particular host species under consideration—clearly wildlife and human populations behave very differently. However, we can safely assume that in many situations the death rate (D_i) of an individual will be a function of the local density, which in turn is a measure of the local competitive effects—this requires a second kernel to calculate the strength of competition.

Movement. Finally, individuals are allowed to move through the environment. The simplest assumption is that individuals randomly and spontaneously move from their current location to a new location given by a local movement kernel. For many species movement may well be more complex, with both the rate of movement and the choice of new location dependent upon the strength of local competition.

Individual-based models account for the spatial interaction between individual hosts distributed on a spatial landscape. They can include a wide variety of complex (more biologically realistic) behavior that often features a spatial component; this can often lead to a huge number of parameters that can be difficult to determine from available data.

Although generic individual-based models such as the one described above are easy to formulate and intuitively appealing, their parameterization may be far more difficult. In the above example, up to five different spatial kernels require estimation. Understanding how errors and biases in these estimates affect the epidemiological dynamics is a computationally intensive problem, but one that cannot be neglected if models are to be robust. In the two model examples given below, only one kernel is required, because there is no demography or movement of individuals, which greatly simplifies both the parameter estimation and the simulation of the spatial epidemics.

One potential difficulty with the simulation of individual-based models is the number of interaction terms that must be considered when the population size becomes large. In principle, these models must consider the probability of each infectious individual infecting each susceptible individual—with the number of permutations being vast. Consider trying to run a stochastic spatial individual-based model with a population of N individuals; the number of infectious-susceptible combinations grows proportional to N^2, whereas the time step between events decreases like N^{-1} (see Chapter 6); therefore, the computational time needed to simulate an epidemic grows proportional to N^3. This means that it becomes increasingly prohibitive to model very large populations. Some approximations and computational shortcuts can be used to dramatically reduce the simulation time and are worth considering in any practical application; these are outlined in Box 7.3, although the techniques given in Section 6.3.5 may also be beneficial.

This is online program 7.5

7.5.1. Application: Spatial Spread of Citrus Tristeza Virus

The model of spatial spread used in this example bridges the gap between cellular automata and individual-based models. We focus on the spread of Citrus Tristeza virus (CTV) between trees in an orchard. Individual trees in an orchard are automatically in a lattice-like distribution as used in cellular automata models. However, transmission is modeled by a transmission kernel and is not confined to simple nearest-neighbor interactions.

Box 7.3 Short-Cuts for Individual-Based Models

Two main shortcuts are used to improve the speed of individual based models:

Discrete time. One difficulty with individual-based models is that as the population size becomes larger, the time interval been events decreases, which in turn increases the number of iterations that must be performed to advance the model by a given time period. By discretizing time and allowing multiple events to occur in each time-step, a great improvement in speed is achieved—this is effectively the τ-leap model discussed in Box 6.5. This approximation can be justified for two reasons. First, in a short time-step the chance of two events occurring in a local neighborhood is small, therefore it is likely that all events that occur in one time-step are independent. Second, most biological systems have a natural time frame (often one day), and at finer (or continuous) timescales the standard assumption of constant parameters may not be true. For example, human populations have a clear daily cycle, and unless the continuous time model differentiates between night and day, a discrete-time model with a daily time step may be equally justifiable.

In such discrete-time models, it is necessary to convert the event rates into probabilities:

$$\mathbb{P}(\text{event}) = 1 - \exp(-\text{Rate} \times \text{time step}).$$

Discrete space. By overlaying space with a grid of sites (squares) and knowing which individuals belong in each square, substantial computational saving can be made in terms of the interaction between susceptible and infectious individuals. If the infection process is very localized, then the disease is unlikely to spread to very distant squares, so considering all susceptible individuals within a distant square is often a fruitless exercise. Instead, it is more efficient to first consider whether infection spreads from the infected individual to any host in the susceptible square.

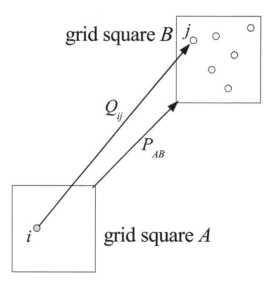

Pictorial example of the mechanism behind using gridded data. We first decide whether infected individual i is likely to have transmitted infection into square B, and only if this probably is realized do we consider the individual hosts within square B.

As an example, consider an infectious individual i in square A and X_B susceptible individuals in square B. An upper bound on the chance that at least one susceptible individual is infected is given by:

$$\mathbb{P}(A \text{ to } B) = 1 - (1 - P_{AB})^{X_B},$$

where P_{AB} is the greatest possible probability of infection between individuals in squares A and B, often calculated from the shortest distance between squares A and B.

Only if this probability of infection is realized is it worth considering the X_B individuals within square B. Of course, when considering these individuals it must be taken into account that the site-level probability has already been realized.

Once it has been determined that infection into square B is realized ($RAND < \mathbb{P}(A \text{ to } B)$), the pseudo-code below shows how individual-level infection should be treated:

$s = 1$
loop $j = 1 : X_B$
$P = 1 - s(1 - P_{AB})^{X_B+1-j}$
$R = RAND$
    if $R < \dfrac{P_{AB}}{P}$
       $s = 0$
       $Q_{ij} = \mathbb{P}\,(j$th susceptible individual in square B is infected
             by infectious individual i in square A)
       if $R < \dfrac{Q_{ij}}{P}$
          Infect the jth individual in square B.
       end if
   end if
end loop

Here the first "if" statement determines whether the individual j in square B would be infected, assuming that the infectious probability is the maximum value P_{AB} and ensuring that at least one individual within the square would be infected if P_{AB} were the infectious probability for all individuals. This statement is therefore the individual-based equivalent of the square-level condition $RAND < \mathbb{P}(A \text{ to } B)$. The second if statement accounts for the fact that P_{AB} is an overestimate of the true transmission probability. This grid-based method confers great time savings because large numbers of susceptibles in distant squares can be dismissed simultaneously. The most efficient square size is generally found by experimentation. If the grid size is too small, then a large number of squares will need to be considered; however, if the grid size is too big, there is a high chance that at least one individual within it is infected, forcing us to consider all individuals within the square. In principle, a hierarchy of grids could be used to overcome this problem, but it is not clear that the computational efficiency would justify the greater complexity.

Citrus Tristeza virus is an infectious disease of citrus trees that is spread by the brown citrus aphid. It has a worldwide distribution, from Asia to Africa to America, and leads to a reduced fruit crop and eventual loss of the tree, depending upon the strain of virus. In fruit-growing areas, CTV can therefore have severe economic consequences and so in Florida ring-culling has been implemented, removing all citrus trees within 1,900 feet of identified infected trees.

In a series of papers, Gibson and coworkers (Gibson and Austin 1996; Gibson 1997a, b) have examined the spread of CTV using sequential data from detailed studies of its spread in an orchard (Marcus et al. 1984); Figure 7.14 shows an example of this data illustrating the spread of infection.

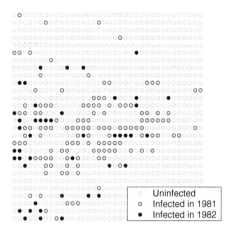

Figure 7.14. Position of infected and susceptible trees in an orchard. Empty circles are susceptible trees, gray-filled circles are trees infected with CTV before the end of 1981, and black-filled circles are trees infected in 1982. Data from Marcus et al. (1984).

In general, the work of Gibson has focused on obtaining accurate parameter values for the spread of CTV infection, which is a highly complex procedure because there are multiple potential sources for each new case. Although detailed parameterization is clearly a prerequisite before model simulation can begin, here we focus primarily on the results from models. CTV is vector transmitted and trees eventually die from infection, however over relatively short spatial and temporal scales we can simply consider the disease as an SI infection (ignoring deaths) and assume that transmission occurs from tree to tree. The proposed model assumes that the rate at which tree i is infected by tree j (assuming i is susceptible and j is infectious) is proportional to a distance kernel $K(d_{ij})$, where d_{ij} is the distance between the two trees. Therefore, the rate that tree i becomes infected (or the force of infection to tree i) is given by the sum over all the rates of infection from all infectious sources:

$$Rate = \lambda_i \propto \sum_{j \in \text{infectious}} K(d_{ij}). \qquad (7.19)$$

Gibson argues that a power-law decay ($K(d) = d^{-2\alpha}$) is the most appropriate form for the kernel, and estimates that α is around 1.3 (Gibson 1997b). Figure 7.15, top graph, shows the instantaneous rate of infection (equal to the force of infection) for all susceptible trees at the start of 1982 using equation (7.19) with the proportionality constant derived to generate the observed number of new infections (45 cases) within one year. Due to the low value of the exponent in this transmission kernel, an infected tree could potentially infect many others outside the sampled orchard and, in fact, the average dispersal distance is infinite (see Section 7.4.2; Shaw 1995). Therefore, it is clear that the scale of an orchard can have an impact on the transmission potential from an infected tree. The bottom-left graph of Figure 7.15 shows how the total force of infection from a single infected tree in the center of a large square orchard scales with the size (number of trees) of the orchard, using the transmission kernel estimated for CTV. This result agrees with our hypothesis stated in Section 7.2.1.1—for plant diseases, R_0 will increase with the number of hosts.

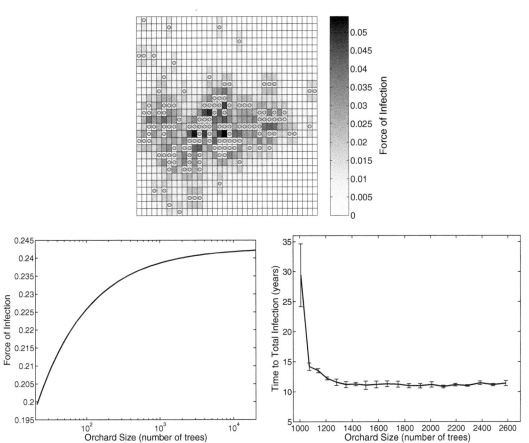

Figure 7.15. Graphs showing the transmission potential and dynamics of CTV, using the power-law kernel estimated by Gibson (1997b). The top graph shows the instantaneous force of infection to every susceptible tree at the end of 1981, with circles marking the position of infected trees. Note that the distance between trees is not equal; this is not a square lattice, there is greater spacing between the rows than the columns. The bottom-left graph shows the effect of orchard size on the force of infection from one infected tree in the center of the orchard. Finally, the bottom-right graph extends this force of infection result, by simulating the full epidemic dynamics across the whole orchard starting with the trees infected in 1981. The graph shows the average time for CTV to infect the entire orchard. The CTV data came from an orchard of size $28 \times 36 = 1{,}008$ trees; larger orchards were simulated by embedding the true orchard in the center of a larger grid (size $(28 + n) \times (36 + n)$ for n between 0 and 19).

The effect of orchard size is even more dramatic when the dynamics are iterated forward. Here orchard size plays two roles: (1) it increases the per capita force of infection, and (2) it increases the number of trees that will become infected in the next generation, hence speeding up the epidemic process still further. This secondary factor is promoted by the SI nature of infection such that trees remain permanently infected. The bottom-right graph of Figure 7.15 shows the average time (from the end of 1981) to infect an entire orchard of varying sizes. Despite the greater number of trees that need to be infected, larger orchards are predicted to be overwhelmed with infection quicker than smaller ones—although the time taken quickly saturates.

Although these simulations offer an understanding of CTV spread, several factors would need to be included if detailed predictions were required. The assumption of SI dynamics is clearly an approximation; infected trees do not instantaneously become infectious and infectious trees are eventually killed by the pathogen—therefore, $SEIR$-type dynamics may be a better assumption. The dynamics of the aphid vector have been ignored; the density of vectors is likely to vary throughout the year, producing some seasonal effects (see Chapter 5). Finally, the extrapolation from the small (28×36) sampled orchard to very large orchards assumes that the dispersal of the aphid vector is not influenced by the presence of trees; it is plausible to think that in a small orchard local transmission might be increased, because the aphids do not frequently leave the orchard, to regions with no fruit trees. However, despite these simplifying assumptions, the research by Gibson and coworkers have illustrated the difficulties in parameterizing transmission kernels from spatial data, but how sophisticated statistical techniques may provide a mechanism for teasing this information from the vast combinatorial array of possible infection scenarios.

7.5.2. Application: Spread of Foot-and-Mouth Disease in the United Kingdom

As a second example of individual-based modeling, we consider the model of Keeling et al. (2001b, 2003) for the spatial spread of foot-and-mouth disease in the United Kingdom. This model framework considers farms as the individual unit, classifying each farm as susceptible, exposed, infectious, or recovered. It is therefore assumed that once one animal in a farm is infected, the rest of the livestock on the farm soon contract the disease— this is a reasonable assumption given the highly transmissible nature of this virus. This model is far simpler than the general form of the individual-based model outlined above because farms do not move, and during the course of an epidemic farms will be neither created nor destroyed—except as part of the control measures. Therefore, the models of the foot-and-mouth epidemic are most closely related to those of plant epidemics, with the spatial dynamics being governed by a single transmission kernel.

The rate λ_i at which susceptible farm i becomes infected is given by:

$$\lambda_i = \text{Sus}_i \sum_{j \in \text{infectious}} \text{Trans}_j\, K(d_{ij}), \qquad (7.20)$$

$$\text{Sus}_i = \sum_{l \in \text{species}} N_{i,l}\, s_l \qquad \text{Trans}_j = \sum_{l \in \text{species}} N_{j,l}\, t_l,$$

This is online program 7.6

where $N_{i,l}$ is the number of livestock of type l on farm i and s and t are species-specific susceptibility and transmissibility. Therefore, the farm-level susceptibility and transmissibility (Sus and $Trans$) are assumed to be the sum of the animal-level susceptibility and transmissibility for each species of livestock. The parameter values for s_{sheep}, s_{cattle}, t_{sheep}, and t_{cattle} were estimated by fitting to the spatial and temporal pattern of the 2001 UK epidemic, whereas the transmission kernel is derived from contact-tracing data that attempts to identify a source for each new reported case.

Figure 7.16 shows examples of the spatial and temporal dynamics of a simulated foot-and-mouth disease (FMD) outbreak in the United Kingdom if only minimal control measures are used (movement restrictions, which determine the kernel, and culling of animals on farms reporting infection). The spatial location of farms and the localized nature of the transmission kernel play a key role in determining the regions of the most severe outbreaks. However, despite the aggregation of cases, a huge prolonged epidemic

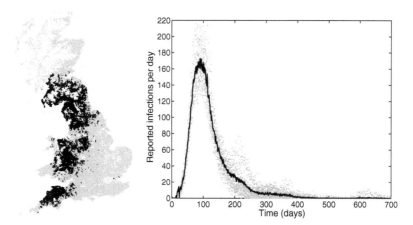

Figure 7.16. Results from the individual-based foot-and-mouth disease model, equation (7.20), using transmission parameters from Keeling et al. (2001b). Limited control measures are used—culling only animals on infected farms, together with movement restrictions—hence the epidemic is far larger, longer, and more widespread than in 2001. The left-hand figure shows the location of all livestock farms within the United Kingdom (gray dots) and those farms infected with FMD in a typical simulation (black dots). The right-hand graph gives the number of reported cases per day from ten simulations (gray dots) together with the average (black line). ($s_{sheep} = 1.0$, $t_{sheep} = 5.1 \times 10^{-7}$, $s_{cow} = 10.5$, $t_{cow} = 7.7 \times 10^{-7}$, $s_{pig} = t_{pig} = 0$, farms are latent for 5 days, then infectious for 5 days before they are culled). Simulation results kindly provided by M.J. Tildesley.

still ensues, lasting between one and two years and infecting more than 16,000 farms. The size of the outbreak is attributable to two main factors. First, foot-and-mouth disease has a relatively short latent and infectious period, and measuring the basic reproductive ratio as the number of secondary *farms* infected shows that $R_0 \approx 2.5$. This leads to a rapid doubling time of around a week in a totally susceptible population (Woolhouse et al. 2001a). In addition, animals (and therefore farms) are infectious for 4 to 5 days before clinic signs become apparent; therefore, a great deal of transmission may occur before it is realized that a farm has become infected (Fraser et al. 2004). This means that the removal of livestock from farms that are diagnosed with FMD is insufficient to control the disease.

In practice, additional reactive culling was performed as a preventative measure in an attempt to both remove farms that are likely to be infected and to remove susceptible farms that are likely to become infected in the future (Keeling et al. 2001b; Ferguson et al. 2001b). These culls took two main forms: Dangerous Contacts (DCs) found through the tracing of movements of vehicles, livestock, and people from an infected farm; and Contiguous Premises (CPs), defined as farms that share a common boundary with an infected farm. Unfortunately, neither of these culling practices is simple enough to be described in terms of a mechanistic formulation suitable for this book. Instead, we consider the likely success of ring culling (Ferguson et al. 2001a), removing the livestock on all farms within a given radius of an infected premise within 1 to 2 days. Figure 7.17 shows the impact of variously sized ring culls; although larger ring culls are always predicted to decrease the number of cases, the total number of farms lost (infected plus culls) is minimized for some intermediate culling policy of around 3.3 kilometers. Finally, the epidemic duration is minimized by large ring culls; therefore, a ring cull of 4 or 5 kilometers may be preferable if a long epidemic has severe economic or political consequences.

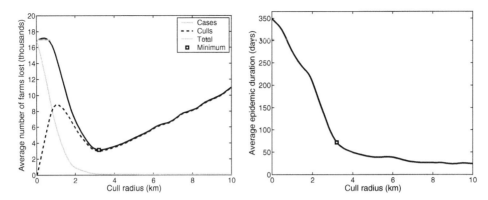

Figure 7.17. The simulated impact of ring-culling using the individual-based foot-and-mouth disease model, equation (7.18); with all other parameters as in Figure 7.16. The left-hand figure gives the average total loss of farms during an epidemic, composed of both reported infections and farms removed as part of a ring cull. There is a clear trade-off between minimizing the number of cases and not losing too many farms as part of the cull. For the same simulations the right-hand graph gives the average duration of the epidemic, with the cull size that minimized loss of farms marked. Results are from 1,250 simulations per radius value considered, and are smoothed using a local spline. Simulation results kindly provided by M. J. Tildesley.

7.6. NETWORKS

Networks provide a unified way to think about the interaction between individuals or populations, and are especially useful when each individual is in direct contact with only a small proportion of the population (Garnett and Anderson 1996; Morris 1997; Dunyak et al. 1998; Potterat et al. 1999; Klovdahl 2001; Rothenberg 2001; Sander et al. 2002; Potterat et al. 2002; Halloran et al. 2002; Liljeros et al. 2003; Rothenberg 2003; Szendroi and Csanyi 2004; Doherty et al. 2005; Keeling and Eames 2005). Networks tend to be very powerful tools for understanding the transmission of infection in human populations due to either social contacts (for airborne infections) or sexual contacts (for sexually transmitted diseases). In either case we expect that each individual will be in contact with only a small proportion of the population, and that the number of contacts will be highly heterogeneous—networks provide a simple means of capturing such interactions. We therefore see that the primary advantage of network models is their ability to capture complex individual-level structure in a simple framework.

To specify all the connections within a network, we can form a matrix from all the interaction strengths ρ_{ij}, which we expect to be sparse with the majority of values being zero. Usually, for simplicity, two individuals (or populations) are either assumed to be connected with a fixed interaction strength or unconnected (and therefore have an interaction strength of zero). In such cases, the network of contacts is specified by a graph matrix G, where G_{ij} is 1 if individuals i and j are connected, or 0 otherwise. For the remainder of this section we will exclusively consider such networks where all the interaction strengths are identical; understanding networks with variable strength connections remains a challenge for the future (Sander et al. 2002). When G is symmetric ($G_{ij} = G_{ji}$) we define the network as undirected and infection can pass in both directions across a contact—this is the standard assumption for the vast majority of infectious diseases. However, there are a few special cases where a network is directed and infection can pass only one way across a contact; examples of such directional transmission include infections transmitted through blood products, infections of livestock transmitted through artifical insemination, and transmission to populations (e.g., farms) through the movement of individuals (livestock).

Networks provide a robust means to consider the individual nature of disease transmission. Two individuals are linked if they have sufficient contact to allow the infection to pass between them.

A matrix G can be used to completely specify the network, indexing all possible transmission links between individuals.

7.6.1. Network Types

Several types of networks are commonly used within the epidemiological literature (as well as by statistical physicists). Although many theoretical approaches to networks use the terms nodes and edges, we will generally refer to individuals and contacts. Below we describe the basic nature of these networks in terms of a few fundamental properties: the way the network is constructed, the heterogeneity in the number of contacts, the clustering of contacts, and the average path length (in terms of the number of steps it takes on average to link two randomly chosen individuals). The five common network types discussed below are illustrated in Figure 7.18.

7.6.1.1. Random Networks

Random networks ignore the actual spatial position of individuals and, as the name suggests, connections are formed at random (Islam et al. 1996; Andersson 1998; Diekmann et al. 1998; Newman et al. 2002; Neal 2003). In one of the most analytically tractable versions of the random network, each individual has the same number of contacts. The random network is therefore characterized by a lack of heterogeneity in the number of contacts and a lack of clustering. The average path length in a random network is low because a large number of "long-distance" contacts exist, effectively spanning the population. A range of analytical techniques can be applied to understanding the dynamics of diseases spreading through such networks (Diekmann et al. 1998; Keeling 1999); of greatest importance, these models show that the initial growth rate of a disease in a network is reduced compared to the random-mixing equivalent:

$$\text{Initial random network growth rate} = \tau(\overline{n} - 2),$$

$$\text{Initial random-mixing growth rate} = \beta = \tau\overline{n},$$

where τ is the transmission rate across a contact and $\overline{n}$ is the average number of contacts (or effective number of contacts in the random-mixing model). This reduction is due to the development of strong negative correlations between susceptible and infected individuals during the early phase of the epidemic.

7.6.1.2. Lattices

As explained in Section 7.3, lattices are associated with a regular grid of contacts and each individual has a fixed number of contacts (usually either 4 or 8) (Bak et al. 1990;

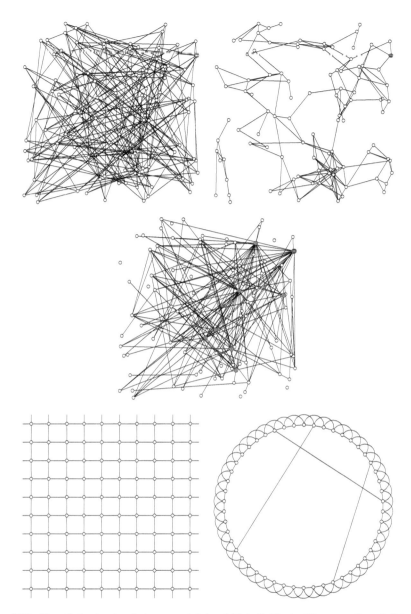

Figure 7.18. Five distinct network types containing 100 individuals. These are from left to right: Random, Spatial (top row), Scale-free (middle row), Lattice, and Small-World (bottom row). The Random, Spatial, and Scale-Free networks all use the same position of individuals—although for the Random and Scale-Free network, the position of the individuals is irrelevant for forming connections. In all five graphs, the average number of contacts per individual is approximately 4. For the scale-free network, individuals with high numbers of contacts are shaded gray.

Rhodes and Anderson 1997). In contrast to random networks, lattices possess far stronger clustering because contacts are localized in space. This higher level of clustering further reduces the initial growth rate of a disease in a lattice compared to the random network, although exact analytical results are no longer available. Given that all the connections are local, the average path length is very long because the only way to transverse the lattice from one side to the other is by steps of a single grid size.

7.6.1.3. Small World Networks

Small world networks are based upon a lattice structure, with a small number of "long-range" connections added. Figure 7.18 shows the classical one-dimensional small-world model (Watts and Strogatz 1998) where each individual is connected to its four nearest neighbors together with three long-range contacts across the entire population. Locally (from the perspective of an individual), small world networks look very much like lattices; they are highly clustered and have little heterogeneity in the number of neighbours—therefore, transmission of infection is predominantly localized so that the strong saturation effects and wavelike spread observed in the lattice models still occur. However, the presence of the few long-range connections provides shortcuts across the network, vastly reducing the average path length and allowing a spreading infection to jump to new susceptible areas (Newman and Watts 1999; Moore and Newman 2000). In practice, it may be difficult to estimate the number of long-range contacts, but small-world networks have highlighted their profound importance for disease dynamics (Boots and Sasaki 1999; Kuperman and Abramson 2001).

7.6.1.4. Spatial Networks

Spatial networks are one of the most flexible forms of networks, and are related to the individual-based models discussed in Section 7.5. In spatial networks, a kernel is often used to calculate the probability of any two individuals being connected depending on the distance between them (Watts 1999; Read and Keeling 2003; Keeling 2005a). By changing the distribution of individuals and the connection kernel, it is possible to generate a wide variety of networks from highly clustered lattices to small world arrangements and globally connected random networks. Spatial networks generally show a reasonably high degree of heterogeneity, with the number of neighbors often being approximately Poisson distributed. In addition, when the connection kernel preferentially links nearby individuals, we can regain the spatial wavelike spread of infection that characterizes lattice models.

7.6.1.5. Scale-Free Networks

In the vast majority of networks that have been studied, the number of contacts per individual is very heterogeneous, with most individuals having a relatively small number and a few have many contacts (Albert et al. 1999; Barabási and Albert 1999; Jeong et al. 2000; Lilijeros et al. 2001). Because the most connected individuals are likely to be disproportionately important in disease transmission (see Chapter 3), networks that can capture this heterogeneity are therefore vital in understanding the spread of real infections—scale-free networks incorporate these heterogeneities. Scale-free networks are generally created dynamically, adding new individuals to a network one at a time with a connection mechanism that mimics the natural formation of social contacts. Each new individual that is added to the population connects preferentially with individuals that already have a large number of contacts; in a social setting, this corresponds to everyone wanting to be friends of the most popular people. The resultant network has a power-law distribution for the probability of having a given number of contacts, $\mathbb{P}(\text{contacts} > n) \approx n^{-\alpha}$. This power-law property was first observed for the World Wide Web connections and has also been recorded in power grid networks and graphs of actor collaborations. The same type of heterogeneities are likely to be present in the social contacts that permit the spread of infection.

Many different types of network structure are possible. These differ in the amount of heterogeneity, clustering, and average path length, thus reflecting the different transmission routes for various infections.

7.6.2. Simulation of Epidemics on Networks

Networks have many similarities with individual-based spatial models (see Section 7.5 and Section 6.3.5), in that spatial interactions can be defined in terms of a kernel. However, in networks, contacts tend to be of equal strength and limited in number. This can be used to considerable advantage in simulations:

$$Rate(\text{Infected individual } j \text{ recovers}) = \gamma,$$

$$Rate(\text{Susceptible individual } i \text{ infected}) = \tau \times \text{number of infectious contacts}$$

$$= \tau \sum_j G_{ji} I_j = \lambda_i,$$

where τ is the rate of transmission across a contact and I_j is one if individual j is infectious or zero otherwise. One immediate implication of the network structure is that the force of infection λ_i depends on the state of only a few individuals. This means that the force of infection does not need to be calculated anew at every iteration, providing huge computational savings. Instead, we can store the force of infection for each individual; when an individual first becomes infectious, the force of infection of all its contacts is increased by τ, and when an individual recovers the force of infection for all its contacts is likewise increased by τ. Hence, each event impinges on the state of only its neighborhood of contacts. Further computational savings can be achieved if the contacts of each individual are stored in a list (such that $C_1^i, C_2^i, \ldots C_{n_i}^i$ are the n_i contacts of individual i) because loops and summations need to be over only the n_i neighbors, which is generally much faster than summing over the total number of individuals. For example:

$$Rate(\text{Susceptible individual } i \text{ infected}) = \tau \sum_{j=1}^{n_i} I_{C_j^i}.$$

This is online program 7.7

Figure 7.19 shows examples of the epidemic dynamics on the network types illustrated in Figure 7.18, giving both the individual epidemic curves (gray lines) and the average of all major epidemics (black line). Clearly the random network (which in many ways is closest to the nonspatial mass-action models) generates the fastest epidemic growth rate and has the highest proportion of infectious individuals at the maximum. Surprisingly, there appears to be little difference between the dynamics of epidemics on the Spatial and Scale-Free networks; this may be attributable to the variance in the number of contacts that is present in both networks. For infinitely large population sizes, Scale-Free

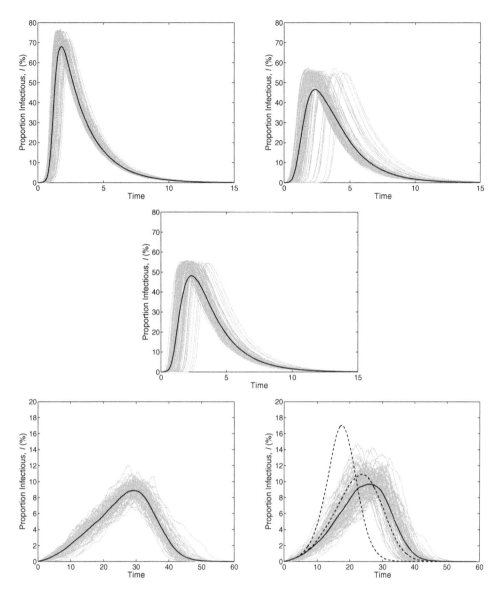

Figure 7.19. Typical epidemics of the five network types described above. These are from left to right: Random, Spatial (top row), Scale-Free (middle row), Lattice, and Small-World (bottom row). Each graph shows 100 epidemic curves (gray), together with the average for all major epidemics (black) for a single example of each network type—therefore, all variablity within each graph is due to the stochastic nature of transmission and not variation in the network. All five networks contain 10,000 individuals, although all individuals are not necessarily interconnected as part of a giant component. For the Spatial and Scale-Free networks, approximately 88% and 74% are part of the giant component and can therefore potentially become infected; for these networks, the proportion of infectious individuals have been rescaled as a fraction of the giant component. For all other types, the entire network is interconnected. In all networks, the average number of contacts per individual is approximately 4, although the Scale-Free network has considerable heterogeneity, with one individual having 85 contacts. For consistency, the Small-World network is formed from a two-dimensional lattice (Watts 1999) (not a one-dimensional circle as shown in Figure 7.18) with 10 additional random "long-range" contacts. The dashed lines show the effect on the mean epidemic of increasing the number of long-range contacts to 20 and 100. ($\tau = 1, \gamma = 0.5$.)

networks theoretically have $R_0 = \infty$ because there will be some individuals with arbitrarily large numbers of contacts (Albert et al. 2000; May and Lloyd 2001). However, in any practical scenario, the maximum number of contacts is always limited and the success of the epidemic depends on encountering these core individuals early.

In contrast, the lattice and two-dimensional small-world networks show much slower epidemic growth rates. In fact, as predicted in Section 7.3, the lattice network leads to an expanding wave of infection and hence an (almost) linear initial increase in the number of cases. As long-range contacts are added to this basic lattice, forming a Small-World network (Watts and Strogatz 1998), the infection is able to spread to new susceptible areas of the network and therefore the infection grows more rapidly.

In general, networks display slower epidemic dynamics compared to randomly mixed models. As a consequence, networks that are most like the random-mixing models— with short-average path length (Small-World, Random, and Scale-free) and little clustering (Random, and Scale-Free)—show the fastest epidemic growth rates for a given average number of contacts per individual.

7.7. WHICH MODEL TO USE?

With such a vast array of possible spatial models, it may be daunting to try to choose between them. As with all model choices, the type of model required will reflect the problem being addressed, the availability of data, and the form of results required. At some fundamental level, *all* spatial models can ultimately be expressed as both metapopulation models or individual-based models with carefully constructed interaction terms. However, the following guide may help discriminate between the model classes:

- *All individuals interact at random.* A spatial model is not required.
- *The population is naturally separable into groups, with strong (random) interaction within each group.* This is the classical metapopulation ideal, and is frequently the preferred model for human disease dynamics at a national scale where the population can be generally grouped by town or city.
- *The population is densely distributed across the entire space.* Here we may safely treat the population as continuous and deterministic; therefore, either PDE or integro-differential models can be appropriate.
- *The environment and distribution of hosts is approximately uniform and a qualitative understanding of spatial effects is required.* In such situations, the approximate lattice-based models may be suitable. It should be noted, however, that such models are unlikely to provide an accurate prediction of the quantitative behavior of any real problem.
- *Hosts have a low density or patchy distribution and stochasticity effects are important.* Here we must resort to individual-based modeling; this has the extra advantage that greater behavioral complexity can be easily included. Note that parameters are generally defined at the individual level, such that aggregate population-level data is difficult to use.
- *Hosts have few contacts to whom they can pass infection.* In such cases, networks are the preferred modeling tool. Again, this is an individual-based approach, such that individual-level data is required for parameterization.

Although the above points provide a general guide to the types of models that can be used, compromises are frequently involved. Often individual-based models will seem to be the preferred model type because they are the most flexible and can most closely mimic reality. However, in a great many scenarios there is insufficient information to parameterized such models—for example, only aggregate data may be available. Individual-based models can also be extremely slow compared to other techniques and it may be difficult to assess the robustness of the model to the wealth of factors that can be included. We therefore see that model choice is a skill in itself; ideally several model formats should be tested, their output compared to the available data, and the predictions of all the models scrutinized in terms of the different elements that have been ignored. The variety of models used during the 2001 foot-and-mouth disease outbreak in the United Kingdom illustrate the advantages of this approach where widely different model assumptions all produced similar control recommendations (Keeling et al. 2001b; Ferguson et al. 2001a,b; Morris et al. 2001; Keeling 2005b). However, the use of multiple models to investigate the use of vaccination in the case of a smallpox outbreak produced conflicting recommendations (Meltzer et al. 2001; Kaplan et al. 2002; Halloran et al. 2002; Bozzette et al. 2003)—highlighting the sensitivity of this problem to model structure and assumptions (Ferguson et al. 2003b). Obviously, such a comprehensive approach is rarely possible for a single researcher (or research team); instead, the merits and assumptions of each approach must be weighed against the available data and the required detail and accuracy of model predictions.

7.8. APPROXIMATIONS

In general, spatial models tend to be computationally intensive, such that most results can be considered as in silico experiments, rather than definitive answers. Researchers are therefore beginning to consider other means of modeling spatial epidemics, such that some of the robustness and understanding that comes from the standard differential equation models can be regained. This approach can be compared to the various approximations that have been used to understand stochastic systems (Chapter 6). Here we briefly consider two forms of approximation, both based on the idea of modeling pairs of hosts and hence capturing the correlations that develop when two individuals interact.

Although more mathematically involved than the standard simulation models described in Sections 7.2 to 7.6, these pair-wise models offer the chance to understand the processes involved in disease transmission in a spatial environment. In both approximation approaches given below, negative correlations between susceptible and infectious individuals are of paramount importance for the dynamic behavior, reducing the growth of epidemics by effectively reducing the transmission. This is a universal feature that we have observed in all the spatial models of this chapter; however, it is only through the use of analytical approximation methods (such as those illustrated below) that these effects become fully apparent.

7.8.1. Pair-Wise Models for Networks

The simplest form of a pair-based approximation model is used to capture disease spread through a network of contacts (see Section 7.6). In its most basic form, this pair-wise model assumes an equal number of contacts per individual and no clustering; this approximation therefore corresponds most closely to the random network (Keeling et al. 1997b; Keeling 1999; Bauch and Rand 2000), although adaptations that capture clustering or heterogeneities are possible (Keeling 1999; Eames and Keeling 2002; Eames and Keeling 2004). To explain the formulation of pair-wise models, we go back to the original SIR (or SIS) models discussed in Chapter 2:

Rate of new infection $=$ transmission rate $\times$ number of susceptibles $\times$

number of contacts $\times$ probability contact is infectious.

In the standard random-mixing (frequency-dependent) models this was approximated by:

Rate of new infection $\approx \tau \times X \times n \times Y/N = \beta XY/N$,

where τ is the transmission rate between contacts, n is the average number of contacts, and Y/N is an approximation for the probability that a contact is infectious. This latter term is an approximation because it neglects all spatial correlations between connected susceptible and infectious individuals.

If we know the number of susceptible-infected pairs (i.e., the number of susceptible individuals in contact with an infectious individual in the network), which we label $[XY]$, then the calculation of the infection dynamics is exact:

Rate of new infection $= \tau[XY]$.

This calculation is exact because $[XY]$ takes into account all local spatial correlations. (We can think of the mean-field models as approximating $[XY]$ by nXY/N.) However, if we wish to use this pair-wise technique predictively, we need to model how $[XY]$ varies over time. We therefore develop a differential equation for the dynamics of $[XY]$ pairs. For the SIR equations:

$$\frac{d[XY]}{dt} = \tau[X\overleftarrow{X}Y] + \gamma[YY] - \tau[\overleftarrow{XY}] - \tau[\overrightarrow{Y}XY] - \gamma[XY]. \qquad (7.21)$$

Here an arrow signifies the direction of transmission and triples ($[ABC]$ represents an A connected to a B connected to a C in the network) are spaced such that the pair in question is more clearly identified. Five events can lead to changes in the number of XY pairs; in the order they appear in equation (7.21) these are: creation of an XY pair by infection of an XX pair, creation of an XY pair by recovery of an infected individual in a YY pair, loss of an XY pair due to the susceptible being infected by the infectious individual within the pair, loss of an XY pair due to the susceptible being infected by an infectious individual outside the pair, and loss of an XY pair due to recovery of the infected individual.

To close the dynamics, we need to know the number of triples—in particular $[XXY]$ and $[YXY]$. We could formulate an equation for the number of triples, which would contain expressions involving the number of quads. However, it is simplest to perform a moment-closure approximation, by approximating the number of triples in terms of the number of pairs and singles. If all individuals within the contact network have exactly n contacts, then the triple approximation becomes:

$$[ABC] \approx \frac{(n-1)}{n} \frac{[AB][BC]}{[B]}.$$

This approximation therefore ignores any correlation that may have developed between the ends of the triple—that is, A and C are correlated only by the fact that they are both connected to B. Therefore, if triples also form triangles, such that A and C are connected in the network, then this approximation is likely to be flawed, although still better than the mean-field approximation that ignores all correlations.

Box 7.4 SIS Pair-Wise Equations

Although equation (7.22) provides the most intuitive description of the pair-wise network approximation for the SIS model, it is informative to show the full set of equations, where the triple approximation has been included and the dynamics are written in terms of just two state variables:

$$\frac{dX}{dt} = \gamma(N - X) - \tau[XY],$$

$$\frac{d[XY]}{dt} = \tau\frac{(n-1)}{n}\frac{(nX - [XY])[XY]}{X} + \gamma(nN - nX - [XY]) -$$

$$\tau[XY] - \tau\frac{(n-1)}{n}\frac{[XY]^2}{X} - \gamma[XY].$$

We are now in a position to formulate a pair-wise equation for the dynamics of a disease on a random (unclustered) network where each individual has exactly n contacts. For a disease with SIS dynamics and no births or deaths:

$$\frac{dX}{dt} = \gamma Y - \tau[\overleftarrow{XY}]$$

$$\frac{d[XY]}{dt} = \tau[\overleftarrow{XX}Y] + \gamma[YY] - \tau[\overleftarrow{XY}] - \tau[\overrightarrow{Y}\,XY] - \gamma[XY].$$

(7.22)

This is online program 7.8

Noting that

$$[ABC] \approx \frac{(n-1)}{n}\frac{[AB][BC]}{[B]}, \qquad Y = N - X, \qquad [XX] = nX - [XY].$$

Figure 7.20 shows a comparison between stochastic SIS epidemics on a random network and results from the deterministic pair-wise model. Clearly there is excellent agreement between the two approaches, even though the pair-wise model is deterministic and requires only two equations, whereas the network simulation is stochastic. However, the advantage of the pair-wise model is in its analytical tractability and its comparison to the standard differential equations (Chapter 2) for disease dynamics.

Network-based pair-wise models provide a deterministic approximation for the dynamics of pairs of connected individuals within a network, such as the number of connected susceptible-infectious pairs, and can therefore account for the buildup of local correlations within the network.

Network-based pair-wise models are most accurate when dealing with networks with low levels of clustering, such as Random or Scale-Free networks.

Network-based pair-wise models approximate the buildup of local correlations within the network by explicitly modeling connected pairs of individuals.

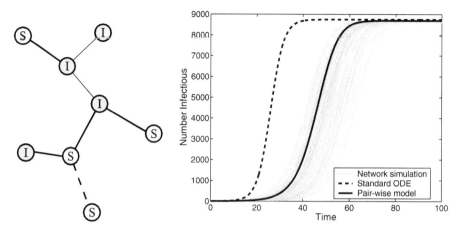

Figure 7.20. The left-hand figure shows a caricature of a network, so that pairs of connected individuals can be readily seen. This particular example has four $[SI]$ pairs (and hence also four $[IS]$ pairs), indicated with thick black contacts; four $[II]$ pairs, because these can be counted in both directions, shown as thin contact lines; and two $[SS]$ pairs, again counting in both directions. The right-hand graph compares the results for a stochastic SIS epidemic on a Random network (see Section 7.5, Figure 7.19) with the corresponding pair-wise equation (7.22) and the standard ODE model (see Chapter 2). Clearly, the correlations in the pair-wise model capture the reduced epidemic growth rate seen in the full network model. ($N = 10,000$, $n = 4$, $\tau = 0.1$, $\gamma = 0.05$.)

7.8.2. Pair-Wise Models for Spatial Processes

A similar approach can be used for approximating the dynamics of full individual-based spatial models (see Section 7.5). In the network-based pair-wise models, a contact was either present or absent, so that all XY pairs had the same strength of transmission. However, for individual-based models, the transmission strength is a function of distance, determined by a transmission kernel; therefore, all XY pairs must be indexed by the distance between them. We therefore write $[XY](\underline{d})$ as the density of XY pairs separated by a vector distance $\underline{d}$; however, in many situations it is easier to consider this pair-wise quantity to be composed of the mean densities plus the spatial covariance (the covariance between two distinct spatial points):

$$[XY](\underline{d}) = X \times Y + C_{XY}(\underline{d}),$$

where X and Y can now be thought of as densities, or more formally probability densities, for an individual to exist at a particular point in space. For such models, the calculation of the rates of change is somewhat more involved due to the nature of individual-based spatial transmission:

$$\text{Rate of new infection } = \tau \int K(\underline{r})[XY](\underline{r})d\underline{r},$$

$$= \beta XY + \tau \int K(\underline{r})C_{XY}(\underline{r})d\underline{r},$$

where K is again the transmission kernel that measures how infection risk decreases with distance, and β ($= \tau \int K(\underline{r})d\underline{r}$) is again an aggregate measure of transmission. For the rate of change of pairs we have:

$$\frac{d[XY](\underline{d})}{dt} = \tau[XX](\underline{d})\frac{\int K(\underline{r})[XY](\underline{r})[XY](\underline{r}-\underline{d})d\underline{r}}{X^2Y},$$

$$-\tau[XY](\underline{d})\frac{\int K(\underline{r})[XY](\underline{r})[YY](\underline{r}-\underline{d})d\underline{r}}{XY^2}, \tag{7.23}$$

$$-\tau[XY](\underline{d})K(\underline{d}) + \gamma[YY](\underline{d}) - \gamma[XY](\underline{d}).$$

Here the first term corresponds to the infection of one susceptible (of a XX pair separated by a distance $\underline{d}$) by an infectious individual who is distance $\underline{r}$ from the target susceptible and therefore distance $\underline{r}-\underline{d}$ from the other susceptible within the pair. This term is composed of three pair-wise components, because all three individuals involved in the event have associated correlations—this differs from the network approach where all triples are "linear"; here all triples form triangles. The second term refers to the loss of an XY pair due to the susceptible being infected from outside the pair. The other three terms do not involve any external interactions and are: infection within an XY pair, recovery within a YY pair, and recovery within an XY pair.

Pair-wise models for spatial processes provide a deterministic approximation for individual-based spatial models.

Although the solution of complex integro-differential equations such as equation (7.23) is an involved process, for some spatial kernels semi-analytic results are possible (Bolker and Pacala 1997; Bolker 1999; Dieckmann et al. 2000) and hence a more comprehensive and robust understanding of the general effects of spatial interaction are possible.

7.9. FUTURE DIRECTIONS

It has only been over the past decade that spatial models have been used in epidemiological applications, and in many areas the implications of spatial structure have yet to be understood. Two main problems stand out as being of both theoretical and applied importance.

The first is the theoretical question of scale—which has applied implications in terms of what is computationally feasible. It is clear from the work in this chapter that spatial structure occurs and operates at a variety of scales; however, it is not clear whether it is necessary to simulate all of these scales to derive accurate predictions of the epidemiological dynamics. In particular, is it always necessary to model the underlying network structure of social contacts to create reliable epidemic models at the national (or international) level? This is one aspect of a more general issue: When can complex fine-scale heterogeneities be absorbed into parameterization and when must they be modeled explicitly? The approximation methods outlined above may be able to provide some insights and lead to general rules about when spatial structure is important.

Second, many difficulties still exist with parameterizing spatial models, either from epidemiological or behavioral data. This is exemplified by three different problems. First, we have very little information about the network of social contacts through which most airborne infections spread, although this has recently become an area of intense focus. Understanding the role of transmission network structure is fundamental to all infectious disease modeling, and may validate or refute the standard (pseudo) mass-action assumption in particular circumstances. Second, at a larger scale, the 2001 foot-and-mouth epidemic in the United Kingdom highlighted the difficulties with assessing the degree of spread between farms. In reality, this spatial spread is the combination of several factors (such as the movement of vehicles or wind-borne transmission), but due to the complexities of parameterization this spread was estimated and modeled as a single transmission kernel. Finally, metapopulation models—treating each town or city as a subpopulation—appear to be an ideal tool for modeling national epidemic patterns; however, it is difficult to assess the level of coupling between communities, especially when many of the links may be sporadic and social.

Spatial models will continue to be an area of high research activity for many years. The importance of local spatial interaction is only recently being appreciated, in terms of both understanding disease dynamics and local control of infection. We can expect to see spatial models increasingly used in public health scenarios, where control usually operates on a regional basis and where preventing infection reaching new populations is a key control aim.

7.10. SUMMARY

Seven different model formulations have been described in this chapter (Metapopulations, Coupled lattices, Cellular automata, Reaction-Diffusion, Integro-Differential, Individual-Based and Network), each with its own merits and disadvantages. The techniques needed to simulate these models and the data needed to parameterize them differ greatly; however, they share a common theme in that local interactions (transmission) generally dominate longer-range interactions, leading to a clustering of cases.

➤ The type of spatial model used is dependent on the host organism, our degree of knowledge about its behavior, and the scale we wish to consider.

➤ **Metapopulations** provide a powerful framework for modeling disease dynamics for hosts that can be naturally partitioned into spatial subunits.

➤ The force of infection within a subpopulation can be modeled as a weighted sum of the prevalence in all subpopulations.

➤ With stochastic metapopulations, the speed of the spread of infection between subpopulations is reduced compared to the equivalent deterministic model.

➤ For plants and other sessile hosts, coupling (or the strength of spatial interaction) generally decreases with distance, mimicking the effects of wind—or vector—dispersal. Adding an extra subpopulation generally increases R_0 because more pathogens can be intercepted by the additional hosts.

➤ Metapopulation models of animal diseases usually capture the transmission of infection by the permanent immigration and emigration of hosts. In these models, R_0 is generally independent of the coupling because each host transmits infection at a constant rate.

➤ The spread of human diseases in metapopulations is best captured by the rapid commuter movements of individuals from their home subpopulation to another subpopulation and back again—requiring us to model both the current location and home location of individuals. When commuter movements are of short duration, this can be approximated by simple coupling. In these models, R_0 is independent of the coupling.

➤ The correlation between disease prevalence in two subpopulations increases sigmoidally with the strength of interaction between the populations. In general, the change from largely independent dynamics to synchrony occurs for interaction strengths from 10^{-3} to 0.1.

➤ Without interaction between the subpopulations, a spatially segregated metapopulation suffers a faster rate of stochastic extinction than its randomly mixed counter-part. When interaction between the subpopulations is included, the level of local (subpopulation-scale) and global (metapopulation-wide) extinctions is an emergent property of the dynamics and cannot be easily predicted from the parameters.

➤ **Levins' metapopulation** models ignore the internal dynamics within each subpopulation, and instead classify each subpopulation as either infected or disease free. Despite differences between the equilibrium-level results of Levins and full metapopulation models, the Levins model still remains a useful and simple tool for studying invasion dynamics.

➤ **Coupled lattice models** are specialized metapopulation models, where subpopulation are arranged on a grid and coupling is generally to the nearest neighbors only.

➤ The wave speed of an invading epidemic in a coupled-lattice model increases almost linearly with the initial growth rate of the infection, $\beta - \gamma - \nu$; increases nonlinearly with the level of coupling, ρ; and is slightly more rapid in deterministic compared to stochastic models.

➤ **Cellular automata** operate on a lattice of sites, with each site generally assumed to hold a single host. Interaction is usually stochastic and with the neighboring (four or eight) lattice sites.

➤ In many locally coupled spatial models (such as cellular automata), the depletion of the locally available susceptible population can reduce the early growth rate of the epidemic and the speed of the invading wave front.

➤ The forest-fire model typifies many stochastic spatial cellular automata models. The fact that transmission is faster than recovery, which is faster than births, which is faster than imports of infection, leads to power-law relationships for the frequency of epidemic sizes.

➤ **Reaction-diffusion models**, which use a PDE formulism, assume local transmission of infection and rely on spatial diffusion of hosts to spread the infection.

➤ For PDE models (in two dimensions with equal diffusion in all directions), infection spreads as a growing circular wave of near constant velocity.

➤ With **integro-differential equations**, the spatial spread of infection is via a transmission kernel that defines how transmission risk decays with distance. The shape of the tail of the transmission kernel determines the eventual spatial pattern of invasion, from wavelike spread, to scattered local foci, to highly probable extremely long-range jumps.

➤ **Individual-based models** account for the spatial interaction between individual hosts distributed on a spatial landscape. They can include a wide variety of complex (more biologically realistic) behavior that often features a spatial component; this can lead to a huge number of parameters that can be difficult to determine from available data.

➤ **Networks** provide a robust means to consider the individual nature of disease transmission. Two individuals are linked if they have sufficient contact to allow the infection to pass between them.

➤ Many different types of network structure are possible. These differ in the amount of heterogeneity, clustering, and average path length, thus reflecting the different transmission routes for various infections.

➤ In general, networks display slower epidemic dynamics compared to randomly mixed models. As a consequence, networks that are most like the random-mixing models—with short average-path length (Small-World, Random, and Scale-Free) and little clustering (Random and Scale-Free)—show the fastest epidemic growth rates for a given average number of contacts per individual.

Chapter Eight

Controlling Infectious Diseases

One of the important uses of epidemiological models is to provide some basic guidelines for public health practitioners. Models have two primary uses in applied settings. First, as we have seen in previous chapters, statistical data analyses and model fitting permit basic epidemiological characteristics of pathogens to be uncovered. This, in conjunction with empirical observations, enables epidemiologists to develop a picture of the kind of pathogen they are faced with, such as its transmission potential, routes of transmission, and latent and infectious periods. The second use of epidemiological models is to provide a means of comparing the effectiveness of different potential management strategies.

The most straightforward objective is simply to minimize *transmission* within a population, with the ultimate aim of reducing it to zero. Alternatively, we may wish to minimize the occurrence of *disease* (or severe illness). Although for many infectious diseases these two objectives may amount to the same control outcomes, for others this change in emphasis can have dramatic implications for control strategies. For example, infectious diseases such as rubella are associated with minimal health risks when affecting the young or the old, whereas infection in pregnant women can have very serious consequences (Behrman and Krliegman 1998). This age-dependent severity has a profound impact on which control strategies are optimal (see later for a full exposition).

In reality, a range of constraints and trade-offs may substantially influence the choice of practical control strategy, and therefore their inclusion in any modeling analysis may be important. These limitations may be simply logistical, in terms of the number of units of vaccine that can be administered in a given time frame, or epidemiological such as adverse reactions to a vaccine. Frequently, epidemiological models need to be coupled to economic considerations, such that control strategies can be judged through holistic cost-benefit analyses (see, for example, Michael et al. 2004). Control of livestock diseases is a scenario when cost-benefit analysis can play a vital role in choosing between cheap, weak controls that lead to a prolonged epidemic, or expensive but more effective controls that lead to a shorter outbreak. For human diseases, cost-benefit analysis may still be applied, but its interpretation is more subjective (see, for example, Hay and Ward 2005).

When attempting to model epidemics and control for public-health applications, there is the compelling urge to make models as sophisticated as possible, including many details of the host and pathogen biology. Although this strategy may be beneficial when such details are known or there exist adequate data to parameterize the model, it may lead to a false sense of accuracy when reliable information is not available. In general, we believe it is better to start with simple models that can provide a generic understanding and then investigate systematically the effects of adding more complexity or detail—this is the approach advocated throughout this book. In addition, it is also important that any detailed model is accompanied by a sensitivity analysis of the assumptions and parameters, without which it is difficult to ascertain the reliability of any predictions.

Despite the assertion that simple models are required to generate a deeper understanding of many issues associated with control, it will become obvious that the models in this chapter are more complex and parameter-rich than those in previous chapters. In general, as we move toward models that are more applied in nature, we encounter a proliferation of parameters necessary to capture the many aspects of transmission and control that are thought to be important. This has two main implications: (1) it is more difficult to obtain a general understanding of the model behavior across all parameter space; and (2) careful, statistically rigorous, parameterization of the model from detailed data becomes ever more important. Finally, even though the models in this chapter are among the most complex in the book, they are still relatively simple compared to some of the sophisticated public health and veterinary models that are used.

In this chapter, we review some mathematical models for different types of control strategies available to decision makers, with some of their inherent limitations and the general principles that emerge from their analyses. We start by considering alternative aspects of vaccination, then move on to controlling infections by reducing transmission opportunities via contact tracing and quarantine methods. Finally, we discuss two case studies, one that highlights the need for a combination of vaccination and contact reduction, and another that demonstrates the importance of seasonality and density-dependent factors in determining the optimal culling of wildlife populations.

8.1. VACCINATION

Perhaps the best documented example of infectious disease management has been the exploration of the levels of prophylactic vaccination necessary for eradication. Since the pioneering work of Edward Jenner on smallpox (Fenner et al. 1988), the process of protecting individuals from infection by immunization has become routine, with substantial historical successes in reducing both mortality and morbidity. Typically, vaccines contain antigens, which are either the whole- or broken-cell protein envelopes from the virus or bacterium causing a specific disease. When efficacious, the presence of such antigens illicits an immune response in the host, intended to be similar to the consequences of actual infection. The assumption (and hope) is that the vaccine provides long-lasting immunity to the infection, preventing both *transmission* and *disease*.

Two forms of random vaccination are possible: The most common for human diseases is pediatric vaccination to reduce the prevalence of an endemic disease; the alternative is random vaccination of the entire population in the face of an epidemic—such a policy of mass-vaccination may be applied in case of any potential smallpox outbreak (see Section 8.1.2).

8.1.1. Pediatric Vaccination

For many potentially dangerous human infections (such as measles, mumps, rubella, whooping cough, polio, etc.), there has been much focus on vaccinating newborns or very young infants. The mathematical treatment of this practice is wonderfully straightforward and requires making a single addition to the $S(E)IR$ equations. Conventionally, the parameter p is used to denote the fraction of newborns (or infants who have lost any maternally derived immunity) who are successfully vaccinated and are therefore "born" into the immune class. This term, p, is the product of the actual vaccination *coverage* (the percentage of newborns who receive the required number of vaccine doses) and the vaccine *efficacy* (the probability that they successfully develop immunity). When incorporated into the SIR system, we get the following set of modified equations:

$$\frac{dS}{dt} = \nu(1-p) - \beta IS - \mu S,$$

$$\frac{dI}{dt} = \beta SI - (\gamma + \mu)I, \qquad (8.1)$$

$$\frac{dR}{dt} = \gamma I + \nu p - \mu R.$$

This is online program 8.1

This modification can be dynamically explored using a simple (linear) change of variables: $S = S'(1-p)$, $I = I'(1-p)$, and $R = R'(1-p) + \frac{\nu}{\mu}p$ (Earn et al. 2000). These substitutions give rise to a new set of ODEs:

$$\frac{(1-p)dS'}{dt} = \nu(1-p) - (\beta I'(1-p) + \mu)S'(1-p),$$

$$\frac{(1-p)dI'}{dt} = \beta S'I'(1-p)^2 - (\gamma + \mu)I'(1-p), \qquad (8.2)$$

$$\frac{(1-p)dR'}{dt} = \gamma I'(1-p) + \nu p - \mu R'(1-p) - \nu p.$$

Clearly, these equations can be simplified by cancelling out the terms $(1-p)$ on both sides. This gives:

$$\frac{dS'}{dt} = \nu - (\beta(1-p)I' + \mu)S',$$

$$\frac{dI'}{dt} = \beta(1-p)S'I' - (\gamma + \mu)I', \qquad (8.3)$$

$$\frac{dR'}{dt} = \gamma I' - \mu R'.$$

After some thought, it becomes obvious that these equations (8.3) are identical to the basic SIR equations with a single important modification: The transmission rate β is replaced with $\beta(1-p)$. Note that if, instead of vaccination, we were attempting to deal with the dynamical consequences of a systematic change in the per capita birth rates (from ν to ν', for instance), then we would instead replace β with $\beta\frac{\nu'}{\nu}$. These observations simply translate into the following general conclusion:

A system either subject to constant long-term vaccination of a fraction p of newborns against an infection with a basic reproductive ratio R_0, or with a modified per capita birth rate of ν', is dynamically identical to a system with $R'_0 = (1-p)\frac{\nu'}{\nu}R_0$.

Although this vaccination result is simple, it is also very powerful. In order to eradicate a pathogen by long-term pediatric vaccination, we need to ensure that the fraction of susceptible individuals in the population is sufficiently small to prevent the spread of the infection (i.e., $dI/dt \le 0$). This is effectively the threshold theorem of Kermack and McKendrick (1927) and means we need to ensure $R'_0 = (1-p)R_0 < 1$, which translates into vaccinating a critical proportion of the newborns

$$p_c = 1 - 1/R_0. \qquad (8.4)$$

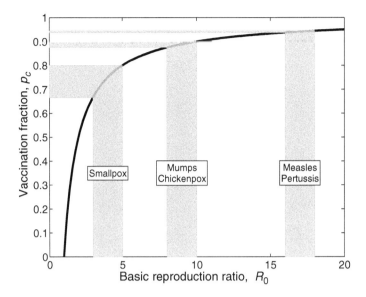

Figure 8.1. The critical fraction of newborns that must be vaccinated to eradicate an infection with a specific basic reproductive ratio (R_0). The figure demonstrates that all incoming susceptibles need not be vaccinated to ensure the infection is not endemic. The shaded regions show the range of p_c for the estimated R_0 of different infections.

This vaccination threshold make good intuitive sense, demonstrating that greater action is required for infectious diseases with a larger basic reproductive ratio. The relationship between p_c and R_0 is plotted in Figure 8.1. It demonstrates that for diseases with very high transmission potential, such as measles and pertussis (R_0 between 16 and 18; Anderson and May 1982), the vaccinated fraction of newborns needed for eradication is somewhere between 93% and 95%. For mumps and chickenpox, on the other hand, the threshold vaccination level is lower, ranging from 87.5% to 90%. For smallpox, p_c was below 80%. To date, smallpox is the only high-profile example of global eradication of a potentially fatal infection. In recent years, there has been a tendency to credit its elimination to a policy of surveillance, containment, and ring vaccination (intensive immunization of individuals in the vicinity of a known case—more below) in the mid- to late-1970s. This misses the point, however, that background vaccination is likely to have sufficiently reduced transmission within the population to such a level that ring vaccination was able to successfully target the remaining susceptibles (Arita et al. 1986).

In order to eradicate an infection, not all individuals need to be vaccinated, as long as a critical proportion (determined by the reproductive ratio of the infection) have been afforded protection. This phenomenon is referred to as "herd immunity" (Fine 1993).

Vaccinating at the critical level p_c does not instantly lead to eradication of the disease. The level of immunity within the population requires time to build up and at the critical level it may take a few generations before the required herd immunity is achieved. Thus, from a public health perspective, p_c acts as a lower bound on what should be achieved, with higher levels of vaccination leading to a more rapid elimination of the disease.

However, the converse is also true. Vaccination is still a worthwhile control measure even when the critical level cannot be achieved. In such cases, vaccination reduces the prevalence of infection:

$$I^* = \frac{\nu(1-p)}{(\gamma + \mu)} - \frac{\mu}{\beta}. \tag{8.5}$$

Hence, the equilibrium fraction of infecteds decreases linearly with increasing vaccination, until eradication is achieved. Thus, even limited vaccination provides protection at the population level, as well as direct protection for those individuals vaccinated. Comparing equation (8.5) with the unvaccinated equilibrium ($I^* = \nu/(\gamma + \mu) - \mu/\beta$; see Chapter 2), we see that $\mu p/\beta$ unvaccinated individuals are saved from infection due to the herd-immunity effects.

Additionally, the stochastic nature of disease transmission can contribute to the beneficial effects of limited vaccination. Consider equation (8.1); in terms of the transmission process, a vaccinated population is similar to a smaller population (of size $(1-p)N$) with a reduced basic reproductive ratio. The effective reduction in R_0 leads not only to lower prevalence, as outlined above; this also enhances the effects of stochasticity and can lead to chance extinction. Thus, even when vaccination does not exceed the deterministic threshold, eradication may occur; the chain of transmission can be broken by chance (see Section 6.3.3.3). However, because the level of control is below the critical threshold, the disease can re-invade after a stochastic extinction, leading to subsequent epidemics.

These principles are illustrated in Figure 8.2. The straight gray line shows how the number of infecteds within the population can be reduced even when vaccinating below the eradication threshold. The black lines are the average time to extinction, taken as the time from the onset of vaccination (when the population is assumed to be at the unvaccinated equilibrium) until the infection undergoes stochastic extinction and is eradicated from the population. This extinction time illustrates three important points: First, when vaccinating at birth, it always takes time to eradicate an infectious disease, even when the vaccination is well above the threshold p_c. Second, eradication can occur below the vaccination threshold due to either stochastic effects or the deep trough in the number of infecteds that accompanies the sudden onset of vaccination (the "honeymoon period," as termed by McLean 1995). Finally, population size plays an overwhelming role; it is only for the very highest levels of vaccination that the large (one million individuals) population is driven extinct as rapidly as the unvaccinated small (100,000 individuals) population.

One of the obvious consequences of the reduced frequency of disease transmission is that those (unvaccinated) individuals who eventually contract the infection are likely to be older than in the absence of vaccination. It is straightforward to derive an expression for the mean age at infection in the SIR model with vaccination (A') as a function of the mean age at infection for an unvaccinated population (A):

$$A' = A/(1-p).$$

As we discussed in Chapter 5 an important implication of this result is that we can produce a single bifurcation diagram that summarizes the *dynamics* of an infectious disease as vaccination or demographic rates systematically vary. This means that we may think of a system under vaccination as dynamically equivalent to a system with a proportionately reduced R_0. As mentioned in Chapter 5, this argument may be somewhat complicated by

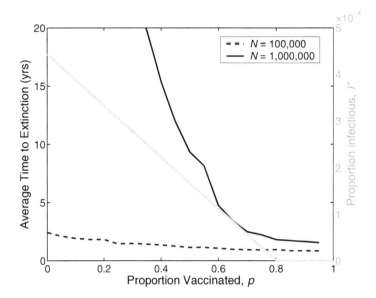

Figure 8.2. The effects of vaccinating a proportion p of the population at birth in a stochastic model. We consider host population sizes of 10^5 and 10^6, and start the simulations with the population at the unvaccinated equilibrium point. The gray line shows the mean prevalence of infection after vaccination, as predicted by equation (8.5). The black curve shows the average time to "extinction" (also called the "first passage time") measured from the start of vaccination. (Model parameters were $\mu = 0.02$ per year, $R_0 = 5$, $1/\gamma = 5$ days.)

changes in the strength of seasonality in transmission because the mean age at infection increases following immunization.

8.1.2. Wildlife Vaccination

There are many instances, especially for wildlife diseases or perhaps when vaccine boosters are necessary, where control by vaccination means targeting the entire susceptible pool and not just the newborns. This can occur through distributing feed containing vaccine (e.g., to control rabies in foxes), or administering vaccines (e.g., to control distemper in domestic dog populations). In such cases, we model the random vaccination of any member of the population (irrespective of disease status), although it is only the vaccination of susceptible individuals that has any effect. The vaccination parameter, v, now necessarily becomes a *rate* rather than a *fraction* and we are concerned with the proportion of the susceptible population immunized per unit time. The changes in the mathematical equations describing this scenario are small:

$$\frac{dS}{dt} = \mu - (\beta I + \mu - v)S,$$

$$\frac{dI}{dt} = \beta S I - (\gamma + \mu)I, \qquad (8.6)$$

$$\frac{dR}{dt} = \gamma I + v S - \mu R.$$

This is online program 8.2

The dynamical consequences of this introduction are qualitatively minor. The system still possesses two equilibria, one that is disease free and another with the infection endemic. The primary difference is that we now need to evaluate the vaccination *rate* required to eliminate the infection. After some algebra, we find the critical rate of vaccination, $v_c = \mu(R_0 - 1)$. This criterion is clearly different in structure to that derived from equation (8.2). Here, v_c increases linearly with R_0 rather than the previously concave relationship. In actuality, however, these two thresholds (p_c and v_c) lead to the same fraction of the population needing to be vaccinated in order to eliminate the infection. At the critical threshold, a fraction $v_c S^*$ are vaccinated daily; substituting for the values gives $\frac{1}{R_0} \times \mu(R_0 - 1)$, which simplifies to $\mu(1 - 1/R_0) = \mu p_c$, as derived in Section 8.1.1. Therefore, these two vaccination schemes are equivalent in terms of the *numbers* of susceptible hosts who need to be immunized. The key practical difference, however, lies in the fact that wildlife vaccination assumes that the fraction of the population susceptible to infection (given by $1/R_0$) cannot be unambiguously identified and therefore vaccination effort is spread across the entire population—even though it is effective only for susceptible animals. For this reason, regulating an infectious disease by reducing the recruitment of individuals susceptible to it (equation (8.1)) may be perhaps easier than attempting to immunize the susceptible population (equation (8.6)).

8.1.3. Random Mass Vaccination

For rare, nonendemic pathogens, continual vaccination at birth is not a cost-effective control measure. Instead, a mass-vaccination program may be initiated whenever there is increased risk of an epidemic. In such situations there is a "race" between the exponential increase of the epidemic, and the logistical constraints upon mass-vaccination. For most human diseases it is possible (and more efficient) to record who has been vaccinated, and only immunize those who have not received the vaccine—an even more refined approach would not vaccinate those individuals who have recovered from the disease because they are already protected. We take as our most simple model:

$$\frac{dS}{dt} = -\beta SI - u,$$

$$\frac{dI}{dt} = \beta SI - \gamma I, \qquad (8.7)$$

$$\frac{dR}{dt} = gI,$$

$$\frac{dV}{dt} = u,$$

where demographics have been ignored because we are primarily interested in the short-term response to an emerging epidemic or pandemic. We obviously insist that vaccination stops once the number of susceptibles reaches zero. Two extremes of this model can be considered. When u is small, vaccination will have little impact on the epidemic and a proportion R_∞ of the population will be infected ($R_\infty = 1 - \exp(-R_0 R_\infty)$; Chapter 2). At the other extreme, when u is large we can use the approximation $S(t) \approx max(S(0) - ut, 0)$, which assumes that the level of susceptibles is decreased by vaccination although the impact of infection on the level of susceptibles is insignificant and ignored. This is a reasonable assumption if the rate of vaccination is sufficient to control the outbreak. Under these assumptions, the number of infectious cases is given by:

$$I(t) \approx \begin{cases} I(0)\exp\left(\left[\beta S(0) - \gamma - \frac{1}{2}\beta ut\right]t\right) & t \le \dfrac{S(0)}{u} \\ I(0)\exp\left(\frac{1}{2}\beta S(0)^2/u - \gamma t\right). & \text{otherwise} \end{cases}$$

Here, the fraction of infecteds follows a Gaussian curve, and the initial disease prevalence at the onset of immunization ($I(0)$) determines the scale of the ensuing epidemic. This conclusion echoes a broad tenet in epidemiology, that the best way to control an epidemic is to hit it hard and hit it early—a strong response leads to the fastest reduction in the susceptible population, which in turn reduces the epidemic, and a rapid response prevents the exponential increase of cases from getting beyond logistical control.

8.1.4. Imperfect Vaccines and Boosting

Despite the effectiveness of vaccines in dramatically reducing the number of new infectious cases (and the severity of illness), the resurgence and epidemic outbreaks of some infectious diseases are considered to be of major public health concern (Orenstein et al. 2004). Among childhood infections, measles is a well-known candidate for such outbreaks and still contributes to over one million deaths annually, mostly among children in developing countries. Clinical studies have proposed several potential explanations, including decreased immunization coverage together with irregularities in the supply of vaccines, incomplete protection conferred by imperfect vaccines, and the loss of vaccine-induced immunity (Garly and Aaby 2003; Janaszek et al. 2003).

To prevent an endemic spread of measles infection, many countries, mostly in the developed world, have revised their vaccination programs to include multiple schedules. The reported clinical data using the strategy of a booster MMR (measles-mumps-rubella) vaccine confirm that these countries have generally succeeded in controlling the spread of infection. Hence, in order to achieve a global eradication, the World Health Organization recommends a booster vaccination program worldwide. The central question to ask is whether this strategy could eventually provide the conditions for global eradication. To address this question, we can develop a framework, modified from the $SEIR$ equations, that would predict the consequences of the introduction of a booster schedule, in terms of the known major factors associated with a vaccination program (McLean and Blower 1993; Gandon et al. 2001, 2003).

The model we present is due to Alexander et al. (2006) and is composed of four distinct classes: Susceptible (S), Vaccinated (S_v), Infectious (I), and Booster vaccinated (or recovered) individuals (V) who are immune for life. It accounts for two major aspects of an imperfect vaccine: (1) incomplete protection, and (2) waning of vaccine-induced immunity. The first may result in the subsequent infection of the pediatric-vaccinated class, perhaps at a lower rate than that of the fully susceptible class. The second leads to an increase in the size of the fully susceptible pool through the loss of vaccine-induced immunity. The model also assumes that, like the natural immunity induced by the infection, the booster vaccine administered to the class of pediatric-vaccinated individuals confers complete protection against the disease. The system can be mathematically expressed by the following system of differential equations:

$$\frac{dS}{dt} = (1-p)\mu - \beta SI - \mu S - \xi S + \delta S_v, \tag{8.8}$$

$$\frac{dS_v}{dt} = p\mu + \xi S - (1-\alpha)\beta S_v I - (\mu + \rho + \delta)S_v, \tag{8.9}$$

$$\frac{dI}{dt} = \beta S I + (1-\alpha)\beta S_v I - (\mu + \gamma)I, \tag{8.10}$$

$$\frac{dV}{dt} = \rho S_v + \gamma I - \mu V, \tag{8.11}$$

where p is the fraction of newborns who receive the pediatric vaccine, α represents the efficacy of the vaccine in terms of reducing the susceptibility of (singly) vaccinated individuals, δ is the waning rate following pediatric vaccination, $1/\gamma$ is the infectious period, μ is the natural death rate, and ρ and ξ are the rates of administration of the booster vaccine to previously vaccinated and susceptible individuals, respectively.

As demonstrated in Chapter 3, by studying the dominant eigenvalues of these equations, we can derive the effective reproductive ratio, r_0, which is defined here as the number of secondary cases resulting from a single index case given the specified vaccination regime (Alexander et al. 2006). We get the following expression:

$$r_0 = \frac{\mu[\delta + (1-p)(\mu+\rho) + (\mu p + \xi)(1-\alpha)]\beta}{(\mu+\gamma)[(\mu+\xi)(\mu+\rho) + \mu\delta]}. \tag{8.12}$$

Naturally, there is significant public health interest to ensure control parameters that would make eradication feasible by reducing r_0 below unity. An increase in μ, δ, or β—which relate to more susceptibles entering the population, a decrease in the mean duration of vaccine-induced immunity, and a higher transmission rate respectively —can all be offset by a higher level of pediatric vaccination. It is useful to rewrite equation (8.12) in terms of the basic reproductive ratio for a population that is wholly susceptible, with no vaccination (R_0). This gives:

$$r_0 = \left(1 - \frac{(\mu p + \xi)(\rho + \mu\alpha)}{(\mu+\rho)(\mu+\xi) + \mu\delta}\right) R_0, \tag{8.13}$$

where, as before, $R_0 = \beta/(\mu+\gamma)$. Clearly, a high value of R_0 requires a high coverage level of pediatric vaccination, p, to prevent the spread of the infectious disease, regardless of the type of vaccine being administered. However, it is practically unfeasible to vaccinate all individuals in the susceptible class (p is always significantly less than 1), particularly in countries where finances play a major role in the number of people who receive the vaccines. Hence, the next best strategy is to determine the critical number needed to be vaccinated and try to achieve this value.

It is instructive to establish the minimum pediatric vaccination level that is required to eliminate the infectious disease in the absence of boosters ($\rho = \xi = 0$)—the equivalent to the standard vaccination model but with waning immunity and partial protection. This is given by:

$$p_c = \left(1 - \frac{1}{R_0}\right)\left(\frac{\mu+\delta}{\mu\alpha}\right), \tag{8.14}$$

such that $r_0 \leq 1$ whenever $p \geq p_c$. Not surprisingly, this threshold reduces to $p_c = 1 - 1/R_0$ for a perfect vaccine ($\alpha = 1$, $\delta = 0$). The most important implication of this result is that eradication may be impossible to achieve once the basic reproductive ratio, R_0, is

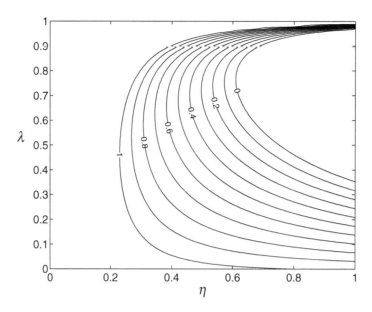

Figure 8.3. Contour plots for various values of critical pediatric vaccination coverage, p_c, as the total rate of booster vaccination, η, and the proportion of booster vaccination given to unvaccinated individuals, λ. These two quantities can be related to those in equations (8.8)–(8.11) as $\xi = \lambda\eta$ and $\rho = (1 - \lambda)\eta$. The critical values of p are found by setting r_0 in equation (8.12) to one. (Parameter values are: $1/\mu = 50$ years, $1/\delta = 20$ years, $\alpha = 0.95$, and $1/\gamma = 14$ days.)

greater than 2. Consider the optimistic case in which the pediatric vaccine provides perfect immunity to infection ($\alpha = 1$), but where protection wanes through time ($\delta > 0$). In this scenario, equation (8.14) means that the critical proportion of the population required to be vaccinated becomes greater than 1 ($p_c \geq 1$), unless the ratio of life expectancy to the period of protection (($\mu + \delta)/\mu$) is less than $R_0/(R_0 - 1)$. As a result, for a pathogen with effective $R_0 = 3$, this result effectively means that eradication requires the period of protection to last for at least 2/3 the duration of life—hence the need for booster vaccination.

We now introduce a new parameter η as the rate of total booster administration, and let $\xi = \lambda\eta$ and $\rho = (1 - \lambda)\eta$ where $0 \leq \lambda \leq 1$. (Note that changing λ but keeping η constant does not imply the vaccination of a constant number of individuals, but a constant vaccination effort partitioned between the two classes.) Let $p_c(\eta, \lambda)$ represent the curve on which $r_0 \equiv 1$ and therefore the minimum level of pediatric vaccination needed to eradicate the infection. Examining the control of an infection such as measles, we set $R_0 = 17$ and show in Figure 8.3 contour plots of $p_c(\eta, \lambda)$ (for feasible ranges of η and λ). For each p_c, there is a critical value η_p (corresponding to a vertical tangent to p_c in Figure 8.3) such that disease control is not feasible if $\eta < \eta_p$. However, for $\eta > \eta_p$, there is a range of λ for which $r_0 < 1$ and the disease can be eradicated. Decreasing the pediatric vaccination coverage p makes the feasible range of λ shrink, with the lower limit of the range showing the greatest change. An important epidemiological consequence of this result is that, for relatively low vaccine coverage (p), a booster program may fail to control the disease if it is mostly targeted to primary vaccinated individuals (λ is too low). The same conclusion can be derived when λ is too high and the booster in effect functions mostly as primary vaccination. More important, the probability of failure of a booster program increases as pediatric vaccination coverage p decreases, leading to a more restricted range of λ for disease control. This highlights the significant role that primary coverage plays in ensuring a successful booster program (Alexander et al. 2006).

In this section, we have considered the impact of imperfect vaccines in reducing the basic reproductive number. We have used a slightly modified SIR framework to evaluate the effect of a booster dose of an imperfect vaccine, to supplement primary vaccination, in reducing R_0. Our findings highlight two important epidemiological messages:

In high incidence areas, eradication with a single dose of imperfect vaccine will be impossible unless the mean period of protection afforded by the vaccine is greater than the average life expectancy multiplied by $1 - 1/R_0$**.**

Eradication by administering a booster vaccination is feasible only if a minimum rate is achieved.

In addition, infections that require a booster vaccine highlight an important role for models in public health. For a perfect vaccine, the obvious public health initiative is to vaccinate as many individuals as possible; models do little to enhance this response other than define the threshold level—although vaccination at levels above the threshold is still desirable because it speeds eradication. In contrast, when booster vaccines are required there is a clear trade-off in the partitioning of resources between primary and booster vaccination campaigns. Models can help to optimize this trade-off.

If such models were to be used in a public health context, then we would potentially require far greater realism. In particular, it may be necessary to include age structure into the model formulation as well as introduce a more refined description of the time scale of waning immunity (Chapter 3).

8.1.5. Pulse Vaccination

The approaches outlined above are based on the immunization of newborns or relatively young individuals and although the details are population and disease specific, theory generally predicts successful eradication if vaccine-induced immunity levels exceed 70–95% (see, for example, Figure 8.1). Achieving such high levels of coverage presents a daunting challenge, however, especially in the face of financial, political, and logistical difficulties.

An alternative approach that has been tried in a variety of locations is pulse vaccination, where children in certain age cohorts are periodically immunized (Agur et al. 1993; Nokes and Swinton 1997; Shulgin et al. 1998). The rationale for this idea is as follows: The prevalence of an infectious disease increases only if $S \times R_0 > 1$, where S is the fraction of the population that is susceptible. Hence, as demonstrated earlier in this chapter, for any specified infectious agent, there is a critical proportion of susceptibles, $S_c = 1/R_0$, below which spread is unlikely. The principle aim of pulse vaccination is to ensure the susceptible fraction is maintained below this level by periodically immunizing a fraction of the susceptible population. These ideas can be best illustrated by example. Assuming a disease has an R_0 of 10, then, in the long run, 10% of the population will be susceptible on average. If a percentage $p_V = 60\%$ of susceptibles are vaccinated in a single pulse, then only 4% of the population will remain susceptible. Given a per capita birth rate (ν) of 2% per year and a constant population size, it will take 3 years for S to reach 10%, thus providing a crude approximation to the period between pulses (Nokes and Anderson 1988). Pulse vaccination has gained in prominence as a result of its highly successful application in the field. Compared to "continual" pediatric vaccination, it has the additional advantage that it is often logistically simpler to implement. A well-publicized example is the spectacular control of poliomyelitis and measles in Central and South America (Sabin, 1991; de Quadros et al. 1996).

The theoretical challenge of pulse vaccination is the a priori determination of the pulse interval for specified values of R_0, the vaccination fraction p_V, and population birth rate μ. Simulation methods used by Agur et al. (1993) demonstrated that the optimal pulsing period was approximately equal to the mean age at infection. To make this more rigorous, Shulgin et al. (1998) examined the following refinement of the SIR system:

$$\frac{dS}{dt} = \mu - \beta SI - \mu S - p_V \sum_{n=0}^{\infty} S(nT^-)\delta(t-nT), \tag{8.15}$$

$$\frac{dI}{dt} = \beta SI - \gamma I - \mu I, \tag{8.16}$$

This is online program 8.3

where T is the interval between pulses and $S(nT^-)$ represents the level of susceptibles in the instant immediately prior to the nth vaccination pulse ($n = 0, 1, \ldots$). $\delta(t)$ is the Dirac delta function which is zero unless $t = 0$; hence, the final term in equation (8.15) ensures that the number of susceptibles is reduced by a fraction p_V at every vaccination pulse. We can now establish the optimal pulsing period by studying the disease-free equilibrium and the conditions required for its stability. Leaving aside the details of this stability analysis, which are complicated by the pulses, we arrive at the optimal gap between pulses to ensure eradication as determined by Shulgin et al. (1998):

$$\frac{(\mu T - p_V)(e^{\mu T} - 1) + \mu p_V T}{\mu T(p_V - 1 + e^{\mu T})} < \frac{1}{R_0}. \tag{8.17}$$

This equation can be solved numerically in order to obtain the threshold value of T_c. In Figure 8.4, we plot the value of T_c for three different R_0 values as p_V is varied. It demonstrates that the pulse interval increases with the vaccination fraction and declining R_0. It also shows that in contrast to the numerically obtained estimates by Agur et al. (1993), the values of T_c can exceed the mean age at infection A. The above approximation has been shown to remain good for the $SEIR$ model (d'Onofrio, 2002).

If, as is the case with childhood infections, transmission varies substantially depending on season (see Chapter 5), then we set $\beta(t) = \beta_0(1 + \beta_1 \cos(2\pi t + \psi_0))$, where ψ_0 represents the phase of seasonal transmission for the susceptible population assuming the year starts after a vaccination pulse. Taking this framework, Shulgin et al. (1998) again used stability analysis to show that T_c satisfies:

$$\frac{((\mu T - p_V)(e^{\mu T} - 1) + \mu p_V T)}{\mu T(p_V - 1 + e^{\mu T})} + \frac{\beta_1 p_V}{T} \frac{(2\pi \sin(\psi_0) - \mu \cos(\psi_0))}{(4\pi^2 + \mu^2)(p_V - 1 + e^{\mu T})} = \frac{1}{R_0}. \tag{8.18}$$

Clearly, the first term in the inequality is simply the unforced case given by inequality (8.17). What we need to examine, therefore, is how much the second term alters the predictions of the constant transmission model. Upon substituting conventional measles parameter values into the above two terms, we find the ratio between the first and the second to be approximately $50:1$. Hence, the optimal timing of pulse vaccination is not substantially altered when seasonality is taken into account. It is interesting to note that T_c is affected by the phase shift, with a maximum at $\psi_0 \cong 3\pi/2$ for parameter values

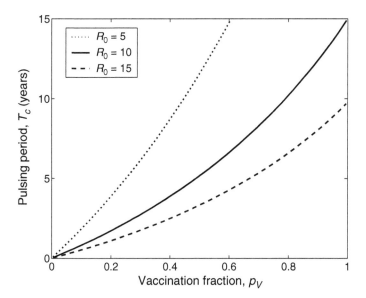

Figure 8.4. The threshold pulsing period (T_c) as a function of the vaccination proportion (p_V) for three different basic reproductive ratios. The curves were generated assuming $1/\mu = 70$ years ($\mu \approx 0.0143$).

representative of measles dynamics. Hence, pulses of vaccination should ideally be applied 3 months after the peak in the seasonal transmission rate $\beta(t)$.

In most countries, pediatric immunization programs are already established and any attempt at pulse vaccination is likely to be *in addition to* constant background vaccination (with probability p) rather than an alternative. Shulgin et al. (1998) demonstrated that the introduction of an additional term to equation (8.15) to take this component into account changes T_c in a very predictable way. The new optimal pulsing period is simply the previously derived value (T_c) divided by $(1 - p)$.

An additional advantage of pulse vaccination becomes apparent when infection is considered in a spatial context (see Section 7.2.3 and Figure 7.5 in Chapter 7). We first note that in England and Wales there was little noticeable decrease in measles persistence following the onset of vaccination at around 60% in the late 1960s—in direct contrast to the predictions from simple mathematical models. It is hypothesized that this surprising observation is due to greater asynchrony of epidemic cycles between communities after vaccination, which in turn means that recolonisation of disease-free populations is enhanced. Pulse vaccination may counter this effect, because the regular decrease in susceptibles due to vaccination may act to synchronize epidemics such that all communities experience troughs at similar times, reducing the chance of recolonization (Earn et al. 1998).

8.1.6. Age-Structured Vaccination

The above models have all ignored the effects of age structure on the vaccination policy. Here, we use the simplest transmission assumption, that the force of infection is independent of age, to consider the optimal age of vaccination.

Consider the scenario when a proportion p of the population are vaccinated at age A_v. The equations for the age-structured population are best considered as two separate sets of identical equations but with different boundary conditions:

$$\frac{\partial S(a)}{\partial t} = -\beta S(a)\widehat{I} - \mu S(a) - \frac{\partial S(a)}{\partial a} \qquad (8.19)$$

$$\frac{\partial I(a)}{\partial t} = \beta S(a)\widehat{I} - \gamma I(a) - \mu I(a) - \frac{\partial I(a)}{\partial a},$$

where $\widehat{I} = \int_0^\infty I(a)da$ is the total prevalence in the population, and the time dependence of the variables is implicit. The boundary conditions are $S(0) = \mu$, $I(0) = 0$ at birth, and $S(A_v+) = (1-p)S(A_v-)$, $I(A_v+) = I(A_v-)$ such that there is an instantaneous reduction in susceptibles at age A_v. If we now look for the equilibrium solutions we find that:

$$S^*(a) = \mu \begin{cases} \exp(-(\beta\widehat{I}^* + \mu)a) & \text{for } a < A_v \\ (1-p)\exp(-(\beta\widehat{I}^* + \mu)a) & \text{for } a > A_v \end{cases} \qquad (8.20)$$

and:

$$\frac{d\widehat{I}}{dt} = \beta \int_0^\infty S(a)da\,\widehat{I} - \gamma\widehat{I} - \mu\widehat{I} \qquad (8.21)$$

$$\Rightarrow \int_0^\infty S(a)da = \frac{\gamma + \mu}{\beta}$$

$$\Rightarrow \frac{1 - p\exp(-[\beta\widehat{I}^* + \mu]A_v)}{\beta\widehat{I}^* + \mu} = \frac{\gamma + \mu}{\beta\mu}. \qquad (8.22)$$

It should be obvious from biological principles, if not from equation (8.22), that prevalence, $\widehat{I}^*$, is minimized when $A_v = 0$ and individuals are vaccinated at birth. In such situations protection is offered for the greatest time span possible, reducing the risk of infection to a minimum. When A_v is nonzero, the equilibrium prevalence is more complex to calculate and must be solved numerically (Figure 8.5). For some infections, such as measles, newborns are afforded protection from maternally derived antibodies, as a result of which any effective vaccination must take place after this period of immunity has lapsed. Thus, current MMR (measles, mumps, and rubella) vaccination schedules in most developed nations take place at 12–15 months. This consideration does, however, raise a difficult problem in the developing world where the mean age at infection tends to be very young, leaving a very small window during which infants can be successfully vaccinated before experiencing substantial exposure to infection (McLean and Anderson 1988a,b).

8.1.6.1. Application: Rubella Vaccination

For some diseases, rubella in humans being a prime example, the number of cases within the population as a whole is largely irrelevant; instead, public health policies are concerned with minimizing the number of cases in the most vulnerable classes of the population. Generally, rubella is a benign infection with mild flu-like symptoms—a healthy person typically recovers from infection and may not even realize they have been sick. However, if caught while pregnant the infection can have severe consequences. Infection in the

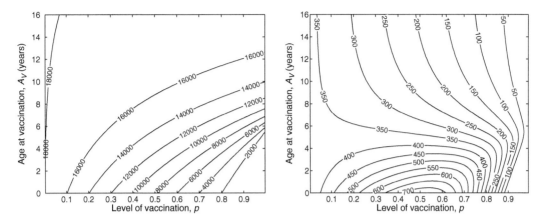

Figure 8.5. The left-hand graph shows the total number of annual cases, $365\gamma \widehat{I}^{*}$, (in a population of one million), which is determined by solving the age-structured vaccination equation (8.22). Clearly, large amounts of early vaccination are necessary to reduce the prevalence. The right-hand figure shows the number of annual cases of rubella (again in a population of one million) in woman of child-bearing age, which are found from equation (8.23). ($1/\mu = 70$ years, $R_0 = 10$, $1/\gamma = 10$ days.)

first trimester may result in miscarriage, stillbirth, or premature birth. Indeed, one out of four babies infected during the first trimester is born with Congenital Rubella Syndrome, which may result in a multitude of birth defects including mental retardation, cataracts, deafness, cardiac complications, or bone lesions (Behrman and Kliegman 1998). From a public health perspective, we are interested in vaccination strategies that minimize C^{*}, the equilibrium number of cases in child-bearing women:

$$C^{*} = \int_{0}^{\infty} \beta S(a)\widehat{I}^{*} P(a)da,$$

where $P(a)$ is the probability that an individual of age a is a pregnant woman.

We can simplify the calculation still further by minimizing the number of cases in women of child-bearing age (say between 16 and 40):

$$C^{*} = \frac{1}{2}\beta\widehat{I}^{*} \int_{16}^{40} S(a)da, \qquad (8.23)$$

where the $\frac{1}{2}$ is due to the fact that women comprise approximately half of the total population size. Again this value has no explicit expression, but Figure 8.5 shows how the number of cases in the at-risk age group changes as the timing and level of vaccination varies.

Three distinct results emerge. When vaccinating before age 5, intermediate levels of vaccination can increase the number of problematical cases. For most levels of vaccination, it is always better to vaccinate late rather than early. However, for the very highest levels of vaccination (greater than 80%), early immunization more readily limits the impact of the disease.

When the severity of an infection is age-dependent, the interplay between the level of vaccination and the average age of infection can lead to counter-intuitive results where, although moderate amounts of vaccination reduce prevalence they can lead to an increase in disease.

Although such models provide an important understanding of the implications of vaccination when there is a strong age-dependence on the severity of infection, several additional elements would need to be considered if this were to be used for planning public health policy. In particular, the model used above assumes random mixing between age groups, however as was shown in Chapter 3, the differential and assortative mixing between age groups can have a significant impact on the age-structured susceptibility profile, $S(a)$. Additionally, immunization against rubella usually involves a booster vaccine, hence models of the type used in Section 8.1.4 become necessary. Finally, there is the question of targeting the vaccine by gender. Again, this is a complex trade-off; vaccinating girls only makes the best use of a limited vaccine supply but vaccinating both boys and girls has the greatest potential to achieve herd immunity.

8.1.7. Targeted Vaccination

There are many ways in which vaccination, and control in general, can be targeted. The above example showed how targeting of particular age classes may be an effective and efficient control strategy. Other methods are generally based upon protecting those elements of the population that are most at risk. We deal with three specific examples: (1) prophylactic vaccination of individuals based upon their at-risk status, (2) vaccination of individuals identified by contact tracing from sources of infection, and (3) ring vaccination around infectious sources, which has been advocated as a means of controlling livestock diseases.

Consider a population that can be decomposed into two classes, high-risk and low-risk individuals; following the work in Chapter 3, we can model these two classes:

$$\frac{dS_H}{dt} = \mu_H - (\beta_{HH} I_H + \beta_{HL} I_L) S_H - \mu S_H,$$

$$\frac{dI_H}{dt} = (\beta_{HH} I_H + \beta_{HL} I_L) S_H - \gamma I_H - \mu I_H,$$

This is online program 8.4

where subscripts denote the high- and low-risk groups, and a similar equation exists for the susceptibles and infecteds in the low-risk group. (Although the risk-structured models of Chapter 3 were focused toward infections with SIS dynamics, here we assume SIR dynamics but note that both models have the same R_0 and invasion criteria.) For most plausible mixing matrices, β, we generally find that if the risk structure is ignored, then the estimated value of R_0 is too low, and consequently we underestimate the level of control needed. Once the true value of R_0 is found, either by creating the correct model or more likely by analysis of case reports, then we can again determine a critical level of vaccination that protects the population. Quite surprisingly, if we simply vaccinate at random, then we retain the familiar threshold for the proportion of the population that must be immunized:

$$p_c = 1 - \frac{1}{R_0}.$$

In general, however, we can improve on this threshold. Mathematically, we need to find a level of vaccination for each class, p_H and p_L, such that the effective reproductive ratio, R, is one and the amount of vaccine administered is minimized. (This framework can be generalized to multiple risk groups although the calculations become increasingly difficult.) Unfortunately, an analytical approach to this problem is complex and does not produce any additional understanding. Instead, a numerical approach is required and

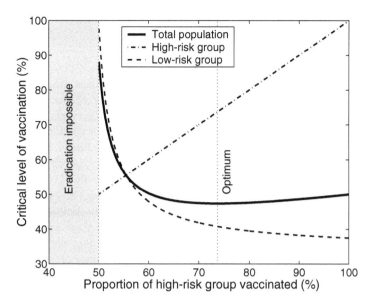

Figure 8.6. The critical level of vaccination needed for eradication, as a percentage of the entire population, for a range of coverage in the high-risk group. In this example, the transmission matrix is $\beta = \begin{pmatrix} 1 & 0.1 \\ 0.1 & 0.2 \end{pmatrix}$ per day, with the proportion of the population in the high-risk and low-risk groups being $N_H = 0.2$ and $N_L = 0.8$, respectively. We take $1/\gamma = 10$ days. For each level of vaccination within the high-risk group, we have searched for the corresponding level of vaccination in the low-risk group that sets the effective reproductive ratio, R, equal to one.

although this cannot produce any rigorous results, it can provide general insights. The most notable is that we can often do much better than the random vaccination threshold when we focus more of our vaccination on the high-risk group. The differences with the random vaccination scenario are most pronounced when a small high-risk group dominates the invasive dynamics. Additionally, although focusing too much vaccination effort on the high-risk group is suboptimal, the penalties are small in comparison to focusing too little effort on this group. Figure 8.6 shows the optimal distribution of vaccine for a given transmission matrix (see Section 3.1.14). For this disease and population structure it is impossible to eradicate the infection unless 50% of the high-risk group are vaccinated. The optimal vaccination policy is to target around 75% of the high-risk group and only 40% of the low-risk group; this bias reflects the relative importance of the two groups in the transmission of infection.

For a general assortative transmission matrix, it is usually better to target vaccination toward the higher-risk groups. Biasing control too much toward high-risk individuals is generally less problematic than biasing control toward low-risk individuals.

Two extreme mixing scenarios are worth further mention, because they bound the standard vaccination approaches. First, consider an infection where the mixing is at random, but proportional to the risk of each class (see the random partnership model, Chapter 3, Section 3.1.1.6). In particular we require the matrix, β, to be the product of an exposure (or susceptibility) vector, σ, and transmission vector, τ:

$$\beta_{XY} = \sigma_X \tau_Y$$

and that greater exposure is associated with greater transmission potential ($\sigma_X > \sigma_Y \Rightarrow \tau_X > \tau_Y$). In essence, we simply need that the higher-risk groups are both more at risk of catching the infection and, once infected, have a greater chance of spreading the disease. This is often the case with sexually transmitted infections where both quantities are proportional to the number of partners; however, it may also hold for airborne infections where transmission and susceptibility could scale with the number of social contacts or the immune status of an individual. In such circumstances, vaccine should initially be exclusively targeted at the high-risk group and only once this group is completely immunized should attention switch to the rest of the population. The intuition behind this optimal policy is as follows: Suppose that at any point in the vaccination campaign we have an extra dose that can be administered; it is always better to vaccinate a high-risk individual because, irrespective of the state of the population, they are both most at risk and have the greatest chance of spreading infection—thus immunization of this individual has the greatest effect.

If individuals interact at random but proportional to the risk of each class, then the optimal strategy is to concentrate all vaccination efforts toward the highest-risk individuals.

An alternative transmission matrix is formed when the mixing is completely assortative, so that transmission occurs only within a class and there is no transmission between classes.

$$\beta_{XY} = \begin{cases} \beta_{XX} \geq 0 & \text{if } X = Y \\ 0 & \text{otherwise.} \end{cases}$$

In this case, it is clear that the highest-risk class should be vaccinated first to the point where infection will be eradicated $p_X = 1 - \frac{\gamma}{\beta_{XX}}$, before vaccinating lower-risk groups. This provides the most rapid decrease in R_0 and also a marginally faster reduction in the number of cases.

When mixing is completely assortative it is optimal to vaccinate each class at its eradication threshold.

These two extreme cases provide an understanding of the general rules for targeting vaccination between risk groups—as mixing changes from random to more assortative, the degree of targeting decreases. However, even for completely assortative mixing the vaccine is still targeted due to the higher eradication threshold in the higher-risk groups. Both of these optimally targeted strategies perform significantly better than random vaccination.

8.2. CONTACT TRACING AND ISOLATION

Isolation or quarantining of individuals provides one of the oldest, yet most effective, means of disease control. Although vaccination has to be targeted toward the large pool of susceptible individuals, isolation is focused toward those individuals who are infected—preventing them from further contact and subsequent transmission. There are also substantial microbiological challenges to the successful use of vaccines, especially when faced with emerging diseases. Vaccination requires that the infectious agent is quickly identified and a safe vaccine is produced and administered. As was seen with the SARS epidemic of 2003, these steps may take a frustratingly long period. In contrast, isolation is effective against any infectious disease even when the etiology is unknown. Finally, whereas there is often a delay between vaccination and immunity, isolation works instantaneously. Therefore, in the majority of outbreak scenarios, the rapid isolation of infected individuals (or those suspected of being infected) is a primary means of control.

The main disadvantage of isolation is that it can be difficult to rapidly identify infectious individuals, especially when symptoms are ambiguous or only emerge sometime after the individual becomes infectious (i.e., when the incubation period is much longer than the latent period). In such cases, contact tracing becomes vital. Contact tracing attempts to identify new cases by tracing all the potential transmission contacts of known infected individuals. Although this is a complex and labor-intensive process, it may be very effective if infectious individuals are rare within the general population (Eames and Keeling 2003).

8.2.1. Simple Isolation

Quarantine can be applied in many ways, but always involves isolating individuals. At the most simple level, infected individuals are quarantined as soon as they are diagnosed. Thus, whereas vaccination acts on susceptible individuals preventing them from becoming infected, quarantine acts by removing infectious individuals from the population, dramatically reducing their risk of transmission. This can be simply modeled by an effective decrease in the infectious period (or an increase in the recovery rate), adding a quarantine class Q to the standard SIR approach:

$$\frac{dS}{dt} = \mu - \beta SI - \mu S,$$

$$\frac{dI}{dt} = \beta SI - \gamma I - d_I I - \mu I,$$

$$\frac{dR}{dt} = \gamma I + \tau Q - \mu R,$$
(8.24)

$$\frac{dQ}{dt} = d_I I - \tau Q,$$

where d_I is the rate at which infectious individuals are detected and "removed" to quarantine in addition to the normal recovery rate, and $1/\tau$ is the average time spent in isolation. We assume that individuals leave the quarantine class, Q, only after they have recovered. This leads to a reproductive ratio of

$$R_Q = \frac{\beta}{\gamma + d_I + \mu},$$

so the critical isolation threshold that ensures $R_Q = 1$ is $d_I^* = \beta - \gamma - \mu$.

Contact tracing and isolation are powerful control methods that can be applied to any disease without needing any detailed knowledge of the infection.

Simple isolation of symptomatic individuals acts to reduce the effective length of the infectious period.

The above calculation assumes that the rates of transmission, recovery, and quarantining are independent of the time since infection. A more realistic formulation takes into account the temporally varying rates that are observed. Let us assume that $\beta(T)$ is the transmission rate and $s(T)$ is the probability that symptoms have been observed at time T since infection. Without quarantine, the basic reproductive ratio can be calculated as the infectivity over time:

$$R_0 = \int_0^\infty \beta(T) \exp(-\mu T) \, dT.$$

When quarantine is applied as soon as symptoms are detected, this reduces to:

$$\int_0^\infty \beta(T) \exp(-\mu T)(1 - s(T)) \, dT.$$

Hence, even when quarantining as fast as possible, if symptoms are displayed only late in the infectious period (the incubation period is much longer than the latent period), it may be impossible to control an infection by this method alone (Fraser et al. 2004).

We now return to equation (8.24) and consider the effects of logistical constraints on the isolation facility (Cooper et al. 2004). In practice, an isolation facility has a maximum capacity, Q_c, that it can accommodate; for example, Cooper et al. (2004) consider an isolation ward in a hospital in the context of an outbreak of Methicillin Resistant Staphylicoccus Aureus (MRSA), although similar considerations could apply locally or nationally for outbreaks of smallpox, SARS, or pandemic influenza. When the isolation facility is operating below capacity ($Q < Q_c$ or $d_I I < \tau Q_c$), then the dynamics are governed by equation (8.24). However, if infection exceeds capacity ($Q = Q_c$ and $d_I I > \tau Q_c$; the isolation facility is full and there is an excess of infection), newly detected infections can enter isolation only when someone leaves, leading to:

$$\frac{dS}{dt} = \mu - \beta S I - \mu S,$$

$$\frac{dI}{dt} = \beta S I - \gamma I - \tau Q_c - \mu I,$$

$$\frac{dR}{dt} = \gamma I + \tau Q_c - \mu R,$$

$$\frac{dQ}{dt} = \tau Q_c - \tau Q_c,$$

$$(8.25)$$

in which case the reproductive ratio depends on the current number of cases and is:

$$R_{Q_c} = \frac{\beta}{\gamma + \tau Q_c / I + \mu} > R_Q.$$

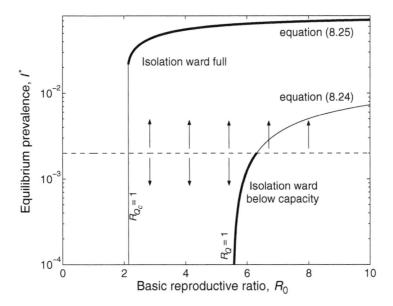

Figure 8.7. The equilibrium behavior of equations (8.24) and (8.25) as the unconstrained basic reproductive ratio $R_0 = \frac{\beta}{\gamma + \mu}$ varies. The dynamics are parameterized to reflect hospital-acquired infections, so that μ measures the rate at which individuals enter and leave the hospital. ($\mu = 0.01$ per day, $d_I = 0.5$ per day, $\gamma = 0.1$ per day, $\tau = 0.1$ per day, $Q_c = 0.01$ measured as a proportion of the hospital population size). Thick lines represent where solutions are stable; the dashed line $d_I I = \tau Q_c$ marks when the isolation ward becomes full and therefore is the boundary between attraction to the two solutions.

Given that there are now two plausible equations, (8.24) and (8.25), two equilibria are now possible (Figure 8.7): one when the isolation ward is below capacity

$$I_Q^* = \mu(R_Q - 1)/\beta,$$

and one when there is an excess of infection

$$\beta(\gamma + \mu)I_{Q_c}^{*}{}^2 + (\beta\tau Q_c + \mu(\gamma + \mu - \beta))I_{Q_c}^* - \mu\tau Q_c.$$

We observe that for some intermediate values of $R_0 (= \beta/(\gamma + \mu))$ the two solutions are both stable. This has a number of far-reaching implications. First, once the quarantine limit is reached the prevalence of infection dramatically increases, jumping from the lower stable solution to the higher stable solution. Second, once the quarantine capacity is reached it may be exceedingly difficult to regain spare capacity because the reproductive ratio, R_{Q_c}, may be much higher. Finally, the size of the isolation facility needed is governed by the prevalence predicted by equation (8.25), and not the normal lower level predicted by equation (8.24)—thus isolation facilities may need to far exceed the usual demand if they are not to succumb to catastrophic failure.

Where isolation or quarantining capacity is limited, bistability can occur. For the same parameters, when the isolation facility is full disease prevalence is high; however, when the facility has spare capacity the incidence of infection can be controlled—both of these solutions are stable.

8.2.2. Contact Tracing to Find Infection

Contact tracing, followed by testing and treatment, is a major weapon in the control of sexually transmitted-infections (STIs). In general, when an individual believes that they are showing symptoms of an STI, they visit their doctor, or a GUM (genitourinary medicine) clinic, where they are tested for the appropriate infection(s). If the test is positive, they are treated and asked for a list of recent (last 6–12 months) sexual partners. These sexual partners, therefore, have an increased likelihood of being infected compared to the general population, and are therefore sought to be tested. If these contacts prove to be infected, then treatment and further contact tracing follows; in this way contact tracing is recursive.

We now define a simple model that captures some of the essential elements of contact tracing for STIs that obey the *SIS* paradigm:

$$\frac{dS}{dt} = -\beta SI + \tau_T T,$$

$$\frac{dI}{dt} = \beta SI - d_I I - cbT, \qquad (8.26)$$

$$\frac{dT}{dt} = d_I I + cbT - \tau_T T,$$

where T is the class of individuals that are being treated and whose contacts are being traced, with $1/\tau_T$ being the average time in the treatment and tracing class. d_I is the rate at which infected individuals seek treatment and therefore plays the role of the recovery rate, c is the rate of contact tracing, and b is the probability that a traced individual is infectious. We are now faced with the difficulty of determining the value of b. Clearly, if the disease is highly infectious such that every contact leads to the transfer of infection, then $b = 1$; however, at the opposite extreme when contacts are no more likely to be infected than the general population, $b = I$. For simplicity, however, we can initially treat cb as a single parameter that defines the effectiveness of the tracing scheme.

Looking at the equilibrium behavior of equation (8.26) we observe that there are two conditions for disease persistence:

$$\tau_T > cb \qquad \text{and} \qquad \beta(\tau_T - cb) > d_I \tau_T.$$

The first of these ensures that contact tracing is not a runaway process and hence places a limit on the probability b; including iterative tracing, the total number of infected individuals that are traced as a result of an initial detection is $n_T = cb/(\tau_T - cb)$. The second criterion ensures that the basic reproductive ratio (in the presence of contact tracing) is greater than one. Given that these conditions are satisfied we find that:

$$I^* = \frac{\beta(\tau_T - cb) - d_I \tau_T}{\beta(d_I + \tau_T - cb)} = \frac{\tau_T}{\beta}\left(\frac{\beta - d_I(1 + n_T)}{\tau_T + d_I(1 + n_T)}\right),$$

and we can calculate the basic reproductive ratio by once again using the eigenvalue approach (see Chapter 3):

$$R_T = \frac{\beta}{d_I} - \frac{cb}{\tau_T - cb} = \frac{\beta}{d_I} - n_T,$$

which is equal to the uncontrolled R_0 (β/d_I) minus the number of individuals traced per infectious case. Therefore, if we wish to use contact tracing to eradicate infection, we need to insure that the total number of individuals traced as a result of each initial detection (n_T) is greater than the basic reproductive ratio of the uncontrolled system minus one ($n_T > \beta/d_I - 1$).

One problem with the above set of equations is that the parameter b is unknown, and will change during the course of an epidemic. Several methods can be used to tackle this problem: The methodology outlined in Section 8.3 below is ideal if individuals are generally contacted before they become infectious; alternatively, for STIs one can construct a network of sexual partnerships (see Chapter 3, Figure 3.5) and explicitly model the disease spread and tracing through this network; finally, it is possible to estimate b in terms of the number of infected-infected pairs within the network. This pairwise approach was adopted by Eames and Keeling (2003). By considering that tracing occurred at a much faster timescale than infection, they were able to show that in a range of scenarios the threshold for eradication was approximately that the proportion of contacts traced has to be equal to $(1 - 1/R_0)$—this has strong resonances to the above result and the threshold for vaccination. Intuitively, this ensures that at the threshold each infection causes only one new case that is not traced; however, it is somewhat surprising that this result holds true under such a wide range of heterogeneities.

Contact tracing (followed by isolation) can control an infectious disease if the tracing scheme detects at least all but one of the secondary cases caused.

8.3. CASE-STUDY: SMALLPOX, CONTACT TRACING, AND ISOLATION

Contact tracing is essentially a localized process, such that a complete model should account for the social structure of contacts (e.g., Halloran et al. 2002; Ferguson et al. 2003b; Ferguson et al. 2005); however, substantial progress can still be made using conventional differential equation models. Here we shall illustrate the principles involved with modeling the control of infection using contact tracing and isolation, using smallpox as an example disease, although the models are sufficiently general to be applicable in other scenarios. As mentioned earlier, a major contributing factor to the elimination of smallpox was the dynamic strategy of surveillance, containment, and vaccination of susceptibles in the immediate locality of an identified case (Fenner et al. 1988). The relative merits of these various control options versus prophylactic preemptive vaccination has been a major source of debate in the attempt to prepare potential target nations against any deliberate exposure to smallpox (Kaplan et al. 2002; Halloran et al. 2002; Ferguson et al. 2003b). The major problem relates to vaccine safety becase there are side effects and risks associated with the smallpox vaccine. Although most people experience mild reactions to receiving the vaccine, other people experience reactions ranging from serious to life-threatening. Historically, in a population of one million vaccinated individuals, approximately 1,000 people experienced reactions that, although not life-threatening, were serious (Fenner et al. 1988), and between 14 and 52 people experienced potentially life-threatening reactions to the vaccine, with 1 or 2 fatalities.

When faced with a potential reintroduction of smallpox, policy makers need to decide whether to preemptively mass-vaccinate (e.g., attempt to immunize all 296 million U.S. citizens) or try to locally control any possible outbreaks using the options summarized

TABLE 8.1.

Alternative options for managing any smallpox attack.

Policy	Benefits	Drawbacks
Quarantine and Isolation Sequester suspect and confirmed cases	Highly effective at reducing transmission	Requires rapid and high levels of compliance
Movement Restrictions Close schools, airports, and other transport	Potentially useful as shown during 2003 SARS epidemic	Very costly and difficult to implement
Ring Vaccination Trace and vaccinate contacts of suspected cases	Optimizes vaccine use and any associated complications	Logistically costly, and needs efficient contact tracing
Targeted Vaccination Immunize specific groups or neighborhoods	Effective locally; no contact tracing needed	Requires high levels of herd immunity
Mass Vaccination Vaccinate entire population under threat	Effective at widespread transmission control	Large numbers need to be rapidly vaccinated; expensive
Prophylactic Vaccination Vaccinate before introduction of disease	Useful to protect "first-responders"; Can prevent rapid spread	High long-term cost; numbers vaccinated

in Table 8.1. In this situation, models can provide useful predictions for choosing among competing options under a range of differing scenarios.

We start by considering an urban population of 1.25 million people, equivalent to a moderately sized city. According to standard theory (Figure 8.1) and assuming an R_0 of 5 (Gani and Leach 2001, Eichner and Dietz 2003), approximately 80% of the population (around 1 million individuals) would need to receive the smallpox vaccine in order to prevent an outbreak. [1] As mentioned above, we may expect serious complications in some 14–52 vaccinated people. The question is whether it would be safer to attempt to use surveillance and containment plus ring vaccination in order to manage any introduction, rather than accept the known complications associated with mass vaccination. Much of the decision in this scenario clearly rests on the perceived likelihood of such an introduction, though the following analyses would be of equal applicability when considering the emergence of novel pathogens.

Consider the simple SIR model, ignoring population demography because it is assumed that the time scale of any outbreak is substantially more rapid than births and background mortality. Although we will often be considering the control of an infection introduced into a totally susceptible population, it is worth noting that for smallpox, some of the

---

[1] In the absence of any precise new information, this argument assumes that any reintroduced smallpox virus is identical to the historically known pathogen. It is well established, however, that at least the former Soviet Union was active in attempting to combine genes of different viruses (notably smallpox and ebola, and smallpox and Venezuelan equine encephalomyelitis; Alibek 1999). Hence, it is plausible that any new strains of smallpox would bear little resemblance to the "speckled monster" seen in previous decades.

population may have residual immunity either due to past infection or earlier vaccination campaigns. To incorporate contact tracing and isolation, we assume that the traditional transmission parameter, β, is formed from the product of the number of contacts per unit time, k, and the probability of transmitting the disease, b. Again we assume that infectious individuals are detected (and isolated) at a rate d_I, and their contacts are sought. We model contact tracing by forcing a fraction q of those who have recently had contact with an infectious individual to be quarantined where they will spend an average $1/\tau_Q$ days. This occurs both for contacts that are not infected (which are quarantined as X_Q) and those who are infected (which are quarantined as Q)—importantly, we assume that these individuals are quarantined before they have a chance to generate any subsequent infection. Because of this latter assumption, contact tracing does not need to be recursive. This framework will subsequently be made more realistic and complex with the addition of delays in reacting to the epidemic and different distributions of latent and infectious periods, but the simple model will allow us to derive some general results. The equations for this model are given by:

$$\frac{dX}{dt} = -\frac{(kbY + qk(1-b)Y)X}{N} + \tau_Q X_Q, \qquad (8.27)$$

$$\frac{dX_Q}{dt} = \frac{qk(1-b)XY}{N} - \tau_Q X_Q, \qquad (8.28)$$

$$\frac{dY}{dt} = \frac{kbY(1-q)X}{N} - d_I Y - \gamma Y, \qquad (8.29)$$

$$\frac{dQ}{dt} = \frac{qkbXY}{N} + d_I Y - \tau_Q Q, \qquad (8.30)$$

$$\frac{dZ}{dt} = \gamma Y + \tau_Q Q. \qquad (8.31)$$

This is online program 8.5

where k is the contact rate and b is the probability of transmitting infection to a contact. The absence of host births in this system means that in the event of an infective introduction, the unstable disease-free equilibrium $(N, 0, 0, 0, 0)$ will give way to a brief outbreak, which eventually dies out with $(X_\infty, 0, 0, 0, Z_\infty)$ (see Chapter 2). Therefore, although the dynamics of this system may not be of particular interest, one important quantity is the number of undetected infectives, Y, that occur during the course of the epidemic. Effective infection management would keep this quantity small. We can calculate the conditions required for equation (8.29) to remain negative, which gives the intuitively appealing condition

$$\frac{X}{N} < \frac{X_c}{N} = \frac{(d_I + \gamma)}{kb(1-q)}. \qquad (8.32)$$

Efficient contact tracing of susceptibles who have come into contact with infectives (given by parameter q) and rapid isolation of infectious individuals (d_I) effectively reduce the likelihood of an outbreak for an infection with $R_0 = kb/\gamma$. In Figure 8.8, we demonstrate how this threshold changes as a function of the control strategy (contact tracing probability q and isolation rate d_I (per year)) and the infection's R_0. The figure demonstrates that, as

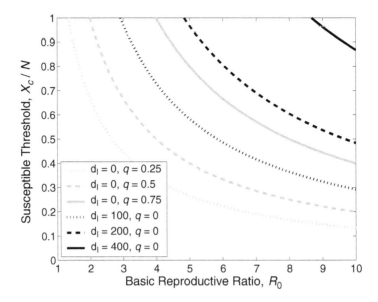

Figure 8.8. The schematic diagram showing the critical threshold fraction of susceptibles necessary for an outbreak as the isolation rate (d_I per year) and quarantine efficiency (q) are varied. Parameter values were $k = 400$ per year and $1/\gamma = 7$ days; b was varied to achieve different R_0 values.

one might expect, $\frac{X_c}{N}$ gets smaller as R_0 increases—an epidemic can occur for a lower level of susceptibles. However, the more useful aspect of the plot is the comparison between the relative effectiveness of contact tracing versus isolating infectives in controlling the infection. We find that rapid detection and quarantining of those infected ($d_I \gg 1$, black lines) is generally a more efficient means of preventing widespread transmission than relying on the successful identification and subsequent isolation of those who recently came into contact with the infection ($q > 0$, gray lines). However, determining the optimal response requires us to estimate the logistical effort associated with these two policies; for example, if contact tracing is much easier than rapid isolation, it may be the preferred strategy despite its weaker impact.

Another important currency for control is the cumulative number of people who contract the infection, especially if infection is associated with serious complications. In Figure 8.9, we plot the cumulative fraction of an initially entirely susceptible population who acquire the infection for various levels of infection management. When q and d_I are small, not surprisingly, a substantial proportion of the population become infected. There is clearly a threshold combination of controls that results in the effective breaking of the chain of transmission. The threshold is derived by studying the necessary conditions for the disease-free equilibrium to become unstable. From equation (8.32), this is given by

$$kb(1 - q) > d_I + \gamma.$$

Now let us explore a hypothetical set of scenarios. Consider the introduction of a smallpox-like infection, with highly detectable symptoms for which an efficacious vaccine was available. The vaccine is assumed to be associated with a small chance of serious

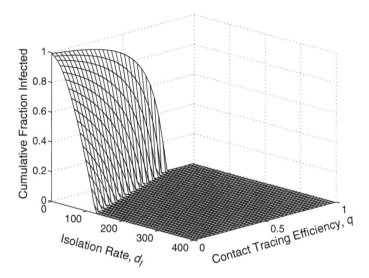

Figure 8.9. The cumulative fraction of the population infected as the isolation rate (d_I) and quarantine efficiency (q) are varied, using equations (8.27) to (8.31). Parameter values were $k = 230$ per year, $1/\gamma = 12$ days, $1/\tau_Q = 21$ days, and $b = 0.65$, such that $R_0 \approx 4.9$.

side effects. Assuming all economic costs and possible logistic obstacles and delays can be overcome, is it "best" to (1) vaccinate a large fraction of the population, (2) attempt to manage any introduction by surveillance and containment, plus additional vaccination? Of course, to some extent the answer depends on what we mean by best. Possible important measures of optimality may include the fewest number of cases, the fewest fatalities, and the shortest duration of outbreak. In order to examine this situation, we use equations (8.27)–(8.31) with the addition of an (asymptomatic and undetectable) exposed class. We assume $R_0 \sim 5$, and that the mean latent and infectious periods are both 12 days (Gani and Leach 2001). Background vaccination levels, p, of 40%, 60%, and 80% are explored by setting $X(0) = N \times (1 - p)$. We assume surveillance and containment to include the daily identification and quarantine for 3 weeks of 30% of infectious cases ($d_I = 120$ per year), along with 50% of susceptible contacts ($q = \frac{1}{2}$). The results of this experiment are presented in Table 8.2.

These findings are informative because they highlight different consequences of alternative courses of action. Clearly, if p_c ($= 1 - 1/R_0$) fraction of the population are vaccinated, then $\frac{dI}{dt} < 0$ and any introduction will be unsuccessful, with a very brief period of public concern. The downside to such a strategy is that vaccinating 80% of the population will present substantial logistical challenges in addition to the great financial cost. In addition, in this example, of the one million people vaccinated, around 15–50 individuals may experience very serious side effects. On the other hand, in the absence of prophylactic vaccination, a policy of contact tracing and isolation in the event of any smallpox introduction will result in no vaccination-induced complications, only 73 cases, 11 infection-induced deaths, and an outbreak that lasts less than 25 weeks. This is based on quarantining 50% of contacts, as well as the isolation of one-third of all infectious individuals per day—the implementation of both of these measures is likely to be expensive

TABLE 8.2.

Outcomes of alternative control measures when 10 infectious individuals are introduced into an entirely susceptible urban population of size 1.25 million. We assume that half of all contacts are traced ($q = 0.5$), and that isolation takes an average of 3 days ($d_I = 0.33$ per day). The probability of infection-induced mortality is assumed to be 15%.

Control Measure	Cases	Deaths	Duration of Epidemic	Peak Number of Infectives	Number Vaccinated
Mass Vaccination					
40%	708,303	106,250	52 weeks	112,571	500,000
60%	403,667	60,550	84 weeks	39,692	750,000
80%	< 10	< 1	2–3 weeks	< 10	1,000,000
Contact Tracing	1,082,040	162,306	71 weeks	67,346	0
Isolation	741,334	111,200	56 weeks	18,958	0
Contact Tracing and Isolation	30	4	18 weeks	10	0

and logistically challenging. Table 8.2 shows that isolating known contacts or infectives alone is not very effective, whereas the two in conjunction can drastically control any epidemic. This is because although each control measure in isolation produces a reduction in R_0, the reduction is not sufficient to bring R_0 close to (or below) one and hence a significant epidemic still occurs.

Although such analyses can be illuminating from a general perspective, in order to make quantitative predictions on infection management, we need to take into account a number of important biological details (Ferguson et al. 2003b; Lloyd-Smith et al. 2003; Wearing et al. 2005). First, we note that the incubation period (time taken from infection to exhibit symptoms) exceeds the latent period, so we split infectious individuals into those who are exhibiting symptoms (I_S) and those without symptoms for whom the incubation period has not elapsed (I_A). Second, we incorporate a more realistic distribution into the latent and infectious classes (see Chapter 3). All periods are assumed to be gamma distributed; the exposed period is subdivided into m_E classes and has an average length of $1/\sigma (= 12$ days), the asymptomatic infectious period (I_A) is divided into m_A classes and has an average length $1/\gamma_A (= 3$ days), and the symptomatic infectious period (I_S) is divided into m_S classes and has an average length $1/\gamma_S (= 9$ days).

We again construct the model by considering the fate of each susceptible-infectious contact, which occur at rate kXY/N. The susceptible involved in such a contact can be classified into one of four different states and these are modeled separately: (1) A proportion $(1 - q)(1 - b)$ are uninfected and never uncovered as part of the contact tracing from the infected contact; these remain in the susceptible class X. (2) A proportion $(1 - q)b$ are infected but never uncovered by the tracing; these progress into the first exposed class W_1 and eventually become infectious. (3) A proportion $q(1 - b)$ are uninfected but detected as part of the contact tracing scheme; these are isolated in class X_Q and are only released a time τ_Q after the initial contacts when it can be certain they will not develop symptoms. (4) The remaining proportion qb are infected and are assumed to be uncovered by tracing before they become infectious. These are labeled Q; although these individuals pass through exposed and infectious stages these do not need to be modeled as they have no effect on the rest on the population because the individual will be isolated. The system of equations describing such a model is given by:

$$\frac{dX}{dt} = -\frac{(b+q(1-b))kX(t)Y(t)}{N} + \frac{q(1-b)kX(t-\tau_Q)Y(t-\tau_Q)}{N},$$

$$\frac{dX_Q}{dt} = \frac{qk(1-b)X(t)Y(t)}{N} - \frac{qk(1-b)X(t-\tau_Q)Y(t-\tau_Q)}{N},$$

$$\frac{dW_1}{dt} = \frac{(1-q)bkX(t)Y(t)}{N} - m_E\sigma W_1(t),$$

$$\frac{dW_i}{dt} = m_E\sigma W_{i-1}(t) - m_E\sigma W_i(t), \quad i=2,\ldots,m_E,$$

$$\frac{dY_{A,1}}{dt} = m_E\sigma W_m(t) - m_A\gamma_A Y_{A,1}(t),$$

$$\frac{dY_{A,i}}{dt} = m_A\gamma_A Y_{A,i-1}(t) - m_A\gamma_A Y_{A,i}(t) \quad i=2,\ldots,m_A, \tag{8.33}$$

$$\frac{dY_{S,1}}{dt} = m_A\gamma_A Y_{A,n_A}(t) - (m_S\gamma_S + d_I)Y_{S,1}(t),$$

$$\frac{dY_{S,i}}{dt} = m_S\gamma_S Y_{S,i-1}(t) - (m_S\gamma_S + d_I)Y_{S,i}(t), \quad i=2,\ldots,m_S,$$

$$\frac{dQ}{dt} = \frac{qbkX(t)Y(t)}{N} + d_I Y_S(t),$$

$$\frac{dZ}{dt} = m_S\gamma_S Y_{S,m_S}(t),$$

where, for convenience, $Y = \sum_{i=1}^{m_A} Y_{A,i} + \sum_{i=1}^{m_S} Y_{S,i}$. Note that here R represents those who recover before they could be isolated or quarantined and therefore those who spend their entire infectious period in the population at large. As the epidemic dies out, $Q + R$ will represent the total number of recovered individuals, because Q keeps track of all those infected individuals who are quarantined or isolated, and effectively removed from the infectious population. Note that we do not adjust N in the denominator of the frequency-dependent mixing term to discount those in quarantine when calculating the contact frequency because we want to assume that the level of mixing remains the same following interventions. Finally, if we wished to accurately represent the known epidemiology of smallpox, we would insist that the contact rate k and the transmission probability b are different for the asymptomatic and symptomatic classes, matching the observation that asymptomatics have more contacts but are less infectious ($k_A > k_S$ and $b_A < b_S$) (Eichner and Dietz 2003). However, due to uncertainties in the parameterization of this extra heterogeneity, we will assume uniform parameter values for simplicity.

As shown in Figure 8.10, the precise levels of isolation of infecteds (d_I) and quarantining (q) required to control the outbreak and the predicted level of disease incidence are affected by whether asymptomatic infectious individuals are successfully traced and isolated, or whether this is achieved only for those with pronounced symptoms. Here we are envisaging the situation where contacts are requested for the past 12 days compared to asking for contacts since the infectious person first started to feel ill. It is not surprising that if the

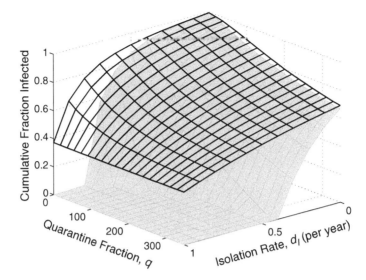

Figure 8.10. The predicted effectiveness of contact tracing and isolation of infectives in a population of 1.25 million susceptibles. The surfaces represent predictions of the model given by equation (8.33) with either contacts with all infected individuals traced and isolated (gray) or only those with pronounced symptoms (black). Although the surface can be calculated from equation (8.33), the black grid requires some modification to the model, separating those in contact with symtomatic and asymtomatic individuals. The figure demonstrates that if the infection is associated with a significant undetectable infectious stage (black grid), effective control is very difficult. (Model parameters $m = 45$, $n_A = 5$, $n_S = 8$, $k = 235$, $b = 0.65$, $1/\sigma = 12$ days, $1/\gamma_A = 3$ days, $1/\gamma_S = 9$ days, $\tau_Q = 20$ days.)

disease is associated with a substantial asymptomatic or undetectable infectious phase, then effective control that relies on the identification of infection is difficult. The figure also demonstrates that quarantining has a greater effect on reducing the cumulative fraction of the population infected than isolation.

Using a slightly more complex infection control model that incorporated a delay between becoming infectious and implementing isolation, Wearing et al. (2005) highlighted an additionally important aspect of such models. (In this model, isolation at rate d_I does not begin until τ_D days after an individual becomes infectious.) They demonstrated that models with exponential waiting times in the infectious period predict more optimistic outcomes of control compared to those with a gamma distribution (see Chapter 3). The explanation for such an observation lies in the process of isolating infectives, which reduces the effective mean infectious period (Wearing et al. 2005). Accounting for the isolation of infecteds at rate D_I after a delay of τ_D days, the average infectious period for the exponentially distributed model ($n_A = n_S = 1$) is

$$\frac{1 - e^{(-\gamma\tau_D)}}{\gamma} + \frac{e^{(-(\gamma+d_I)\tau_D)}}{\gamma + d_I}. \tag{8.34}$$

In contrast, the mean infectious period for a fixed infectious period ($n_A, n_S \to \infty$) is given by

$$\tau_D + \frac{1 - e^{(-d_I(1/\gamma - \tau_D))}}{d_I}. \tag{8.35}$$

From these equations, the isolation of infecteds is predicted to be much more effective when the infectious period is exponentially distributed because it essentially truncates the long tail of the distribution, so that the infectious period of a few individuals is dramatically reduced. This effect is not as pronounced in the gamma-distributed models because there is less variation in the infectious periods. Indeed, as long as the infectious period in the absence of isolation is greater than the delay $(1/\gamma > \tau_D)$, then expression (8.34) is less than expression (8.35) for all $d_I > 0$ (Wearing et al. 2005). In the same way, a longer delay in detecting infected individuals has fewer predicted consequences for the exponentially distributed model because during this time many individuals will have naturally left the infectious class. Under the assumption of a gamma-distributed infectious period, most individuals are infectious for a minimum period of time so early detection is more important. Although the predicted difference between the exponential and gamma-distributed models depends on the duration of the infectious period and the fraction of contacts traced (q), it is generally true that models with an exponentially distributed infectious period will give rise to overly optimistic predictions concerning the effectiveness of isolating infectives.

A further important point to notice from comparing Figures 8.9 and 8.10 is that similar levels of infection management are predicted to result in dramatically reduced infection control. Indeed, irrespective of the assumptions made on latent and infectious period distributions in equation (8.33), the rapid removal of symptomatic infectives $(d_I = 365)$ and quarantine of 80% of contacts is still predicted to result in almost 20% of the population contracting the infection (Figure 8.10). In contrast, the results in Figure 8.9 suggest that eradication is relatively easy. The primary cause for this substantial and very important difference between the two model results is the introduction of the effective delays in implementing controls. The take-home message from Table 8.2 is that effective isolation and quarantine can successfully contain any introduced infection, whereas the results shown in Figure 8.10 demonstrate that in reality this is very difficult, perhaps even unlikely. This work points toward the need for additional (and possibly targeted) vaccination in order to suppress an outbreak. This topic, along with the spatial consequences of such actions, are discussed in the next section.

8.4. CASE-STUDY: FOOT-AND-MOUTH DISEASE, SPATIAL SPREAD, AND LOCAL CONTROL

One of the best demonstrations of the usefulness of mathematical models was during the high-profile 2001 epidemic of foot-and-mouth disease (FMD) in the United Kingdom. As with the smallpox case study, modeling is useful due to the complex trade-off between culling and control: Too little culling and the disease is not controlled; too much culling and although the disease is eliminated, many more livestock may be lost due to culling. As discussed in Chapters 4 and 7, this infection is made more complex by the presence of more than one host (sheep, cattle, and pigs, though pig farms were largely unscathed in the 2001 outbreak) and the spatial aspect of transmission, combining both localized farm-farm interactions and longer-range transmission due to movements of machinery and personnel. To model the 2001 outbreak, Keeling et al. (2001b) developed a stochastic, spatially explicit farm-based model. This framework contains the inherent assumption that transmission within a farm is rapid, hence it is reasonable to categorize each farm as susceptible or infected. Once fully parameterized, the model was used to investigate different options for managing the epidemic.

After the first reported case on February 20, 2001, the FMD virus spread rapidly across the country, with over 600 farms infected within the first 5 weeks of the epidemic. To control the outbreak, the UK Ministry of Agriculture, Fisheries and Food (MAFF) instigated a national ban on livestock movement, together with a requirement for all livestock on infected farms (called infected premises, or IPs) to be culled within 24 hours of the infection being reported. In addition, the livestock on at-risk farms (dangerous contacts, DCs, and contiguous premises, CPs) were also to be culled within 48 hours of FMD detection on the associated IP. Conceptually, we can equate control of dangerous contacts with contact tracing, whereas control of contiguous premises is comparable with local measures such as ring culling. Eventually, in October 2001, the epidemic was declared over. More that 2,000 farms had been infected and over 8,000 culled as part of the control. Given the catastrophic economic cost of implementing measures to control the epidemic, as well as its consequences for UK industries (such as farming and tourism), the newly formed Department of Environment, Food and Rural Affairs (DEFRA) was interested in developing protocols for future management strategies. Mathematical models played an important role in the formulation of DEFRA's contingency plan, published in 2004, which now contains the provision for localized vaccination against a future epidemic.

Box 8.1 Individual-Based FMD Model

As described in Chapter 7 (see also Keeling et al. 2001b; Keeling et al. 2003; and Tildesley et al. 2006), the transmission of foot-and-mouth disease can be modeled using a spatial transmission kernel, $K(d)$, which is a function of the distance d between the infecting and susceptible farm. The rate λ_i at which susceptible farm i becomes infected is given by:

$$\lambda_i = Sus_i \sum_{j \in \text{infectious}} Trans_j \, K(d_{ij}),$$

$$Sus_i = \sum_{l \in \text{species}} N_{i,l} \, s_l, \qquad Trans_j = \sum_{l \in \text{species}} N_{j,l} \, t_l$$

where $N_{i,l}$ is the number of livestock of type l on farm i and s_l and t_l are species-specific susceptibility and transmissibility for cattle, sheep, and pigs. The model is iterated forward daily by performing the following two steps:

1. Each susceptible farm is infected with probability $1 - \exp(-\lambda_i)$.

2. The time since infection for each infected farm is increased (by one day) and the status of the farm is updated appropriately. Farms that have been infected for between 0 and 5 days are assumed to be exposed, those infected for more than 5 days are infectious, whereas those infected for 9 days are symptomatic.

In addition to the epidemiological dynamics, a range of culling strategies also need to be modeled. The processes involved are detailed below.

IP Culling

Infected premises (IPs) should have their livestock culled as quickly as possible following detection of the disease, so as to reduce the infectious period by as much as possible. This process is modeled by removing the farm (by setting $N_{i,l} = 0$) τ_{IP} (usually equal to 1) days after the disease is identified; in general, this is between 10 and 15 days after infection.

DC Culling

Dangerous contacts (DCs) are conceptually equivalent to traced contacts; they are farms that are assumed to be at a high risk of infection due to their connections with an identified IP. If farm i is an infected premise, then the probability that farm j is a dangerous contact is modeled as

$$\begin{cases} 1 - f \exp(-F\,Sus_j\,Trans_i\,K(d_{ij})) & \text{if } j \text{ has been infected by } i \\ 1 - \exp(-F\,Sus_j\,Trans_i\,K(d_{ij})) & \text{otherwise,} \end{cases}$$

where the parameter F accounts for the number of DCs identified and culled per IP, and the parameter f accounts for local knowledge of potential transmission routes, which increases the accuracy of identifying infected farms. After a delay of τ_{DC} (usually equal to 2) days from when an infected farm is diagnosed, all of the livestock on the associated DC that can be identified are culled. In general, an average of one or two DCs are usually identified per infected farm.

CP Culling

Contiguous premise (CP) culling was initiated as an alternative strategy to a fixed-size ring cull, targeted toward farms that share a common boundary. Although in practice CPs are identified using detailed maps of farm boundaries, this information is not available electronically and so some approximate criteria needs to be defined. We construct CPs by assuming that farms tessellate across the landscape, such that the boundary between two farms is given by a perpendicular line halfway between their coordinates (dashed line on the figure below).

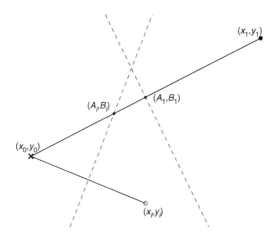

Graphical representation of the methodology for determining contiguous premises in terms of the positions of potential boundaries (shown as dashed lines).

As a simple approximation, farm 1 at position (x_1, y_1) is considered to be contiguous to farm 0 at position (x_0, y_0) if their boundary is closer to farm 0 than any other boundary:

$$\frac{\left(\dfrac{(y_i - y_0)^2 + (x_i - x_0)^2}{(y_i - y_0)} \right)}{(x_1 - x_0) \left(\dfrac{(y_1 - y_0)}{(x_1 - x_0)} + \dfrac{(x_i - x_0)}{(y_i - y_0)} \right)} > 1 \quad \forall i.$$

In the full simulation models (Keeling et al. 2001b; Keeling et al. 2003; and Tildesley et al. 2006) account was also taken of the relative areas of the farms, which biased the position of the boundary and limited the potential for very distant farms to share a boundary.

CP culling then takes place τ_{CP} (usually equal to 2) days after the infected farm is identified by culling all the animals on a proportion p_{CP} of all the contiguous premises of the IP.

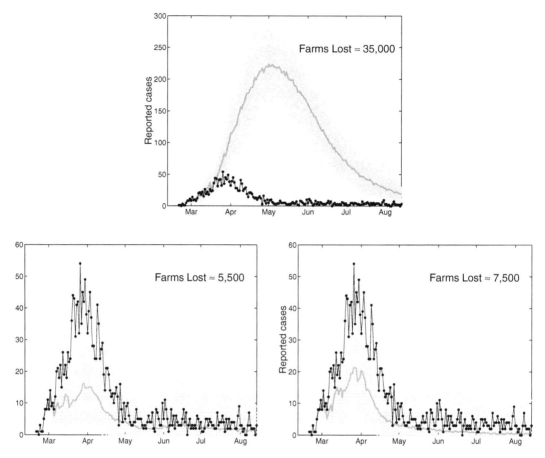

Figure 8.11. The figure demonstrates the effects of varying the culling procedure on the epidemic curve (from Keeling et al. 2001b). Black dots denote actual reported cases, and pale gray dots are the results of 50 simulations, with the thick gray line giving the average of those simulations. Data is shown only until the middle of August for greater clarity. In the top graph, only the IP cull is performed with the delay between reporting and culling following the pattern observed in 2001. The culls in the bottom-left graph follow the observed level, but the delays from reporting to slaughter of the animals on the IP and from reporting to culling of animals on DCs and CPs are assumed to be prompt (24 and 48 hours, respectively) from the beginning of the simulation. In the bottom-right graph, the delays now follow the observed pattern, but the levels of CP and DC culls are constant throughout the simulation and mimic the high levels achieved in the latter part of the 2001 epidemic. The inset values on each graph give the average number of farms losing their livestock either due to infection or control measures.

Modeling demonstrated that culling IPs, CPs, and DCs, as performed in 2001, was effective at controlling the epidemic with minimal consequences (Figure 8.11). Had the cull been restricted to IPs only—just targeting identified sources of infection—then a total of over 35,000 farms are predicted to have been infected. On the other hand, if the 24–48 hour time scale from reporting to cull had been imposed from the very beginning (bottom-left graph), or the higher culling levels achieved later on in the epidemic had been exercised earlier (bottom-right graph), then the total number of farms affected and animals slaughtered could have been substantially reduced. This echoes a general tenet in epidemiology, that it is often a good idea to apply control measures promptly and intensively; however, as shown in Chapter 7, Figure 7.17, which examines ring culling for the same model, if the control measures are too intensive, their consequences can be worse than the epidemic itself.

Planning for any future FMD outbreak has inevitably involved detailed *scenario modeling* involving a range of different strategies. Vaccination remains an important tool in the control of any disease, and could potentially be used against FMD in the future. Vaccination was not implemented during 2001, due to logistical constraints and the modeling predictions that vaccination late in the epidemic would be ineffective (Keeling et al. 2001b; Keeling et al. 2003). The most important aspect of modeling the effects of vaccination is the recognition of three key biological properties of FMD vaccines. First, these vaccines do not provide complete protection, with typical efficacies in the 90–95% range (Barnett and Carabin 2002). It is generally assumed in these models that 90% of vaccinated animals are fully protected, whereas the remaining 10% are fully susceptible, although this is an approximation to the complex reality of immunization. This means that, although all the animals on a farm may be vaccinated, some of the animals may remain susceptible so that the farm can still become infected and act as a source of further cases. Second, as with other vaccines, FMD vaccines do not provide immediate protection; the high potency vaccines may protect within 3–4 days, whereas standard vaccines take considerably longer, around 10 days (Woolhouse 2003). Third, vaccination has no or limited effect on animals that are already infected. Finally, vaccination provides only short-duration protection and repeated vaccination may be required every 6 or 12 months. These traits clearly need to be taken into account when considering future vaccination campaigns.

One option for preventing future epidemics is to maintain a high level of herd immunity to FMD within the UK livestock population, similar to the vaccination policies against childhood diseases (see Section 8.1.1), although requiring repeated vaccination. Such prophylactic vaccination, together with the culling of infected and at-risk Dangerous Contact (DC) farms, has a strong effect on the final epidemic size (Figure 8.12). In general, the vaccination of cattle in 25,000 randomly chosen farms reduces the size of the epidemic to levels similar to those achieved by stringent and expensive IP, DC, and CP culling procedures. If over 80,000 farms can be protected from infection (which requires the vaccination of most cattle in the United Kingdom), then any introduction of the FMD virus is very unlikely to result in further spread. The precise effect of vaccination is determined in part by the strategy for selecting which farms to vaccinate; the preferential immunization of large cattle farms, which are generally at greater risk, is significantly more effective than simple random vaccination. Again, this echoes results from Chapter 3, where control should be targeted toward high-risk individuals—in this case, large cattle farms that have a high susceptibility and transmissibility.

The right-hand graph of Figure 8.12 examines the results of a range of other prophylactic vaccination options. First, we consider vaccination of sheep and pigs as well as cattle. Although in practice this would be more difficult to achieve, it is heuristically useful to explore in terms of the impact of population-level immunity. Not surprisingly, the vaccination of all three species (dashed lines) gives rise to smaller epidemics than vaccinating cattle alone. However, if we consider the effective control achieved as a function of the total numbers of animals vaccinated, then concentrating on cattle alone is much more efficient. This graph also stresses the need for maintaining DC culling (comparing black and gray curves). If DC culling were relaxed (gray curves), because

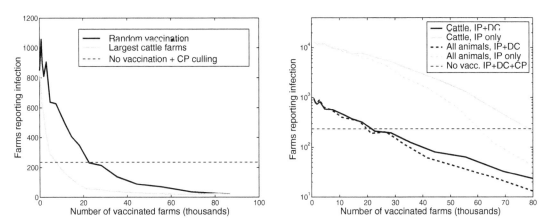

Figure 8.12. The fate of an FMD introduction into five randomly chosen farms in Great Britain after prophylactic vaccination, from Keeling et al. 2003. IP and DC culls (averaging one DC per IP) were used to control the epidemics, and movement restrictions were enforced two weeks after the arrival of the disease. CP culling was not performed due to the strong opposition to this strategy in 2001; however, a comparison is made to the case where 70% CP culling is used instead of vaccination (horizontal dashed line). Results are the average of 500 simulations; in many of which the epidemic fails to take off, we generally assume that only cattle are vaccinated and the efficacy is 90%. The graphs show the effect of changing the number of farms that are vaccinated before the epidemic begins. The left-hand graph shows the average number of farms reporting infection for two different vaccination approaches. The right-hand graph compares the effects of vaccinating just cattle (solid lines) and all animals (dashed lines), and the effects of IP and DC culling (black lines) with IP-only culling (gray lines)—vaccination is at random.

it was felt that vaccination gave sufficient protection, then the expected number of cases increases tenfold compared to a combined DC-culling and vaccination strategy.

As mentioned in Table 8.1, one of the drawbacks of prophylactic vaccination is the need for long-term maintenance of substantial herd immunity, the cumulative cost of which is very large. Instead, reactive vaccination—the immunization of livestock after an outbreak has been detected—may represent a more appealing option. Tildesley et al. (2006) have investigated the possibility of using ring vaccination to control any future epidemic. It is assumed that ring vaccination will be a successful control measure because it is targeted at farms surrounding an infected premise. These farms are known to be at greater risk of infection due to the localized transmission of foot-and-mouth disease (see Section 7.5.2). However, if the farm is already infected, vaccination is assumed to have no effect; therefore, very localized vaccination may often be infective. In agreement with known vaccine behavior (Barnett and Carabin 2002) and adopted UK policy (DEFRA 2004), we assume that a combination of IP culling, DC culling, and ring vaccination are applied and that cattle on farms within a ring surrounding each reported premise are identified for vaccination 2 days after reporting. Due to logistic constraints (in terms of the number of vaccination teams available), only a limited number of cattle can be vaccinated on each day—these are prioritized in the order in which the central infected farm is identified. This logistic constraint generates a trade-off within the model: If the vaccinated ring is too small, too few animals will be protected; in contrast, if the vaccination ring is too large, the constraint on the number of cattle that can be vaccinated per day means that considerable delays may develop.

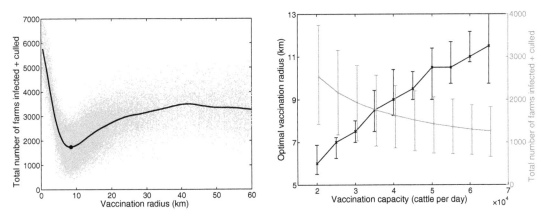

Figure 8.13. Results from detailed stochastic spatial simulations of foot-and-mouth disease in the United Kingdom, taking into account farm heterogeneity and animal composition; parameters are derived by fitting to the behavior of the 2001 epidemic, from Tildesley et al. (2006). Control is a mixture of IP culling within 24 hours of detection, DC culling 24 hours later, and ring vaccination. The left-hand graph shows the total loss of farms (due to either infectious or culling) for various ring sizes when there is the capacity to vaccine 35,000 cattle per day; the gray dots are the results of individual simulations, the black curve is a smoothed fit to the mean, and the large black dot gives the position of the optimal radius. In the right-hand graph, this concept is extended and the optimal ring size and total loss of farms is shown as the vaccination capacity varies. Throughout it is assumed that the vaccine has an efficacy of 90% and that there is a delay of 4 days between vaccination and protection.

The left-hand graph of Figure 8.13 shows how the number of farms lost (either due to infection or DC culling) varies with the size of the vaccination ring. Clearly there exists an optimal vaccination radius that maximizes the number of cattle vaccinated per day but minimizes the delay. This optimal vaccination strategy (combined with IP and DC culling) leads to the loss of far fewer farms than a more intensive cull-based control in which IPs, DCs, and CPs are culled—thus demonstrating the advantages of well-targeted vaccination. The right-hand graph demonstrates how the optimal vaccination radius (gray line) and the number of farms lost (black line) is affected by the vaccination capacity. As expected, when more cattle can be vaccinated per day, the optimal radius can increase and hence the total number of farms lost decreases—however, there is no simple relationship between vaccination capacity and optimal ring size, illustrating the need for complex simulation models that can account for the spatial structure and heterogeneities present in the UK farming landscape.

8.5. CASE-STUDY: SWINE FEVER VIRUS, SEASONAL DYNAMICS, AND PULSED CONTROL

Another arena where mathematical models can be usefully employed is the examination of emerging infectious diseases, such as AIDS, severe acute respiratory syndrome (SARS), or the new influenza A viruses (Morens et al. 2004). These emerging infectious diseases (EIDs) are especially interesting and important because the threat they pose is not confined to humans, with documented examples in animal (Daszak et al. 2000) and plant (Anderson et al. 2004) species. It is now apparent that a substantial proportion of emerging infectious diseases are due to novel pathogens that have managed to cross species boundaries (Antia et al. 2003; Woolhouse et al. 2005). Typical examples include the simian origin of HIV (Hahn et al. 2000), and the identification of viruses related to the SARS coronavirus in small asiatic mammals (Guan et al. 2003). Consequently, the origins of marburg and ebola viruses are currently sought in African wildlife populations (Monath 1999).

Cross-species pathogen transmission also seems to be a common phenomenon between free-living wild animal populations and their domestic counterparts (Daszak et al. 2000), with pathogen transmission generally facilitated by the genetic relatedness of the two host species. An additional aspect of these systems is that many of the free-living wild animal populations that harbor infectious diseases of potential threat are often hunted for recreation and/or population regulation. Here, we follow Choisy and Rohani (2006) and examine a simple general model that accounts for the dynamics of a directly transmitted disease in a harvested/hunted host population.

The general framework we introduce here is intended to mimic the dynamics of a viral or bacterial disease in a harvested population of wild ungulates, in a temperate region. The underlying motivation behind this model is to assess the role of culling wild boar to reduce the transmission of classical swine fever (CSF) to domestic livestock. We consider a naturally regulated host population that experiences infection by a directly transmitted microparasitic disease conferring permanent immunity to those recovered. The general structure of the model is as follows:

$$\frac{dX}{dt} = \varphi(t)N(t)\nu(N) - [\mu_X(N) + \psi(t)q_X H_X + \lambda(t)]X(t), \tag{8.36}$$

$$\frac{dW}{dt} = \lambda(t)X(t) - [\mu_W(N) + \psi(t)q_W H_W + \sigma]W(t), \tag{8.37}$$

$$\frac{dY}{dt} = \sigma W(t) - [\mu_Y(N) + \psi(t)q_Y H_Y + \gamma + m]Y(t), \tag{8.38}$$

$$\frac{dZ}{dt} = \gamma Y(t) - [\mu_Z(N) + \psi(t)q_Z H_Z]Z(t), \tag{8.39}$$

where, as usual, the variables X, W, Y, and Z represent the numbers of susceptible, exposed, infectious, and recovered animals, respectively. We stress that the total population size $N(t)$ $(= X + W + Y + Z)$ is not necessarily constant. The term $\nu(N)$ is the per capita density dependent birth rate, and $\mu_i(N)$ is the per capita density-dependent natural death rate in compartment i ($i \in \{X, W, Y, Z\}$). Parameters q_i ($0 \le q_i \le 1$) and H_i are, respectively, the catchability and the harvest effort in compartment i, reflecting the fact that infected animals may behave differently (Kot 2001). Susceptible individuals get infected at a rate given by the force of infection $\lambda(t) = \beta \frac{Y(t)}{N(t)}$, where β reflects the transmission rate and we are assuming frequency-dependent transmission. Although wildlife diseases are generally modeled using density-dependent transmission, the social structure within the wild boar population means that frequency-dependent transmission is a better assumption, although the truth is likely to lie between the two. We assume that the mean duration of the exposed and infectious periods are given by $1/\sigma$ and $1/\gamma$, respectively—taking the simplest assumption of constant rates rather than explicit distributions. Finally, the disease can induce additional mortality on the infectious individuals, at a rate m.

In most large herbivore species, it is well documented that births and deaths are density dependent (Gaillard et al. 2000). For simplicity, we assume that these two (per capita) rates are linearly related to population density such that

$$v(N) = b - BN(t - \tau), \tag{8.40}$$

$$\mu_i(N) = d_i + D_i N(t), \tag{8.41}$$

where b is the maximum per capita birth rate, d_i is the minimum per capita death rate in compartment i, and B and D determine the strength of the density-dependence in birth and death rates. The time delay τ reflects the duration of gestation between conception and birth.

In most temperate mammal species, births are seasonal (Macdonald 1984, Section 5.3.2.1), and this is accounted for by the function $\varphi(t)$. In wildlife management, the most common practice is to permit hunting only during a specified season, usually short compared to the rest of the year (Xu et al. 2005). This additional form of seasonality in "harvesting" is represented by the function $\psi(t)$. These two seasonality functions—$\varphi(t)$ on the birth rate and $\psi(t)$ on the harvest—are modeled by a simple periodic square function:

$$\varphi(t) = \begin{cases} 1 & \text{if } b_1 < (t \mod 1\text{year}) < b_2 \\ 0 & \text{otherwise}, \end{cases} \tag{8.42}$$

$$\psi(t) = \begin{cases} 1 & \text{if } h_1 < (t \mod 1\text{year}) < h_2 \\ 0 & \text{otherwise}, \end{cases} \tag{8.43}$$

hence, $[b_1, b_2]$ and $[h_1, h_2]$ are the periods of each year corresponding to the birth and harvest seasons, respectively, and $t \mod 1\text{year}$ is the time from the start of the year.

Although this framework is intended to be quite general, it has features that correspond closely to the dynamics of classical swine fever (CSF) disease in wild boar (Paton and Greiser-Wilke 2003). CSF is a viral disease affecting wild and domestic swine worldwide (Paton and Greiser-Wilke 2003). Outbreaks can result in severe losses in pig farms and, in Europe, wild boar populations are suspected as potential reservoirs of the disease. Moreover, wild boar populations experience wide seasonal population fluctuations due to strong density-dependent processes and harvesting. Culling of boar is considered a possibility for preventing spillover of CSF to livestock populations. In Section 8.5.1, we examine the dynamical consequences of such a strategy.

8.5.1. Equilibrium Properties

To ease analytical tractability, we initially make some simplifying assumptions, such as ignoring seasonality, using identical density-dependent terms ($B = D_X = D_W = D_Y = D_Z \equiv D$) and setting the minimum death rates to zero ($d_X = d_W = d_Y = d_Z = 0$). In addition, the maximum birth rate b is set to $2DK$ in order to give a constant carrying capacity K in the absence of harvesting and disease-induced mortality. The harvest effort and host catchability are also assumed to be independent of disease status (i.e., $H_S = H_E = H_I = H_R \equiv H$ and $q_S = q_E = q_I = q_R \equiv q$). In the absence of disease-induced mortality ($m = 0$), we can sum equations (8.36)–(8.39) to obtain a differential equation for the rate of change of the population size,

$$\frac{dN}{dt} = (b - qH)N - 2DN^2. \tag{8.44}$$

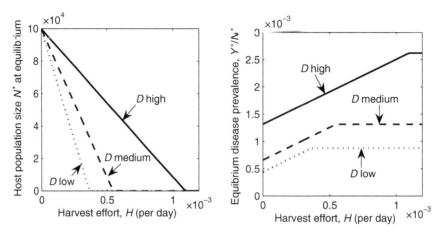

Figure 8.14. Effect of harvest effort H, on host population size N^* (left-hand graph) and disease prevalence Y^*/N^* (right-hand graph) at equilibrium, for different strengths of density dependence: $D \simeq 2.7397 \times 10^{-9}$ per day (full line), $D \simeq 1.3699 \times 10^{-9}$ per day (dashed line), and $D \simeq 9.1324 \times 10^{-10}$ per day (dotted line). The minimum per capita death rates are zero ($d_S = d_E = d_I = d_R = 0$) and the maximum birth rate b is adjusted so that the carrying capacity in the absence of harvesting is $K = 10^5$ individuals, whatever the strength of the density dependence. We thus have $b \simeq 5.4795 \times 10^{-4}$ per day (full line), $b \simeq 2.7397 \times 10^{-4}$ per day (dashed line), and $b \simeq 1.8265 \times 10^{-4}$ per day (dotted line). (Other parameter values are $q_S = q_E = q_I = q_R = 0.5$, $\beta = 5.4794$ per day, $1/\sigma = 8$ days, $1/\gamma = 5$ days and $m = 0$.)

This equation can be set to zero to obtain an equilibrium population abundance,

$$N = K - \frac{q}{2D}H, \qquad (8.45)$$

which requires harvest pressure to be sufficiently small ($qH < b = 2DK$) to ensure nonnegative population equilibrium. Additionally, we can show that harvest effort H has a positive effect on the equilibrium disease prevalence Y^*/N^*:

$$\frac{Y^*}{N^*} \simeq \left(K\frac{D}{\gamma} + \frac{q}{2\gamma}H \right)\left(1 - \frac{1}{R_0} \right). \qquad (8.46)$$

Not surprisingly, the equilibrium disease prevalence depends on the mean duration of the infectious period ($1/\gamma$) and the strength of density dependence (D). Given that the carrying capacity is fixed, the strength of density dependence reflects the population turnover rate, and thus the rate of susceptible recruitment, known to be a major determinant of the dynamics of diseases conferring lifelong immunity to their host (Earn et al. 2000, Chapter 2).

As harvest effort H increases, the equilibrium population size N^* decreases, but somewhat surprisingly the equilibrium disease prevalence Y^*/N^* increases (equation (8.46) and Figure 8.14, left graph). Above a harvest threshold $H^\star$, the equilibrium population size N^* reaches zero (Figure 8.14, left graph) with disease prevalence Y^*/N^* reaching its maximum $(Y^*/N^*)_{\max}$ (Figure 8.14, right graph). From equation (8.45) we can express the threshold harvest value, $H^\star = 2K\frac{D}{q}$. Substituting this into equation (8.46), we get the maximum value of the disease prevalence at equilibrium:

$$\left(\frac{Y^*}{N^*} \right)_{\max} = 2K\frac{D}{\gamma}\left(1 - \frac{1}{R_0} \right). \qquad (8.47)$$

Interestingly, this means that harvesting can increase disease prevalence at equilibrium (Y^*/N^*) from its level in the absence of hunting $(KD(1 - 1/R_0)/\gamma)$ to twice this value.

Now, if we consider the possibility of disease-induced mortality $(m > 0)$, we observe a decrease in the equilibrium population size. If we sum the differential equations (8.36) to (8.39), to give an equation for the population size, N,

$$\frac{dN}{dt} = 2DKN - DN^2 - DN^2 - qHN - mY,$$

and assuming that Y is at the equilibrium value given by equation (8.46), then the population size is reduced to:

$$N^\star \approx K\left(1 - \frac{m}{2\gamma}\left[1 - \frac{1}{R_0}\right]\right) - \frac{q}{2D}\left(1 + \frac{m}{2\gamma}\left[1 - \frac{1}{R_0}\right]\right)H. \tag{8.48}$$

This is an approximation, because we have neglected the effect on the prevalence of infection of this change in the total population size. However, we expect this approximation to be reasonable unless the disease has high mortality. Equation (8.48) shows that, in the absence of harvesting, disease-induced mortality decreases the equilibrium population size by a factor of $m\xi/(2\gamma)$, where for notational convenience we set $\xi = 1 - 1/R_0$. However, in the presence of harvesting, disease-induced mortality mitigates the compensatory effect of density dependence by a factor of $1 + m\xi/(2\gamma)$. By rewriting equation (8.48), we can tease apart all sources of mortality:

$$N^\star = K - \frac{K}{2\gamma}\left[1 - \frac{1}{R_0}\right]m - \frac{q}{2D}H - \frac{q}{4\gamma D}\left[1 - \frac{1}{R_0}\right]mH. \tag{8.49}$$

Thus, the total population size is reduced by the action of three elements: $Km\xi/(2\gamma)$ individuals are lost due to disease mortality, $qH/(2D)$ individuals are lost due to hunting, and $qm\xi H/(4\gamma D)$ individuals are lost due to the synergistic interaction between the disease and harvesting. Put simply, when h individuals are harvested, an additional $\frac{m\xi}{2\gamma}h$ individuals "die" from a harvest-induced increase in disease-induced mortality. Interestingly, the additional proportion of individuals dying from the disease as a side effect of harvest $(m\xi/(2\gamma))$ does not depend on the strength of the density dependence D in the host population.

8.5.2. Dynamical Properties

We now focus on model dynamics to examine the consequences of seasonality on our conclusions. At equilibrium, harvest toll on the host population cannot be more than partially compensated by density-dependent population processes—leading to damped oscillations—whereas with seasonality, it has been shown that over-compensation is possible—leading to sustained oscillations (Kokko and Lindstrm 1998; Jonzén and Lundberg 1999; Boyce et al. 1999; Xu et al. 2005). Thus, we reintroduce the seasonality functions $\varphi(t)$ (birth seasonality) and $\psi(t)$ (harvest seasonality) as well as the gestation time delay, τ. The introduction of these time-dependent terms renders the system analytically intractable, and we will thus rely on numerical simulation to explore model dynamics.

For simplicity, we will consider the infection to be benign $(m = 0)$. It has been documented for a wide variety of vertebrates that density-dependent mortality is much

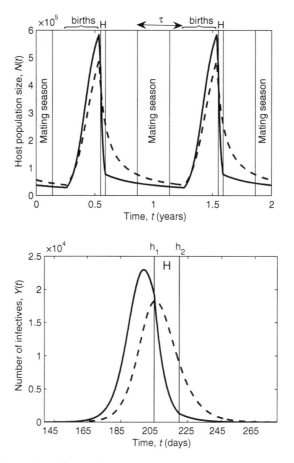

Figure 8.15. The dynamical effects of periodic harvesting and reproduction on the total host population size, $N(t)$, and the number of infected animals, $Y(t)$. The top graph shows 2 years of simulations of the model, after discarding transient dynamics. The periods of mating and harvesting (labeled H) are represented on the graph, whereas the periods of giving birth are indicated above the graph. The arrow represents the delay τ between mating and birth. The dashed curve represents the total host population size, $N(t)$, in the absence of harvesting ($H = 0$), and the full line curve represents the dynamics of the total host population size, $N(t)$, in the presence of harvesting ($H = 0.2$). The harvest season begins on day $h_1 = 205$, and ends on day $h_2 = h_1 + 15$. The lower figure shows the changes in number infected during the epidemic peak in a single year, clearly demonstrating the increase in Y as a result of harvesting (comparing solid and dashed lines). Parameter values are $K = 10^5$, $b \simeq 0.1461$ per day, $d_S = d_E = d_I = d_R = 0$, $B \simeq 4.5662 \times 10^{-7}$, $D_S = D_E = D_I = D_R \simeq 9.1324 \times 10^{-8}$, $\tau = 145$ days, $b_1 = 100$, $b_2 = 200$, $q_S = q_E = q_I = q_R = 0.5$, $\beta = 5.4795$ per day, $1/\sigma = 8$ days, $1/\gamma = 5$ days, and $m = 0$.

stronger in the younger age classes than in adults, where it is often negligible (Hudson 1992; Clutton-Brock et al. 1997; Gaillard et al. 2000). Therefore, we assume a constant mortality rate for the recovered class of individuals, which generally contains the older animals: $D_R > 0$ and $d_R = 0$.

In Figure 8.15 (top graph) we show two years of the total long-term host population dynamics with (full line) and without (dashed line) harvesting. Based on the hunting of wild boar, the harvest season is assumed to begin 5 days after the end of the birth season and to last for 15 days. In the absence of harvesting, the host population dynamics are pronouncedly annual, driven by birth pulses. As can be seen in the window of harvesting in Figure 8.15 (top graph), hunting very quickly and dramatically reduces the number of individuals in the population, with two key consequences. First, the number $N(t - \tau)$ of individuals during the breeding season is reduced; this reduces the competition on the remaining individuals, which in turn increases the per capita birth rate ν that season (see equation (8.40)). Second, the per capita death rate is decreased (equation (8.41)), which means more females survive from mating to giving birth; in fact, N immediately prior to the birth season is very close to its size in the absence of harvesting. The combination of these two effects results in a substantial increase in the number $\nu \times N$ of individuals born. This is a clear case of overcompensation where the effect of hunting increases the host population size after the birth season. Not surprisingly, this also results in a substantial increase in disease prevalence in the population (Figure 8.15, bottom graph).

In wildlife populations subject to age-specific density dependence with an endemic pathogen, harvesting of adults can release intraspecific competitive pressure, thereby increasing the influx of susceptible juveniles. Therefore, paradoxically, attempts to reduce infection prevalence by culling can result in a dramatic increase in diseased individuals.

8.6. FUTURE DIRECTIONS

The refined control of infectious diseases is one of the key strategic uses of epidemiological modeling. To this end, models can have two main inputs. The first is to provide a better generic understanding of infectious disease transmission and control so that informed decisions can be made in the early stages of an outbreak, even before explicit models or parameters are available. The second benefit of modeling comes from creating detailed models that can make reasonably accurate predictions, and therefore allow a refinement of early control decisions and estimates of the magnitude of the epidemic and resources needed for control.

There are many particular areas of epidemiologically modeling where great progress can still be made on understanding and optimally targeting control measures.

1. For heterogeneous populations, it is still an open problem to determine the optimal deployment of control measures (e.g., vaccine, tracing effort) between risk-groups—in particular, a generic understanding of the impact of over- or under-targeting would be highly beneficial.

2. For diseases that affect multiple hosts (especially zoonotic infections), models may allow us to understand the disease behavior in the largely unsampled animal reservoir, and the implications of these dynamics for human cases. When multiple strains of a pathogen are interacting within the community, it may be possible to apply control so as to influence the competitive pressures on the strains, driving the more harmful pathogen extinct. This is particularly important for antibiotic resistant infections, where the act of control is itself a driving force for evolutionary change.

3. The fluctuations driven by seasonal forcing can also potentially be utilized by a control program, and even enhanced by tools such as pulse vaccination. It still remains

an open problem to understand how seasonal forcing and pulsed control interact, and whether this interaction can be used advantageously.

4. Similarly, control methods that enhance the risk of stochastic extinction could be used to great effect if subsequent reintroductions of infection could be controlled. The risk of such stochastic extinctions could be increased by inducing oscillatory dynamics, or by effectively subdividing the population and limiting the interaction between subgroups. It is still an open and difficult modeling exercise to assess how spatial structure or population subdivision interacts with stochasticity.

5. When dealing with spatially structured populations, there is a clear trade-off between local targeting of control measures in infected regions and the prevention of spread to new susceptible areas. Yet again, models can help in assessing the optimal balance of these two conflicting demands on resources.

Finally, one of the biggest challenges facing epidemiological modeling today is to be able to respond quickly to new and emerging infections. This not only requires the rapid development of suitable modeling tools, but also necessitates a range of sophisticated statistical methods that can uncover parameter values (and confidence intervals) from early case reports. Even determining R_0 from the first few generations of transmission is a difficult task, and prone to error. It is therefore vital that parameter estimation matches the continued development of ever more complex and parameter-rich models.

8.7. SUMMARY

In this chapter, we examined the modeling issues surrounding a number of alternative methods of disease management in different scenarios. Using simple models, we explored topics such as widespread pediatric immunization to control childhood infections, the consequences of imperfect vaccines, the role of pulse vaccination, as well as nonvaccine-related control options such as contact tracing and isolation. These models were appropriately modified to explore complexities. We examined age-related aspects of vaccination within the context of rubella, potential delays in implementing quarantine measures, and the importance of alternative assumptions concerning the distributions of latent and infectious periods when thinking about control of deliberate exposure to smallpox. The most complex and detailed model we considered was the stochastic, spatially explicit multi-pathogen framework used to study aspects of foot-and-mouth disease.

The main findings of this chapter can be summarized along the following lines:

➤ A system subject to constant long-term vaccination of a fraction p of newborns against an infection with a basic reproductive ratio R_0, with a modified per capita birth rate of μ', is dynamically equivalent to an unvaccinated population with a birth rate of μ but with $R_0' = (1 - p)\frac{\mu'}{\mu} R_0$.

➤ To control a disease conforming to the assumptions of the SIR model, we can estimate the fraction of recruits into the susceptible class that need to be vaccinated to eradicate an infection: $p_c = 1 - 1/R_0$. All newborns need not be vaccinated as long as a threshold level of "herd immunity" within the population is established. Note that the critical vaccination threshold is a saturating function of R_0.

➤ When a fraction p of infants are vaccinated, the mean age at infection increases by a factor $\frac{1}{1-p}$.

➤ When vaccination is aimed at the susceptible population in general (and not newborns), the critical *rate* of vaccination for effective control is $p_c \geq \mu(R_0 - 1)$. Note that this scales linearly with R_0.

➤ Eradication with an imperfect vaccine, subject to incomplete protection and loss of vaccine-induced immunity, is much more difficult. If the vaccine is 100% protective, eradication requires the mean period of protection afforded by the vaccine to exceed the average life expectancy by a factor $1 - 1/R_0$.

➤ Pulse vaccination of susceptibles is an alternative approach to controlling infections. For any specific infection and host demographic traits, it is possible to derive the relationship between the pulse vaccination level (p_V) and the optimal pulsing period (T_c).

➤ Age-specific considerations point toward a focus on the very young ages in order to exert strongest control. This argument is strongly affected by complicating factors such as transplacentally derived antibodies.

➤ Age-specific complications of infection (as exemplified by rubella) can affect the aims of a public health program. In such situations controls will be aimed at reducing the level of disease (or severe symptoms), rather than simply reducing the total prevalence of infection.

➤ Targeted vaccination can be incorporated into models in order to highlight the differential effort that needs to be concentrated on high-risk groups before attention is turned to lower- risk categories.

➤ In simple models, it is possible to derive levels of isolation and quarantine practices needed for infection control during an epidemic. The addition of realistic complexities highlight the need for additional controls such as targeted and ring vaccination.

➤ The assumptions made concerning the distribution of latent and infectious periods is shown to be very important when making quantitative predictions.

➤ Using foot-and-mouth disease as a case study, the role of a large number of epidemiological and ecological complexities, in addition to alternative control options, are addressed. In particular, we have seen how effective control can be locally targetted when transmission is a spatial process.

➤ Examining the dynamics of swine fever in wild boar populations has highlighted how multiple seasonal factors can interact to produce complex and often counter-intuitive behavior.

Throughout this chapter we have seen how the understanding from simple tractable models can be extended to more applied senarios. Although the addition of extra realism generally involved a substantial increase in the number of parameters (which need to be estimated), such models are vital if modelers are to provide informed and accurate predictions.

References

Abad, F.X., Pinto, R.M., and Bosch, A. (1994) Survival of Enteric Viruses on Environmental Fomites. *App. Env. Micro.* **60** 3704–3710.

Abbey, H. (1952) An Examination of the Reed-Frost Theory of Epidemics. *Human Biol.* **24** 201–233.

Abramson, G., and Kenkre, V.M. (2002) Spatiotemporal patterns in the Hantavirus infection. *Phys. Rev. E* **66**, Art. No. 011912, Part 1.

Abramson, G., and Kenkre, V.M., Yates, T.L., and Parmenter, R.R. (2003) Traveling waves of infection in the hantavirus epidemics. *Bull. Math. Biol.* **65** 519–534.

Acha, P.N., and Szyfres, B. (1989) *Zoonoses and Communicable Diseases Common to Man and Animals.* Washington, D C: Pan American Health Organization.

Ades, A.E., and Nokes, D.J. (1993) Modelling age and time specific incidence from seroprevalence: Toxoplasmosis. *Am. J. Epid.* **137** 1022–1034.

Agur, Z., Cojocaru, L., Mazor, G., Anderson, R.M., and Danon, Y.L. (1993) Pulse Mass Measles Vaccination Across Age Cohorts. *P. N. A. S.* **90** 11698–11702.

Albert, R., Jeong, H., and Barabási, A-L. (1999) Diameter of the world-wide web. *Nature* **401** 130–131.

Albert, R., Jeong, H., and Barabási, A-L. (2000) Error and attack tolerance of complex networks. *Nature* **406** 378–381.

Albina, E. (1997) Epidemiology of porcine reproductive and respiratory syndrome (PRRS): An overview. *Vet. Micro.* **55** 309–316.

Alexander, M.E., Moghadas, S.M., Rohani, P., and Summers, A.R. (2006) Modelling the effect of a booster vaccination on disease epidemiology. *J. Math. Biol.* **52** 290–306.

Alibek, K., and Handelman, S. (1999) *Biohazard: the chilling true story of the largest covert biological weapons program in the world.* New York: Random House.

Allen, A.P., Li, B.L., and Charnov, E.L. (2001) Population fluctuations, power laws and mixtures of lognormal distributions. *Ecol. Lett.* **4** 1–3.

Allen, L.J.S., and Burgin, A.M. (2000) Comparison of deterministic and stochastic SIS and SIR models in discrete time. *Math. BioSci.* **163** 1–33.

Almond, J.W. (1998) Bovine spongiform encephalopathy and new variant Creutzfeldt-Jakob disease. *Brit. Med. Bull.* **54** 749–759.

Altizer, S.M., Hochachka, W.M., and Dhondt, A.A. (2004) Seasonal dynamics of *mycoplasmal conjunctivitis* in eastern North American House Finches. *J. An. Ecol.* **73** 309–322.

Altizer, S., Dobson, A. P., Hosseini, P., Hudson, P. J., Pascual, M., and Rohani, P. (2006) Seasonality and population dynamics: infectious diseases as case studies. *Ecol. Letters* **9** 467–484.

Anderson, P.K., Cunningham, A.A., Patel, N.G., Morales, F.J., Epstein, P.R., and Daszak, P. (2004) Emerging infectious diseases of plants: pathogen pollution, climate change and agrotechnology drivers. *Trends Ecol. and Evol.* **19** 535–544.

Anderson, R.M., and May, R.M. (1979) Population Biology of Infectious Diseases 1. *Nature* **280** 361–367.

Anderson, R.M., and May, R.M. (1982) Directly Transmitted Infectious Diseases: Control by Vaccination. *Science* **215** 1053–1060.

Anderson, R.M., and May, R.M. (1991) *Infectious Diseases of Humans.* Oxford, UK: Oxford University Press.

Anderson, R.M., May, R.M., Ng, T.W., and Rowley, J.T. (1992) Age-Dependent Choice of Sexual Partners and the Transmission Dynamics of HIV in Sub-Saharan Africa. *Phil. Trans. Roy. Soc. Lond. B* **336** 135–155.

Anderson, R.M., Medley, G.F., May, R.M., and Johnson, A.M. (1986) A Preliminary Study of the Transmission Dynamics of the Human Immunodeficiency Virus (HIV), the Causative Agent of AIDS. *IMA J. Math. Appl. Med. Biol.* **3** 229–263.

Andersson, H. (1998) Limit theorems for a random graph epidemic model. *Ann. Appl. Probab.* **8** 1331–1349.

Andersson, H., and Britton, T. (2000) Stochastic epidemics in dynamic populations: quasi-stationarity and extinction. *J. Math. Biol.* **41** 559–580.

Andreasen, V., Levin, S.A., and Lin, J. (1996) A model of influenza A drift evolution. *Z. Angew. Math. Mech.* **76** (S2) 421–424.

Andreasen, V., Lin, J., and Levin, S.A. (1997) The dynamics of co-circulating influenza strains confering partial cross-immunity. *J. Math. Biol.* **35** 825–842.

Anon (1978) Influenza in a boarding school. *Brit. Med. J.* **1**:587.

Ansari, M.A., and Razdan, R.K. (1998) Seasonal prevalence of *Aedes aegypti* in five localities of Delhi, India. *Dengue Bull.* **22** 28–34.

Ansari, S.A., Springthorpe, V.S., and Sattar, S.A. (1991) Survival and Vehicular Spread of Human Rotaviruses—Possible Relation to Seasonality of Outbreaks. *Rev. Inf. Dis.* **13** 448–461.

Antia, R., Regoes, R.R., Koella, J.C., and Bergstrom, C.T. (2003) The role of evolution in the emergence of infectious diseases. *Nature* **426** 658–661.

Arca, M., Perucci, C.A., and Spadea, T. (1992) The Epidemic Dynamics of HIV-1 in Italy—Modeling the interaction between Intravenous-Drug-Users and Heterosexual population. *Stat. Med.* **11** 1657–1684.

Arita, I., Wickett, J., and Fenner, F. (1986) Impact of population density on immunization programmes. *J. Hyg.* **96** 459–66.

Arrowsmith, D.K., and Place, C.M. (1992) *Dynamical Systems: Differential Equations, Maps and Chaotic Behaviour.* Boca Raton, FL: CRC Press.

Artzrouni, M., Brown, T., Feeney, G., Garnett, G., Ghys, P., Grassly, N., Schneider, D., Stanecki, K., Stover, J., Schwartlander, B., Walker, N., Way, P., Yan, P., Zaba, B., Zlotnik, H., Timaeus, I., and Walker, N. (2002) Improved methods and assumptions for estimation of the HIV/AIDS epidemic and its impact: Recommendations of the UNAIDS Reference Group on Estimates, Modelling and Projections. *AIDS* **16** W1–14.

Augustine, D.J. (1998) Modelling Chlamydia-koala interactions: coexistence, population dynamics and conservation implications. *J. Appl. Ecol.* **35** 261–271.

Austin, D.J., and Anderson, R.M. (1999a) Studies of antibiotic resistance within the patient, hospitals and the community using simple mathematical models. *Phil. Trans. Roy. Soc. B* **354** 721–738.

Austin, D.J., and Anderson, R.M. (1999b) Transmission dynamics of epidemic methicillin-resistant Staphylococcus aureus and vancomycin resistant enterococci in England and Wales. *J. Infect. Dis.* **179** 883–891.

Auvert, B., Buonamico, G., Lagarde, E., and Williams, B. (2000) Sexual behavior, heterosexual transmission, and the spread of HIV in sub-Saharan Africa: A simulation study. *Comput. Biomed. Res.* **33** 84–96.

Bailey, N.T.J. (1975) *The Mathematical Theory of Infectious Diseases and its Applications.* London: Charles Griffin and Company.

Bak, P., Chen, K., and Tang, C. (1990) A forest fire model and some thoughts on turbulence. *Phys. Letters A* **147** 297–300.

Baquero F., and Blazquez J. (1997) Evolution of antibiotic resistance. *Trends Ecol. & Evol.* **12** 482–487.

Barabási, A.L., and Albert, R. (1999) Emergence of scaling in random networks. *Science* **286** 509–512.

Barlow, N.D., Kean, J.M., Hickling, G., Livingstone, P.G., and Robson, A.B. (1997) A simulation model for the spread of bovine tuberculosis within New Zealand cattle herds. *Prev. Vet. Med.* **32** 57–75.

Barnett, P.V., and Carabin, H. (2002) A review of emergency foot-and-mouth disease (FMD) vaccines. *Vaccine* **20** 1505–1514.

Barre-Sinioussi, F., Chermann, J.C., Rey, F., Nugeyre, M.T., Chamaret, S., Gruest, J., Dauguet, C., Axler-Blin, C., Vezinet-Brun, F., Rouzioux, C., Rozenbaum, W., and Montagnier, L. (1983) Isolation of a T-lymphotropic retrovirus from a patient at risk for acquired immune deficiency (AIDS). *Science* **220** 868–871.

Bartlett, M.S. (1956) Deterministic and Stochastic Models for recurrent epidemics. *Proc. Third Berkley Symp. on Math. Stats. and Prob.* **4** 81–108.

Bartlett, M.S. (1957) Measles periodicity and community size. *J. R. Statistical Soc. A* **120** 48–70.

Bartlett, M.S. (1960) The critical community size for measles in the U.S. *J. R. Statistical Soc. A* **123** 37–44.

Bascompte J., and Rodriguez-Trelles F. (1998) Eradication thresholds in epidemiology, conservation biology and genetics. *J. Theor. Biol.* **192** 415–418.

Bates, T.W., Thurmond, M.C., and Carpenter, T.E. (2003) Description of an epidemic simulation model for use in evaluating strategies to control an outbreak of foot-and-mouth disease. *Am. J. Vet. Res.* **64** 195–204.

Bauch, C.J., and Rand, D.A. (2000) A moment closure model for sexually transmitted disease transmission through a concurrent partnership network. *Proc. Roy. Soc. Lond. B* **267** 2019–2027.

Bauch, C.T., and Earn, D.J.D. (2003) Transients and attractors in epidemics. *Proc. Roy. Soc. Lond.* **270** 1573–1578.

Beardmore, I., and Beardmore, R. (2003) The global structure of a spatial model of infectious disease. *Proc. Roy. Soc. Lond. A* **459** 1427–1448.

Begon, M., Bennett, M., Bowers, R.G., French, N.P., Hazel, S.M., and Turner, J. (2002) A clarification of transmission terms in host-microparasite models: numbers, densities and areas. *Epidemiol. Infect.* **129** 147–153.

Begon, M., Harper, J.L., and Townsend, C.R. (1996) *Ecol.* Oxford, UK: Blackwell Science.

Begon, M., Hazel, S.M., Baxby, D., Bown, K., Cavanagh, R., Chantrey, J., Jones, T., et al. (1999) Transmission dynamics of a zoonotic pathogen within and between wildlife host species. *Proc. Roy. Soc. Lond. B* **266** 1939–1945.

Behrman, R.E., and Kliegman, R.M. (1998) *Nelson Essentials of Pediatrics*. Philadelphia: Saunders.

Bergelson, J., Dwyer, G., and Emerson, J.J. (2001) Models and data on plant-enemy coevolution. *Annu. Rev. Genet.* **35** 469–499.

Bjørnstad, O.N., Finkenstädt, B.F., and Grenfell, B.T. (2002) Dynamics of measles epidemics: Estimating scaling of transmission rates using a time series SIR model. *Ecol. Monographs* **72** 185–202.

Black, F.L. (1966) Measles endemicity in insular populations: critical community size and its evolutionary implications. *J. Theo. Biol.* **11** 207–211.

Bohannan, B.J.M., and Lenski, R.E. (2000) Linking genetic change to community evolution: insights from studies of bacteria and bacteriophage. *Ecol. Letters* **3** 362–377.

Bohrer, G., Shem-Tov, S., Summer, E., Or, K., and Saltz, D. (2002) The effectiveness of various rabies spatial vaccination patterns in a simulated host population with clumped distribution. *Ecol. Model.* **152** 205–211.

Boily, M.C., and Masse, B. (1997) Mathematical models of disease transmission: A precious tool for the study of sexually transmitted diseases. *Can. J. Pub. Health* **88** 255–265.

Bolker, B.M. (1993) Chaos and Complexity in measles models: a comparative numerical study *IMA J. Math. Appl. Med. Biol.* **10** 83–95.

Bolker, B.M. (1999) Analytic models for the patchy spread of plant disease. *Bull. Math. Biol.* **61** 849–874.

Bolker, B.M., and Grenfell, B.T. (1993) Chaos and biological complexity in measles dynamics. *Proc. Roy. Soc. Lon. B* **251** 75–81.

Bolker, B.M., and Grenfell, B.T. (1995) Space, persistence and dynamics of measles epidemics *Phil. Trans. Roy. Soc. Lond. B* **348** 309–320.

Bolker, B.M., and Grenfell, B.T. (1996) Impact of vaccination on the spatial correlation and persistence of measles dynamics. *PNAS* **93** 12648–12653.

Bolker, B.M., and Pacala, S.W. (1997) Using moment equations to understand stochastically driven spatial pattern formation in ecological systems. *Theo. Pop. Biol.* **52** 179–197.

Bolzoni, L., Dobson, A.P., Gatto, M., and De Leo, G.A. (2006) The Effect of Seasonality on the Epidemics of Wildlife Diseases. *Ecol. Letters* (in press).

Bongaarts, J. (1989) A Model of the Spread of HIV Infection and the Demographic-Impact of AIDS. *Stat. Med.* **8** 103–120.

Boots, M., and Sasaki, A. (1999) 'Small worlds' and the evolution of virulence: infection occurs locally and at a distance. *Proc. Roy. Soc. Lond. B* **266** 1933–1938.

Boots, M., and Sasaki, A. (2002) Parasite-driven extinction in spatially explicit host-parasite systems. *Am. Nat.* **159** 706–713.

Bourne, J., Donnelly, C.A., Cox, D.R., Gettinby, G., McInerney, J.P., Morrison, I., and Woodroffe, R. (2000) Bovine tuberculosis: towards a future control strategy. *Vet. Rec.* **146** 207–210.

Box, G.E.P., and Muller, M.E. (1958) A note on the generation of random normal deviates. *Annals Math. Stat.* **29** 610–611.

Boyce, M.S., Sinclair, A.R.E., and White, G.C. (1999) Seasonal compensation of predation and harvesting. *Oikos* **87** 419–426.

Bozzette, S.A., Boer, R., Bhatnagar, V., Brower, J.L., Keeler, E.B., Morton, S.C., and Stoto, M.A. (2003) A model for a smallpox-vaccination policy. *N. Engl. J. Med.* **348** 416–425.

Brauer, F. (2002) Basic Ideas of Mathematical Epidemiology. *Mathematical Approaches for Emerging and Reemerging Infectious Diseases: Models, Methods, and Theory* (eds. Castillo-Chavez, C., Blower, S., van den Driessche, P., Kirschner, D., and Yakubu, A-A.). Berlin: Springer-Verlag.

Bremermann, H.J., and Thieme, H.R. (1989) A Competitive-Exclusion Principle for Pathogen Virulence. *J. Math. Bio.* **27** 179–190.

Broadfoot, J.D., Rosatte, R.C., and O'Leary, D.T. (2001) Raccoon and skunk population models for urban disease control planning in Ontario, Canada. *Ecol. Appl.* **11** 295–303.

Buceta, J., Escudero, C., de la Rubia, F.J., and Lindenberg, K. (2004) Outbreaks of Hantavirus induced by seasonality. *Phys. Rev. E* **69**, Art. No. 021906, Part 1.

Bunn, C.M., Garner, M.G., and Cannon, R.M. (1998) The 1872 outbreak of foot-and-mouth disease in Australia—why didn't it become established? *Aust. Vet. J.* **76** 262–269.

Burdon, J.J., and Thrall, P.H. (1999) Spatial and temporal patterns in coevolving plant and pathogen associations. *Am. Nat.* **153** S15–S33.

Burdon, J.J., Ericson, L., and Muller, W.J. (1995) Temporal and spatial changes in a metapopulation of the rust pathogen Triphragmium ulmariae and its host, Filipendula ulmaria. *J. Ecol.* **83** 979–989.

Campbell, G.L., Marfin, A.A., Lanciotti, R.S., and Gubler, D.J. (2002) West Nile Virus. *Lancet Infect. Dis.* **2** 519–529.

Cannon, R.M., and Garner, M.G. (1999) Assessing the risk of wind-borne spread of foot-and-mouth disease in Australia. *Environ. Int.* **25** 713–723.

Caraco, T., Glavanakov, S., Chen, G., Flaherty, J.E., Ohsumi, T.K., and Szymanski, B.K. (2002) Stage-structured infection transmission and a spatial epidemic: A model for Lyme disease. *Am. Nat.* **160** 348–359.

Caswell, H. (2000) *Matrix Population Models: Construction, Analysis, and Interpretation.* Sunderland, MA: Sinauer.

Cates, W. et al. (1999) Estimates of the Incidence and Prevalence of Sexually Transmitted Diseases in the United States. *Sex. Trans. Dis.* **26** (suppl):S2–S7.

Cates, W., Rothenberg, R.B., and Blount, J.H. (1996) Syphilis control—The historic context and epidemiologic basis for interrupting sexual transmission of Treponema pallidum. *Sex. Trans. Dis.* **23** 68–75.

Caughey, B., and Chesebro, B. (1997) Prion protein and the transmissible spongiform encephalopathies. *Trends Cell Biol.* **7** 56–62.

Chesson, H.W., and Pinkerton, S.D. (2000) Sexually transmitted diseases and the increased risk for HIV transmission: Implications for cost-effectiveness analyses of sexually transmitted disease prevention interventions. *J. AIDS* **24** 48–56.

Chesson, H.W., Pinkerton, S.D., Irwin, K.L., Rein, D., and Kassler, W.J. (1999) New HIV cases attributable to syphilis in the USA: estimates from a simplified transmission model. *AIDS* **13** 1387–1396.

Choisy, M., and Rohani, P. (2006) Harvesting can increase severity of wildlife disease epidemics. *Proc. Roy. Soc. Lond. Series B* **273** 2025–2034.

Choisy, M., Guegan, J.F., and Rohani, P. (2006) Resonance effects and the dynamics of infectious diseases. *Physica D* **223** 26–35.

Clancy, D. (1996) Carrier-borne epidemic models incorporating population mobility. *Math. BioSci.* **132** 185–204.

Clancy, D., and French, N.P. (2001) A Stochastic model for Disease Transmission in a Managed Herd, motivated by Neospora caninum amoungst dairy cattle. *Math. BioSci.* **170** 113–132.

Clark, J.S., and Bjornstad, O.N. (2004) Population Time Series: Process Variability, Observation Errors, Missing Values, Lags, and Hidden States. *Ecol.* **85** 3140–3150.

Clarkson, J.A., and Fine, P.E.M. (1985) The Efficiency of Measles and Pertussis Notification in England and Wales. *Int. J. Epid.* **14** 153–168.

Cleaveland, S., Laurenson, M.K., and Taylor, L.H. (2001) Diseases of humans and their domestic mammals: pathogen characteristics, host range and the risk of emergence. *Phil. Trans. Roy. Soc. Lond. B* **356** 991–999.

Cliff, A., and Haggett, P. (2004) Time, travel and infection. *Brit. Med. Bull.* **69** 87–99.

Cliff, A., Haggett, P., and Smallman-Raynor, M. (1993) *Measles. An historical geography of a major human viral disease from global expansion to local retreat 1840–1990.* Oxford, UK: Blackwell.

Cliff, A.P., Haggett, P., Ord, J.K., and Versey, G.R. (1981) *Spatial diffusion: an historical geography of epidemics in an island community.* Cambridge, UK: Cambridge University Press.

Clutton-Brock, T.H., Illius, A.W., Wilson, K., Grenfell, B.T., MacColl, A.D.C., and Albon, S.D. (1997) Stability and instability in ungulate populations: an empirical analysis. *Am. Nat.* **149** 195–219.

Coggins, C., and Segal, S. (1998) AIDS and reproductive health. *J. Reprod. Immunol.* **41** 3–15.

Collins, S.D., Wheeler, R.E., and Shannon, R.D. (1942) The occurrence of whooping cough, chickenpox, mumps, measles and german measles in 200,000 surveyed families in 28 large cities. *Special Studies Series*, 1, Washington U.S.P.H.S.

Cooper, B.S., Medley, G.F., Stone, S.P., Kibbler, C.C., Cookson, B.D., Roberts, J.A., Duckworth, G., Lai, R., and Ebrahim, S. (2004) Methicillin-resistant Staphylococcus aureus in hospitals and the community: Stealth dynamics and control catastrophes. *Proc. Natl. Acad. Sci. USA* **101** 10223–10228.

Corbel, M.J. (1997) Brucellosis: an overview. *Emerg. Infect. Dis.* **3** 213–221.

Coulson, T.N., Rohani, P., and Pascual, M. (2004) Skeletons, noise and population growth: the end of an old debate? *Trends Ecol. & Evol.* **19** 360–364.

Courchamp, F., Clutton-Brock, T., and Grenfell, B. (1999) Inverse density dependence and the Allee effect. *Trends Ecol. & Evol.* **14** 405–410

Daley, D.J., and Gani, J. (1999) *Epidemic Modelling.* Cambridge, UK: Cambridge University Press.

Dawes, J.H.P., and Gog, J.R. (2001) The onset of oscillatory dynamics in models of multiple disease strains. *J. Math. Biol.* **45** 471–510.

Day, K.P., and Marsh, K. (1991) Naturally Acquired-Immunity to Plasmodium-Falciparum. *Immunoparasitol. Today* **3** A68–A71.

Daszak, P., Cunningham, A.A., and Hyatt, A.D. (2000) Emerging infectious diseases of wildlife—Threats to biodiversity and human health. *Science* **287** 443–449.

DEFRA (2004) Foot and Mouth Disease Contingency Plan, http://www.defra.gov.uk/footandmouth/contingency/index.htm

de Jong, J.C., Rimmelzwaan, G.F., Fouchier, R.A.M., and Osterhaus, A.D.M.E. (2000) Influenza virus: a master of metamorphosis. *J. Infect.* **40** 218–228.

de Jong, M.C.M., Dieckmann, O., and Heesterbeek, J. (1995) How does transmission of infection depend on population size? *Epidemic Models: Their Structure and Relation to Data* (ed. Mollison, D.),p. 8494. Cambridge, UK: Cambridge University Press.

Delamaza, M.A., and Delamaza, L.M. (1995) A New Computer-Model for Estimating the Impact of Vaccination Protocols and its Application to the Study of Chlamydia-Trachomatis Genital Infections. *Vaccine* **13** 119–127.

de Mendonça, J.R.G. (1999) Precise critical exponents for the basic contact process. *J. Phys. A: Math. Gen.* **32** L467–L473.

de Quadros, C.A., Olive, J.M., Hersh, B.S., Strassburg, M.A., Henderson, D.A., Brandling-Bennett, D., and Alleyne, G.A. (1996) Measles elimination in the Americas. Evolving strategies. *J. Am. Med. Assoc.* **275** 224–229.

Dexter, N. (2003) Stochastic models of foot and mouth disease in feral pigs in the Australian semi-arid rangelands. *J. Appl. Ecol.* **40** 293–306.

Dhooge, A., Govaerts, W., and Kuznetsov, Y.A. (2003) MATCONT: A MATLAB package for numerical bifurcation analysis of ODEs. *ACM Trans. Math. Soft.* **29** 141–164.

Dieckmann, U., Law, R., and Metz, J.A.J. (2000) *The Geometry of Ecological Interactions: Simplifying Spatial Complexity*. Cambridge Studies in Adaptive Dynamics. Cambridge, UK: Cambridge University Press.

Diekmann, O. (1978) Thesholds and Travelling Waves for the Geographical Spread of Infection. *J. Math. Biol.* **6** 109–130.

Dieckmann, O., de Jong, M.C.M., and Metz, J.A.J. (1998) A deterministic epidemic model taking account of repeated contacts between the same individuals. *J. Appl. Prob.* **35** 448–462.

Diekmann, O., and Heesterbeek, J.A.P. (2000) *Mathematical Epidemiology of Infectious Diseases: Model Building, Analysis and Interpretation*. Hoboken, NJ: Wiley.

Diekmann, O., Heesterbeek, J.A.P., and Metz, J.A.J. (1990) On the Definition and the Computation of the Basic Reproduction Ratio R_0 in Models for Infectious-Diseases in Heterogeneous Populations. *J. Math. Biol.* **28** 365–382.

Dietz, K. (1967) Epidemics and Rumours: A Survey. *J. Roy. Stat. Soc. A* **130** 505–528.

Dietz, K. (1976) The Incidence of Infectious Diseases Under the Influence of Seasonal Fluctuations. *Lecture Notes in Biomath.* **11** 1–15.

Dietz, R., Heide-Jørgenson, M-P., and Härkönen, T. (1989) Mass death of harbour seals *Phoca vitulina* in Europe. *Ambio* **18** 258–264.

Dingle, H. (1996) *Migration: The Biology of Life on the Move.* New York: Oxford University Press.

Dobson, A. (2004) Population Dynamics of Pathogens with Multiple Host Species. *Am. Nat.* **164** S64–S78.

Dobson, A., and Meagher, M. (1996) The population dynamics of brucellosis in the Yellowstone National Park. *Ecol.* **77** 1026–1036.

Doedel, E.J., Champneys, A.R., Fairgrieve, T.F., Kuznetsov, Y.A., Standstede, B., and Wang, X. (1998) *AUTO 97: Continuation and Bifurcation Software For Ordinary Differential Equations.*

Doherty, I.A., Padian, N.S., Marlow, C., and Aral, S.O. (2005) Determinants and consequences of sexual networks as they affect the spread of sexually transmitted infections. *J. Inf. Dis.* **191** S42–S54.

Donnelly, C.A., Ferguson, M.N., Ghani, A.C., Woolhouse, M.E.J., Watt, C.J., and Anderson, R.M. (1997) The epidemiology of BSE in cattle herds in Great Britain. I. Epidemiological processes, demography of cattle and approaches to control by culling. *Phil. Trans. Roy. Soc. Lond. B* **352** 781–802.

Donnelly, C.A., Ghani, A.C., Leung, G.M., Hedley, A.J., Fraser, C., Riley, S., Abu-Raddad, L.J., Ho, L.M., Thach, T.Q., Chau, P., Chan, K.P., Lam, T.H., Tse, L.Y., Tsang, T., Liu, S.H., Kong, J.H.B., Lau, E.M.C., Ferguson, N.M., and Anderson R.M. (2003) Epidemiological determinants of spread of causal agent of severe acute respiratory syndrome in Hong Kong. *Lancet* **361** 1761–1766.

d'Onofrio, A. (2002) Stability properties of pulse vaccination strategy in *SEIR* epidemic model. *Math. Biosci.* **179** 57–72.

Downs, A.M., Heisterkamp, S.H., Brunet, J.B., and Hamers, F.F. (1977) Reconstruction and prediction of the HIV/AIDS epidemic among adults in the European Union and in the low prevalence countries of central and eastern Europe. *AIDS* **11** 649–662.

Dubey, J.P. (1988) *Toxoplasmosis of animals and man.* Boca Raton, FL: CRC Press.

Dugaw, C.J., Hastings, A., Preisser, E.L., and Strong, D.R. (2004) Seasonally limited host supply generates microparasite population cycles. *Bull. Math. Biol.* **66** 583–594.

Dunyak, J., Martin, C., and Lampe, R. (1998) Analysis of the influence of social structure on a measles epidemic. *Appl. Math. Comput.* **92** 282–296.

Dushoff, J., Plotkin, J.B., Levin, S.A., and Earn, D.J.D. (2004) Dynamical resonance can account for seasonality of influenza epidemics. *Proc. Natl. Acad. Sci. U.S.A.* **101** 16915–16916.

Eames, K.T.D., and Keeling, M.J. (2002) Modeling dynamic and network heterogeneities in the spread of sexually transmitted disease. *Proc. Natl. Acad. Sci. U.S.A.* **99** 13330–13335.

Eames, K.T.D., and Keeling, M.J. (2003) Contact tracing and disease control. *Proc. Royal. Soc. Lond. Series B* **270** 2565–2571.

Eames, K.T.D., and Keeling, M.J. (2004) Monogamous networks and the spread of sexually transmitted diseases *Math. Biosci.* **189** 115–130.

Earn, D.J., Dushoff, J., and Levin, S.A. (2002) Ecology and evolution of the flu. *Trends Ecol. & Evol.* **17** 334–240.

Earn, D.J.D., Rohani, P., and Grenfell, B.T. (1998) Persistence, chaos and synchrony in ecology and epidemiology. *Proc. Roy. Soc. Lond. B* **265** 7–10.

Earn, D.J.D., Rohani, P., Bolker, B.M., and Grenfell, B.T. (2000) A Simple Model for Complex Dynamical Transitions in Epidemics. *Science* **287** 667–670.

East, M.L., Hofer, H., Cox, J.H., Wulle, U., Wiik, H., and Pitra, C. (2001) Regular exposure to rabies virus and lack of symptomatic disease in Serengeti spotted hyenas. *Proc. Natl. Acad. Sci. U.S.A.* **98** 15026–15031.

Ebert, D., and Bull, J.J. (2003) Challenging the trade-off model for the evolution of virulence: is virulence management feasible? *Trends Microbiol.* **11** 15–20.

Edmunds, W.J., O'Callaghan, C.J., and Nokes, D.J. (1997) Who mixes with whom? A method to determine the contact patterns of adults that may lead to the spread of airborne infections. *Proc. Roy. Soc. Lond. B* **1384** 949–957.

Eichner, M., and Dietz, K. (2003) Transmission Potential of Smallpox: Estimates Based on Detailed Data from an Outbreak. *Am. J. Epidemiol.* **158** 110–117.

Eidson, M., Komar, N., Sorhage, F., Nelson, R., Talbot, T., Mostashari, F., and McLean, R. (2001) Crow Deaths as a Sentinel Surveillance System for West Nile Virus in the Northern United States, 1999. *Emerg. Inf. Dis.* **7** 615–620.

Elbers, A.R.W., Stegeman, A., Moser, H., Ekker, H.M., Smak, J.A., Pluimers, F.H. (1999) The classical swine fever epidemic 1997–1998 in the Netherlands: descriptive epidemiology. *Prev. Vet. Med.* **42** 157–184.

Ericson, L., Burdon, J.J., and Muller, W.J. (1999) Spatial and temporal dynamics of epidemics of the rust fungus Uromyces valerianae on populations of its host Valeriana salina. *J. Ecol.* **87** 649–658.

Erlander, S., and Stewart, N.F. (1990) *The gravity model in transportation analysis—theory and extensions.* The Netherlands: International Science Publishers.

Escriu, F., Fraile, A., and Garc-a-Arenala, F. (2002) The Evolution of Virulence in a plant virus. *Evolut.* **57** 75765.

Farr, W. (1840) Progress of Epidemic. *Second Report of the Registrar General of England* 91–98.

Farrington, C.P., and Grant, A.D. (1999) The distribution of time to extinction in subcritical branching processes: applications to outbreaks of infectious disease. *J. Appl. Prob.* **36** 771–779.

Fenner, F., Henderson, D.A., Arita, I., Jezek, Z., and Ladnyi, I.D. (1988) *Smallpox and Its Eradication.* Geneva: World Health Organization.

Ferguson, N.M., Anderson, R.M., and Garnett, G.P. (1996a) Mass vaccination to control chickenpox: The influence of zoster. *Proc. Natl. Adac. Sci. USA* **93** 7231–7235.

Ferguson, N.M., Anderson, R.M., and Gupta, S. (1999a) The effect of antibody-dependent enhancement on the transmission dynamics and persistence of multiple-strain pathogens. *Proc. Natl. Adac. Sci. USA* **96** 790–794.

Ferguson, N.M., Anderson, R.M., and May, R.M. (1997a) Scale, persistence and synchronicity: measles as a paradigm of a spatially-structured ecosystem. *Spatial Ecology: the Role of Space in Population Dynamics and Interspecific Interactions* (eds. Tilman, D. and Kareiva, P.), Princeton, NJ: Princeton University Press.

Ferguson, N.M., Cummings, D.A., Cauchemez, S., Fraser, C., Riley, S., Meeyai, A., Iamsirithaworn, S., and Burke, D.S. (2005) Strategies for containing an emerging influenza pandemic in Southeast Asia. *Nature* **437** 209–214.

Ferguson, N.M., Cummings, D.A., Fraser, C., Cajka, J.C., Cooley, P.C., and Burke, D.S. (2006) Strategies for mitigating an influenza pandemic. *Nature* (in press).

Ferguson, N.M., Donnelly, C.A., and Anderson, R.M. (2001a) The foot-and-mouth epidemic in Great Britain: Pattern of spread and impact of interventions. *Science* **292** 1155–1160.

Ferguson, N.M., Donnelly, C.A., and Anderson, R.M. (2001b) Transmission intensity and impact of control policies on the foot and mouth epidemic in Great Britain. *Nature* **413** 542–548.

Ferguson, N.M., Donnelly, C.A., Woolhouse, M.E.J., and Anderson, R.M. (1997b) The epidemiology of BSE in cattle herds in Great Britain. I. Model construction and analysis of transmission dynamics. *Phil. Trans. Roy. Soc. Lond. B* **352** 803–838.

Ferguson, N.M., Donnelly, C.A., Woolhouse, M.E.J., and Anderson, R.M. (1999b) Estimation of the basic reproductive number of BSE: the intensity of transmission in British cattle. *Proc. Roy. Soc. Lond. B* **266** 23–32.

Ferguson, N.M., Galvani, A.P., and Bush, R.M. (2003a) Ecological and immunological determinants of influenza evolution. *Nature* **422** 428–433.

Ferguson, N.M., Keeling, M.J., Edmunds, W.J., Gant, R., Grenfell, B.T., Anderson, R.M., and Leach, S. (2003b) Planning for smallpox outbreaks. *Nature* **425** 681–685.

Ferguson, N.M., Nokes, D.J., and Anderson, R.M. (1996b) Dynamical complexity in age-structured models of the transmission of measles virus. *Math. BioSci.* **138** 101–130.

Ferrari, M.J., Bjornstad, O.N., and Dobson, A. (2005) Estimation and inference of R_0 of an infectious pathogen by a removal method. *Math. Biosci.* **198** 14–26.

Fine, P.E. (1993) Herd immunity: history, theory, practice. *Epidemiol. Rev.* **15** 265–302.

Fine, P.E.M., and Clarkson, J.A. (1982) Measles in England and Wales. I: An Analysis of Factors Underlying Seasonal Patterns *Int. J. Epidemiol.* **11** 5–14.

Finkenstädt, B., and Grenfell, B. (2000) Time series modelling of childhood diseases: a dynamical systems approach. *J. R. Statist. Soc. C., Applied Statist.* **49** 187–205.

Finkenstädt, B., and Grenfell, B. (1998) Empirical determinants of measles metapopulation dynamics in England and Wales. *Proc. Roy. Soc. Lond. B* **265** 211–220.

Finkenstädt, B., Keeling, M.J., and Grenfell, B. (1998) Patterns of density dependence in measles dynamics. *Proc. Roy. Soc. Lond. B* **265** 753–762.

Fitch, W.M., Bush, R.M., Bender, C.A., and Cox, N.J. (1997) Long term trends in the evolution of H(3)HA1 human influenza type A. *Proc. Natl. Adac. Sci. USA* **94** 7712–7718.

Foley, J.E., Foley, P., and Pedersen, N.C. (1999) The persistence of a *SIS* disease in a metapopulation. *J. Appl. Ecol.* **36** 555–563.

Folstad, I., Nilssen, F.I., Halvorsen, A.C., and Andersen, O. (1991) Parasite avoidance: the cause of post-calving migrations in Rangifer? *Canadian J. Zool.* **69** 2423–2429.

Frank, S.A., and Jeffrey, J.S. (2001) The probability of severe disease in zoonotic and commensal infections. *Proc. Roy. Soc. Lond. B* **268** 53–60.

Fraser, C., Riley, S., Anderson, R.M., and Ferguson, N.M. (2004) Factors that make an infectious disease outbreak controllable. *Proc. Natl. Adac. Sci. USA* **101** 6146–6151.

Fulford, G.R., Roberts, M.G., and Heesterbeek, J.A.P. (2002) The metapopulation dynamics of an infectious disease: Tuberculosis in possums. *Theor. Pop. Biol.* **61** 15–29.

Furniss, P.R., and Hahn, B.D. (1981) A Mathematical-model of an Anthrax epizootic in the Kruger National-Park. *Appl. Math. Model.* **5** 130–136.

Gaillard, J.M., Festa-Bianchet, M., Yoccoz, N.G., Loison, A., and Toïgo, C. (2000) Temporal variation in fitness component and population dynamics of large herbivores. *Ann. Rev. Ecol. Syst.* **31** 367–393.

Gamerman, D. (1997) *Markov Chain Monte Carlo: Stochastic Simulation for Bayesian Inference.* Chapman & Hall.

Gandon, S., Mackinnon, M.J., Nee, S., and Read, A.F. (2001) Imperfect vaccines and the evolution of pathogen virulence. *Nature* **414** 751–756.

Gandon, S., Mackinnon, M.J., Nee, S., and Read, A.F. (2003) Imperfect vaccination: some epidemiological and evolutionary consequences. *Proc. Roy. Soc. Lond. B* **270** 1129–1136.

Gani, R., and Leach, S. (2001) Transmission potential of smallpox in contemporary populations. *Nat.* **414** 748–751.

Garly, M.A., and Aaby, P. (2003) The challenge of improving the efficacy of measles vaccine. *Acta Tropica* **85** 1–17.

Garner, M.G., and Lack, M.B. (1995) An Evaluation of Alternative Control Strategies for Foot-and-Mouth-Disease in Australia—A Regional Approach. *Prev. Vet. Med.* **23** 9–32.

Garnett, G.P., and Anderson, R.M. (1993a) Contact Tracing and the Estimation of Sexual Mixing Patterns: The Epidemiology of Gonococcal Infections. *Sex. Trans. Dis.* **20** 181–191.

Garnett, G.P., and Anderson, R.M. (1993b) Factors Controlling the Spread of HIV in Heterosexual Communities in Developing-Countries—Patterns of Mixing between Different Age and Sexual-Activity Classes. *Phil. Trans. Roy. Soc. Lond. B* **342** 137–159.

Garnett, G.P., and Anderson, R.M. (1996) Sexually transmitted diseases and sexual behavior: Insights from mathematical models. *J. Infect. Dis.* **174** S150–S161.

Garnett, G.P., and Bowden, F.J. (2000) Epidemiology and control of curable sexually transmitted diseases—Opportunities and problems. *Sex. Trans. Dis.* **27** 588–599.

Garnett, G.P., Mertz, K.J., Finelli, L., Levine, W.C., and St Louis, M.E. (1999) The transmission dynamics of gonorrhoea: modelling the reported behaviour of infected patients from Newark, New Jersey. *Phil. Trans. Roy. Soc. Lond. B* **354** 787–797.

Ghani, A.C., Ferguson, N.M., Donnelly, C.A., and Anderson, R.M. (2000) Predicted vCJD mortality in Great Britain—Modelling the latest data puts a ceiling on the likely number of vCJD cases. *Nature* **406** 583–584.

Ghani, A.C., Ferguson, N.M., Donnelly, C.A., and Anderson, R.M. (2003a) Factors determining the pattern of the variant Creutzfeldt-Jakob disease (vCJD) epidemic in the UK. *Proc. Roy. Soc. Lond. B* **270** 689–698.

Ghani, A.C., Ferguson, N.M., Donnelly, C.A., and Anderson, R.M. (2003b) Short-term projections for variant Creutzfeldt-Jakob disease onsets. *Stat. Meth. Med. Res.* **12** 191–201.

Gibbs, M.J., Armstrong, J.S., and Gibbs, A.J. (2001) Recombination in the hemagglutinin gene of the 1918 "Spanish flu." *Science* **293** 1842–1845.

Gibson, G.J. (1997a) Investigating Mechanisms of Spatiotemporal Epidemic Spread Using Stochastic Models. *Phytopathol.* **87** 139–146.

Gibson, G.J. (1997b) Markov Chain Monte Carlo Methods for Fitting Spatiotemporal Stochastic Models in Plant Epidemiology. *Appl. Statist.* **46** 215–233.

Gibson, G.J., and Austin, E.J. (1996) Fitting and testing spatio-temporal stochastic models with application in plant epidemiology. *Plant Pathol.* **45** 172–184.

Gibson, G.J., Gilligan, C.A., and Kleczkowski, A. (1999) Predicting variability in biological control of a plant-pathogen system using stochastic models. *Proc. Roy. Soc. Lond. B* **266** 1743–1753.

Gibson, M.A., and Bruck J. (2000) Efficient Exact Stochastic Simulation of Chemical Systems with Many Species and Many Channels. *J. Phys. Chem.* **104** 1876–1889.

Gillespie, D.T. (1976) General method for numerically simulating stochastic evolution of coupled chemical-reactions. *J. Comput. Phys.* **22** 403–434.

Gillespie, D.T. (1977) Exact stochastic simulation of coupled chemical reactions. *J. Phys. Chem.* **81** 2340–2361.

Gillespie, D.T. (2001) Approximate accelerated stochastic simulation of chemically reacting systems. *J. Chem. Phys.* **115** 1716–1733.

Gillespie, D.T. and Petzold L.R. (2003) Improved leap-size selection for accelerated stochastic simulation. *J. Chem. Phys.* **119** 8229–8234.

Goddard, L.B., Roth, A.E., Reisen, W.K., and Scott, T.W. (2002) Vector Competence of California Mosquitoes for West Nile Virus. *Emerg. Inf. Dis.* **8** 1385–1391.

Gog, J., Woodroffe, R., and Swinton, J. (2002) Disease in endangered metapopulations: the importance of alternative hosts. *Proc. Roy. Soc. Lond. B* **269** 671–676.

Gog, J.R., and Grenfell, B.T. (2002) Dynamics and selection of many-strain pathogens. *Proc. Natl. Acad. Sci. USA* **99** 17209–17214.

Gog, J.R., and Swinton, J. (2002) A status based approach to multiple strain dynamics. *J. Math. Biol.* **44** 169–184.

Gomes, M.G.M., Medley, G.F., and Nokes, D.J. (2002) On the determinants of population structure in antigenically diverse pathogens. *Proc. Roy. Soc. Lond. B* **269** 227–233.

Goodchild, A.V., and Clifton-Hadley, R.S. (2001) Cattle-to-cattle transmission of *Mycobacterium bovis*. *Tuberculosis* **81** 23–41.

Grais, R.F., Ellis, J.H., and Glass, G.E. (2003) Assessing the impact of airline travel on the geographic spread of pandemic influenza. *Eur. J. Epidemiol.* **18** 1065–1072.

Grassly, N.C., Fraser, C., and Garnett, G.P. (2005) Host immunity and synchronized epidemics of syphilis across the United States. *Nature* **433** 417–421.

Grassly, N.C., and Fraser, C. (2006) Seasonality in Infectious Diseases. *Proc. Roy. Soc. Lond. B* **273** 2541–2550.

Green, D.G., and Sadedin, S. (2005) Interactions matter—complexity in landscapes and ecosystems. *Ecol. Compl.* **2** 117–130.

Greenman, J., Kamo, M., and Boots, M. (2004) External forcing of ecological and epidemiological systems: a resonance approach. *Physica D* **190** 136–151.

Gremillion-Smith, C., and Woolf, A. (1988) Epizootiology of skunk rabies in North America. *J. Wildlife Dis.* **24** 620–626.

Grenfell, B.T. (1992) Chance and chaos in measles dynamics. *J. R. Stat. Soc. B* **54** 383–398.

Grenfell, B.T., Bjørnstad, O.N., and Kappey J. (2001) Travelling waves and spatial hierarchies in measles epidemics. *Nature* **414** 716–723.

Grenfell, B.T., and Anderson, R.M. (1985) The estimation of age related rates of infection from case notifications and serological data. *J. Hyg.* **95** 419–36.

Grenfell, B.T., and Bolker, B.M. (1998) Cities and villages: infection hierarchies in a measles metapopulation. *Ecol. Lett.* **1** 63–70.

Grenfell, B.T., Bolker, B.M., and Kleczkowski, A. (1995) Seasonality and extinction in chaotic metapopulations. *Proc. R. Soc. Lond. B* **259** 97–103.

Grenfell, B.T., and Dobson, A.P. (1995) *Ecology of Infectious Disease in Natural Populations.* Cambridge, UK: Cambridge University Press.

Grenfell, B.T., Kleczkowski, A., Ellner, S., and Bolker, B.M. (1994) Measles as a case study in nonlinear forecasting and chaos. *Phil. Trans. R. Soc. Lond. A* **348** 515–530.

Grossman, Z. (1980) Oscillatory phenomenon in a model of infectious diseases. *Theo. Pop. Biol.* **18** 204–243.

Guan, Y., Zheng, B.J., He, Y.Q., Liu, X.L., Zhuang, Z.X., Cheung, C.L., Luo, S.W., Li, P.H., Zhang, L.J., Guan, Y.J., Butt, K.M., Wong, K.L., Chan, K.W., Lim, W., Shortridge, K.F., Yuen, K.Y., Peiris, J.S.M., and Poon, L.L.M. (2003) Isolation and characterization of viruses related to the SARS coronavirus from animals in Southern China. *Science* **302** 276–278.

Gudelj, I., White, K.A.J., and Britton, N.F. (2004) The effects of spatial movement and group interactions on disease dynamics of social animals. *Bull. Math. Biol.* **66** 91–108.

Gupta, S., and Maiden, M.C.J. (2001) Exploring the evolution of diversity in pathogen populations. *Trends Microbiol.* **9** 181–185.

Gupta, S., Anderson, R.M., and May, R.M. (1989) Networks of sexual contacts: implications for the pattern of spread of HIV. *AIDS* **3** 807–817.

Gupta, S., Ferguson, N., and Anderson, R. (1998) Chaos, Persistence and Evolution of Strain Structure in Antigenically Diverse Infectious Agents. *Nature* **280** 912–915.

Gupta, S., Trenhome, K., Anderson, R.M., and Day, K.P. (1994) Antigenic diversity and the transmission dynamics of *Plasmodium falciparum*. *Science* **263** 961–963.

Hagenaars, T.J., Donnelly, C.A., and Ferguson, N.M. (2004) Spatial heterogeneity and the persistence of infectious diseases. *Theor. Pop. Biol.* **229** 349–359.

Hagenaars, T.J., Ferguson, N.M., Donnelly, C.A., and Anderson, R.M. (2001) Persistence patterns of scrapie in a sheep flock. *Epidemiol. Infect.* **127** 157–167.

Hahn, B.D., and Furniss, P.R. (1983) A Deterministic Model of an Anthrax Epizootic—Threshold Results. *Ecol. Model.* **20** 233–241.

Hahn, B.H., Shaw, G.M., De Cock, K.M., and Sharp, P.M. (2000) AIDS as a zoonosis: scientific and public health implications. *Science* **287** 607–614.

Halloran, M.E., Longini, I.M., Nizam, A., and Yang, Y. (2002) Containing bioterrorist smallpox. *Science* **298** 1428–1432.

Hamer, W.H. (1897) Age-Incidence in Relation with Cycles of Disease-Prevalence. *Trans. Epidemiol. Soc. London* **XVI** 64–77.

Hamer, W.H. (1906) Epidemic diseases in England—the evidence of variability and of persistency of type. *Lancet* **1** 733–739.

Hamilton, J. (1994) *Time Series Analysis*. Princeton, NJ: Princeton University Press.

Hanski, I. (1999) Habitat connectivity, habitat continuity, and metapopulations in dynamic landscapes. *Oikos* **87** 209–219.

Hanski, I.A., and Gaggiotti, O.E. (eds.) (2004) *Ecology, Genetics, and Evolution of Metapopulations*. Elsevier Academic Press.

Hanski, I., and Gilpin, M. (1991) Metapopulation Dynamics—Brief-history and Conceptual Domain. *Biol. J. Linnean Soc.* **42** 3–16.

Hanski, I.A., and Gilpin, M.E. (eds.) (1997) *Metapopulation Biology. Ecology, Genetics and Evolution*. Elsevier Academic Press.

Harris, T.E. (1974) Contact interactions on a lattice. *Ann. Probab.* **2** 969–988.

Hassell, M. P. (1978) *The dynamics of arthropod predator-prey systems*. Princeton, NJ: Princeton University Press.

Hassell, M.P., Comins, H.N., and May, R.M. (1991) Spatial Structure and Chaos in Insect Population Dynamics. *Nature* **353** 255–258.

Hay, J.W., and Ward, J.L. (2005) Economic considerations for pertussis booster vaccination in adolescents. *J. Ped. Infect. Dis.* **24** S127–33.

Hedrich, A. (1933) *Epidemic Studies: The Monthly Variation of Measles Susceptibles in Baltimore, Maryland from 1901 to 1928*. Thesis, Johns Hopkins University.

Heesterbeek, J.A.P. (2002) A brief history of R_0 and a recipe for its calculation. *Acta Biotheor.* **50** 189–204.

Herbert, J., and Isham, V. (2000) Stochastic host-parasite interaction models. *J. Math. Biol.* **40** 343–371.

Hess, G. (1996) Disease in metapopulations: implications for conservation. *Ecology* **77** 1617–1632.

Hethcote, H. (2000) The Mathematics of Infectious Diseases. *SIAM Review* **42** 599–653.

Hethcote, H., and Tudor, D.W. (1980) Integral equation models for endemic infectious diseases. *J. Math. Biol.* **9** 37–47.

Hilton, D.A., Ghani, A.C., Conyers, L., Edwards, P., McCardle, L., Ritchie, D., Penney, M., Hegazy, D., and Ironside, J.W. (2004) Prevalence of lymphoreticular prion protein accumulation in UK tissue samples. *J. Path.* **203** 733–739.

Holt, R.D. (1977) Predation, Apparent Competition, and Structure of Prey Communities. *Theo. Pop. Biol.* **12** 197–229.

Hosseini, P.R., Dhondt, A.A., and Dobson, A.P. (2004) Seasonality and wildlife disease: how seasonal birth, aggregation and variability in immunity affect the dynamics of *Mycoplasma gallisepticum* in house finches. *Proc. Roy. Soc. Lond. B* **271** 2569–2577.

Huang, Y., and Rohani, P. (2005) The dynamical implications of disease interference: Correlations and coexistence. *Theo. Pop. Biol.* **68** 205–215.

Huang, Y., and Rohani, P. (2006) Age-structured effects determine interference between childhood infections. *Proc. Roy. Soc. Lond. B* **273** 1229–1237.

Hubalek Z., and Halouzka J. (1999) West Nile fever—a reemerging mosquito-borne viral disease in Europe. *Emerging Infect. Dis.* **5** 643–650.

Hudson, P.J. (1992) *Grouse in Space in Time.* Game Conservancy, Hampshire, UK.

Hudson, P.J., Dobson, A.P., and Newborn, D. (1998) Prevention of population cycles by parasite removal. *Science* **282** 2256–2258.

Hudson, P.J., Rizzoli, A., Grenfell, B.T., Heesterbeek, H., and Dobson, A.P. (2001) *The Ecology of Wildlife Diseases.* Oxford, UK: Oxford University Press.

Ireland, J.M., Norman, R.A., and Greenman, J.V. (2004) The effect of seasonal host birth rates on population dynamics: the importance of resonance. *J. Theo. Biol.* **231** 229–238.

Islam, M.N., O'Shaughnessy, C.D., and Smith, B. (1996) A random graph model for the final-size distribution of household infections. *Stat. Med.* **15** 837–843.

Jacquez, J.A., and O'Neill, P. (1991) Reproduction Numbers and Thresholds in Stochastic Epidemic Models. 1. Homogeneous Populations. *Math. BioSci.* **107** 161–186.

Jacquez, J.A., and Simon, C.P. (1993) The Stochastic *SI* model with Recruitment and Deaths. 1 Comparison with the closed *SIS* model. *Math. BioSci.* **117** 77–125.

Jacquez, J.A., Simon, C.P., Koopman, J., Sattenspiel, L., and Perry, T. (1988) Modeling and Analyzing HIV Transmission—The Effect of Contact Patterns. *Math. Biosci.* **92** 119–199.

Janaszek,W., Gay, N.J., and Gut, W. (2003) Measles vaccine efficacy during an epidemic in 1998 in the highly vaccinated population in Poland. *Vaccine* **21** 473–478.

Jeong, H., Tombar, B., Albert, R., Oltvai, Z.N., and Barabási, A.L. (2000) The large-scale organization of metabolic networks. *Nature* **407** 651–654.

Johnson, A.M., Wadsworth, J., Wellings K., and Field, J. (1994) *Sexual Attitudes and Lifestyles.* Oxford, UK: Blackwell Scientific Publications.

Jonzén, N., and Lundberg, P. (1999) Temporally structured density-dependence and population management. *Annales Zoologici Fennici* **36** 39–44.

Kallen, A., Arcuri, P., and Murray, J.D. (1985) A Simple-model for the Spatial Spread and Control of Rabies. *J. Theo. Biol.* **116** 377–393.

Kao, R.R. (2003) The impact of local heterogeneity on alternative control strategies for foot-and-mouth disease. *Proc. Roy. Soc. Lond. B* **270** 2557–2564.

Kaplan, E.H., Craft, D.L., and Wein, L.M. (2002) Emergency response to a smallpox attack: The case for mass vaccination. *Proc. Natl. Acad. Sci. USA* **99** 10935–10940.

Katok, A., and Hasselblatt, B. (1996) *Introduction to the Modern Theory of Dynamical Systems.* Cambridge, UK: Cambridge University Press.

Keeling, M.J. (1997) Modelling the persistence of measles. *Trends MicroBiol.* **5** 513–518.

Keeling, M.J. (1999) The effects of local spatial structure on epidemiological invasions. *Proc. Roy. Soc. Lond. B* **266** 859–867.

Keeling, M.J. (2000a) Simple Stochastic Models and their Power-law Type Behaviour. *Theo. Pop. Biol.* **58** 21–31.

Keeling, M.J. (2000b) Metapopulation moments: coupling, stochasticity and persistence. *J. Animal Ecol.* **69** 725–736.

Keeling, M.J. (2005) The implications of network structure for epidemic dynamics. *Theo. Pop. Biol.* **67** 1–8.

Keeling, M.J., and Eames, K.T.D. (2005) Networks and Epidemic Models. *Interface* **2** 295–307.

Keeling, M.J., and Gilligan, C.A. (2000) Bubonic plague: a metapopulation model of a zoonosis. *Proc. Roy. Soc. Lond. B* **267** 2219–2230.

Keeling, M.J., and Grenfell, B.T. (1997a) Disease Extinction and Community Size: Modeling the Persistence of Measles. *Science* **275** 65–67.

Keeling, M.J., and Grenfell, B.T. (1997b) Impact of Variability in Infection Period on the Persistence and Spatial Spread of Infectious Diseases. *Math. BioSci.* **147** 206–227.

Keeling, M.J., and Grenfell, B.T. (1999) Stochastic Dynamics and a power law for measles variability. *Phil. Trans. Roy. Soc. Lond. B* **354** 769–776.

Keeling, M.J., and Grenfell, B.T. (2002) Understanding the Persistence of Measles: Reconciling Theory, Simulation and Observation. *Proc. Roy. Soc. Lond. B* **269** 335–343.

Keeling, M.J., and Rohani, P. (2002) Estimating Spatial Coupling in Epidemiological Systems: a Mechanistic Approach. *Ecol. Letters* **5** 20–29.

Keeling, M.J., Bjørnstad, O.N., and Grenfell, B.T. (2004a) Metapopulation Dynamics of Infectious Diseases. (*Ecology, Genetics and Evolution of Metapopulations*, (eds. I. Hanski and O.E. Gaggiotti,) Elsevier Academic Press, 415–446.

Keeling, M.J., Brooks, S.P., and Gilligan, C.A. (2004b) Using conservation of pattern to estimate spatial parameters from a single snapshot. *Proc. Natl. Acad. Sci. USA* **101** 9155–9160.

Keeling, M.J., Mezic, I., Hendry, R.J., McGlade, J., and Rand, D.A. (1997a) Characteristic length scales of spatial models in ecology via fluctuation analysis. *Phil. Trans. Roy. Soc. Lond. B* **352** 1589–1601.

Keeling, M.J., Rand, D.A., and Morris, A.J. (1997b) Correlation models for childhood epidemics. *Proc. Roy. Soc. Lond. B* **264** 1149–1156.

Keeling, M.J., Rohani, P., and Grenfell, B.T. (2001a) Seasonally Forced Disease Dynamics Explained by Switching Between Attractors. *Physica D* **148** 317–335.

Keeling, M.J., Wilson, H.B., and Pacala, S.W. (2000) Re-interpreting Space, Time-lags and Functional Responses in Ecological Models. *Science* **290** 1758–1761.

Keeling, M.J., Woolhouse, M.E.J., Shaw, D.J., Matthews, L., Chase-Topping, M., Haydon, D.T., Cornell, S.J., Kappey, J., Wilesmith, J., and Grenfell, B.T. (2001b) Dynamics of the 2001 UK Foot and Mouth Epidemic: Stochastic Dispersal in a Heterogeneous Landscape. *Science* **294** 813–817.

Keeling, M.J., Woolhouse, M.E.J., May, R.M., Davies, G., and Grenfell, B.T. (2003) Modelling Vaccination Strategies against Foot-and-Mouth Disease. *Nature* **421** 136–142.

Kelly, M., and Meentemeyer, R.K. (2002) Landscape dynamics of the spread of sudden oak death. *Photogramm. Eng. Rem.* **S68** 1001–1009.

Kendall, D.G. (1949) Stochastic processes and population growth. *J. Roy. Stat. Soc. B* **11** 230–264.

Kermack, W.O., and McKendrick, A.G. (1927) A Contribution to the Mathematical Theory of Epidemics. *Proc. Roy. Soc. Lond. A* **115** 700–721.

Kitala, P.M., McDermott, J.J., Coleman, P.G., and Dye, C. (2002) Comparison of vaccination strategies for the control of dog rabies in Machakos District, Kenya. *Epidemiol. Infect.* **129** 215–222.

Kleczkowski, A., Gilligan, C.A., and Bailey, D.J. (1997) Scaling and spatial dynamics in plant-pathogen systems: From individuals to populations. *Proc. Roy. Soc. Lond. B* **264** 979–984.

Klovdahl, A.S. (1985) Social Networks and the Spread of Infectious Diseases: The AIDS Example. *Soc. Sci. Med.* **21** 1203–1216.

Klovdahl, A.S. (2001) Networks and Pathogens. *Sex. Transm. Dis.* **28** 25–28.

Koelle, K., and Pascual, M. (2004) Disentangling extrinsic from intrinsic factors in disease dynamics: A nonlinear time series approach with an application to cholera. *Am. Nat.* **163** 901–913.

Koelle, K., Rodo, X., Pascual, M., Yunus, M., and Mostafa, G. (2005) Refractory periods and climate forcing in cholera dynamics. *Nature* **436** 696–700.

Kokko, H., and Lindstrm, J. (1998) Seasonal density dependence, timing of mortality, and sustainable harvesting. *Ecol. Mod.* **110** 293–304.

Kolata, G. (1987) Mathematical-Model Predicts AIDS Spread. *Science* **235** 1464–1465.

Komar, N., Panella, N.A., Burns, J.E., Dusza, S.W., Mascarenhas, T.M., and Talbot, T.O. (2001) Serologic Evidence for West Nile Virus Infection in Mosquitoes, Birds, Horses and Humans, Staten Island, New York, 2000. *Emerg. Inf. Dis.* **7** 621–625.

Koopman, J. (2004) Modeling infection transmission. *Annu. Rev. Public Health* **25** 303–326.

Koopman, J.S., Jacquez, J.A., Welch, G.W., Simon, C.P., Foxman, B., Pollock, S.M., Barth-Jones, D., Adams, A.L., and Lange, K. (1997) The role of early HIV infection in the spread of HIV through populations. *J. Acq. Immun. Defic. Synd. Hum. R.* **14** 249–258.

Kot, M. (2000) *Elements of Mathematical Biology.* Cambridge, UK: Cambridge University Press.

Krebs, J.R. (1997) *Bovine tuberculosis in cattle and badgers*: report to the Rt. Hon. Dr. Jack Cunningham MP by the Independent Scientific Review Group. London: DEFRA.

Kretzschmar, M. (2002) Mathematical epidemiology of Chlamydia trachomatis infections. *Neth. J. Med.* **60** 35–41.

Kretzschmar, M., van Duynhoven, Y.T.H.P., and Severijnen, A.J. (1996) Modeling prevention strategies for gonorrhea and Chlamydia using stochastic network simulations. *Am. J. Epid.* **144** 306–317.

Kretzschmar, M., Welte, R., van den Hoek, A., and Postma, M.J. (2001) Comparative model-based analysis of screening programs for Chlamydia trachomatis infections. *Am. J. Epid.* **153** 90–101.

Kuperman, M., and Abramson, G. (2001) Small world effect in an epidemiological model. *Phys. Rev. Lett.* **86** 2909–2912.

Kuznetsov, Y.A. (1994) *Elements of Applied Bifurcation Theory.* New York: Springer-Verlag.

Kuznetsov, Y.A., and Piccardi, C. (1994) Bifurcation Analysis of Periodic $SEIR$ and SIR Epidemic Models. *J. Math. Biol.* **32** 109–121.

Lack, D. (1948) Natural Selection and family size in the starling. *Evolution* **2** 95–110.

Laing, J.S., and Hay, M. (1902) Whooping-cough: its prevalence and mortality in Aberdeen. *Public Health* **14** 584–598.

Langlois, J.P., Fahrig, L., Merriam, G., and Artsob, H. (2001) Landscape structure influences—continental distribution of hantavirus in deer mice. *Landsc. Ecol.* **16** 255–266.

Levin, S.A., and Durrett, R. (1996) From individuals to epidemics. *Proc. Roy. Soc. Lond. B* **351** 1615–1621.

Levin, S.A., Grenfell, B., Hastings, A., and Perelson, A.S. (1997) Mathematical and Computational Challenges in Population Biology and Ecosystems Science. *Science* **275** 334–343.

Levins, R. (1969) Some demographic and genetic consequences of environmental heterogeneity for biological control. *Bull. Ent. Soc. Am.* **15** 237–240.

Lewis, M.A. (2000) Spread rate for a nonlinear stochastic invasion. *J. Math. Biol.* **41** 430–454.

Liljeros, F., Edling, C.R., and Amaral, L.A.N. (2003) Sexual networks: implications for the transmission of sexually transmitted infections. *Microbes Infect.* **5** 189–196.

Liljeros, F., Edling, C.R., Amaral, L.A.N., Stanley, H.E., and Åberg, Y. (2001) The web of human sexual contacts. *Nature* **411** 907–908.

Lipsitch, M., Bergstrom, C.T., and Levin B.R. (2000) The epidemiology of antibiotic resistance in hospitals: Paradoxes and prescriptions. *Proc. Natl. Acad. Sci. USA* **97** 1938–1943.

Lloyd, A.L. (2001) Destabilization of epidemic models with the inclusion of realistic distributions of infectious periods. *Proc. Roy. Soc. Lond. B* **268** 985–993.

Lloyd, H.G. (1983) Past and present distribution of red and grey squirrels. *Mammal Rev.* **13** 69–80.

Lloyd-Smith, J.O., Galvani, A.P., and Getz, W.M. (2003) Curtailing transmission of severe acute respiratory syndrome within a community and its hospital *Proc. Roy. Soc. Lond. B* **270** 1979–1989.

Lloyd-Smith, J.O., Cross, P.C., Briggs, C.J., Daugherty, M., Getz, W.M., Latto, H., Sanchez, M.S., et al. (2005) Should we expect population thresholds for wildlife disease? *Trends Ecol. & Evol.* **20** 511–519.

Lloyd-Smith, J.O., Schreiber, S.J., Kopp, P.E., and Getz, W.M. (2005) Superspreading and the effect of individual variation on disease emergence. *Nature* **438** 355–359.

Loehle, C. (1995) Social barriers to pathogen transmission in wild animal populations. *Ecology* **76** 326–335.

London, W.P., and Yorke, J.A. (1973) Recurrent outbreaks of measles, chickenpox and mumps. I. Seasonal variations in contact rates. *Am. J. Epidem.* **98** 453–468.

Longini, I.M., Clark, W.S., Byers, R.H., Ward, J.W., Darrow, W.W., Lemp, G.F., and Hethcote, H.W. (1989) Statistical-Analysis of the Stages of HIV Infection Using a Markov Model. *Stats. in Med.* **8** 831–843.

Longini, I.M., Nizam, A., Xu, S.F., Ungchusak, K., Hanshaoworakul, W., Cummings, D.A.T., and Halloran, M.E., (2005) Containing pandemic influenza at the source. *Science* **309** 1083–1087.

Lopez, L.F., Coutinho, F.A.B., Burattini, M.N., and Massad, E. (1999) Modelling the spread of infections when the contact rate among individuals is short ranged: Propagation of epidemic waves. *Math. Comput. Model.* **29** 55–69.

Lupton, R. (1993) *Statistics in theory and practise.* Princeton, NJ: Princeton University Press.

Macdonald, D. (1984) *The encyclopedia of mammals.* London: Allen and Unwin.

MacKenzie, K., and Bishop, S.C. (2001) Developing stochastic epidemiological models to quantify the dynamics of infectious diseases in domestic livestock. *J. An. Sci.* **79** 2047–2056.

Marcus, R. (1991) Deterministic and Stochastic Logistic-Models for Describing Increase of Plant Diseases. *Crop Prot.* **10** 155–159.

MacKinnon, K. (1978) Competition between red and grey squirrels. *Mammal Rev.* **8** 185–190.

Maddison, A.C., Holt, J., and Jeger, M.J. (1996) Spatial dynamics of a monocyclic disease in a perennial crop. *Ecol. Model.* **88** 45–52.

Mahul, O., and Durand, B. (2000) Simulated economic consequences of foot-and-mouth disease epidemics and their public control in France. *Prev. Vet. Med.* **47** 23–38.

Mangen, M.J.J., Nielen, M., and Burrell, A.M. (2002) Simulated effect of pig-population density on epidemic size and choice of control strategy for classical swine fever epidemics in The Netherlands. *Prev. Vet. Med.* **56** 141–163.

Mann, N.H., Cook, A., Millard, A., Bailey, S., and Clokie, M. (2003) Marine ecosystems: Bacterial photosynthesis genes in a virus. *Nature* **424** 741.

Marcus, R., Svetlana, F., Talpaz, H., Salomon, R., and Bar-Joseph, M. (1984) On the spatial distribution of citrus tristeza virus disease. *Phytoparasitica* **12** 45–52.

May, R.M. (1973) *Stability and Complexity in Model Ecoosystems.* Princeton, NJ: Princeton University Press.

May, R.M., and Anderson, R.M. (1979) Population Biology of Infectious Diseases 2. *Nature* **280** 455–461.

May, R.M., and Anderson, R.M. (1983) Epidemiology and genetics in the coevolution of parasites and hsots. *Proc. Roy. Soc. Lond. Series B* **219** 281–313.

May, R.M., and Anderson, R.M. (1987) Transmission dynamics of HIV infection. *Nature* **326** 137–142.

May, R.M., and Anderson, R.M. (1989) Transmission dynamics of human immunodeficiency virus (HIV). *Phil. Trans. Roy. Soc. Lond. B* **321** 565–607.

May, R.M., and Lloyd, A.L. (2001) Infection dynamics on scale-free networks. *Phys. Rev. E* **64**, Art. No. 066112.

McCallum, H., and Dobson, A. (2002) Disease, habitat fragmentation and conservation. *Proc. Roy. Soc. Lond. B* **269** 2041–2049.

McKane, A.J., and Newman T.J. (2005) Predator-prey cycles from resonant amplification of demographic stochasticity. *Phys. Rev. Letters* **94** Art No. 218102.

McKenzie, F.E., Killeen, G.F., Beier, J.C., and Bossert, W.H. (2001) Seasonality, parasite diversity, and local extinctions in Plasmodium falciparum malaria. *Ecology* **82** 2673–2681.

McKusick, M.L., Horstman, W., and Coates, T.J. (1985) AIDS and sexual behaviour reported by gay men in San Francisco. *Am. J. Public Health* **75** 493–496.

McLean, A.R., and Anderson, R.M. (1988a) Measles in Developing Countries Part I. Epidemiological Parameters and Patterns. *Epidem. & Inf.* **100** 111–133.

McLean, A.R., and Anderson, R.M. (1988b) Measles in Developing Countries Part II. The Predicted Impact of Mass Vaccination. *Epidem. & Inf.* **100** 419–442 (1988).

McLean, A.R. (1998) Vaccines and their impact on the control of disease. *British Med. Bull.* **54** 545–556.

McLean A.R. (1995) After the honeymoon in measles control. *Lancet* **345** 272.

McLean, A.R., and Blower, S.M. (1993) Imperfect vaccines and herd immunity to HIV. *Proc. Roy. Soc. Lond. B* **253** 9–13.

McManus, T.J., and Envoy, M. (1987) Some aspects of male homosexual behaviour in the United Kingdom. *Brit. J. Sex. Med.* **14** 110–120.

Mead, R. (1974) A test for spatial pattern at several scales using data from a grid of contiguous quadrats. *Biometrics* **30** 295–307.

Medley, G.F., Lindop, N.A., Edmunds, W.J., and Nokes D.J. (2001) Explaining HBV endemicity: heterogeneity, catastrophic dynamics and control. *Nature Medicine* **7** 619–624.

Meltzer, M.I., Damon, I., LeDuc, J.W., and Millar, J.D. (2001) Modeling potential responses to smallpox as a bioterrorist weapon. *Emerg. Inf. Dis.* **7** 959–969.

Mertens, T.E., and LowBeer, D. (1996) HIV and AIDS: Where is the epidemic going? *Bull. World Health Organ.* **74** 121–129.

Metz, J.A.J. (1978) The epidemic in a closed population with all susceptibles equally vunerable; some results for large susceptible populations and small initial infections. *Acta Biotheoretica* **27** 75–123.

Michael, E., Grenfell, B.T., Isham, V.S., Denham, D.A., and Bundy, D.A.P. (1998) Modelling variability in lymphatic filariasis: macrofilarial dynamics in the Brugia pahangi cat model. *Proc. Roy. Soc. Lond. B* **265** 155–165.

Michael, E., Malecela-Lazaro, M.N., Simonsen, P.E., Pedersen, E.M., Barker, G., Kumar, A., Kazura, J.W. (2004) Mathematical modelling and the control of lymphatic filariasis. *Lancet Infect. Dis.* **4** 223–234

Miller, E., Vurdien, J.E. and White, J.M. (1992) The Epidemiology of Pertussis in England and Wales. *Communicable Disease Report* **2** R152–R155.

Mills, J.N., Ksiazek, T.G., Peters, C.J., and Childs, J.E. (1999) Long-term studies of hantavirus reservoir populations in the southwestern United States: A synthesis. *Emerg. Infect. Dis.* **5** 135–142.

Mollison, D. (1991) Dependence of epidemic and population velocities on basic parameters. *Math. Biosci.* **107** 255–287.

Mollison, D., and Ud Din, S. (1993) Deterministic and stochastic models for the seasonal variability of measles transmission. *Math. Biosci.* **117** 155–177.

Mollison, D., Isham, V., and Grenfell B. (1994) Epidemics—Models and Data. *J. Roy. Stat. Soc. A* **157** 115–149.

Monath, T.P. (1999) Ecology of Marburg and Ebola viruses: speculations and directions for future research. *J. Inf. Dis.* **179** S127–S138.

Montgomery, S.S.J., and Montgomery, W.I. (1988) Cyclic and non-cyclic dynamics in populations of the helminth parasites of wood mice, Apodemus sylvaticus. *J. Helminthol.* **62** 978–990.

Moore, C., and Newman, M.E.J. (2000) Epidemics and percolation in small-world networks. *Phys. Rev. E* **61** 5678–5682.

Moore, D.A. (1999) Spatial diffusion of raccoon rabies in Pennsylvania, USA. *Prev. Vet. Med.* **40** 19–32.

Morens, D.M., Folkers, G.K., and Fauci, A.S. (2004) The challenge of emerging and re-emerging infectious diseases. *Nature* **430** 242–249.

Morris, N. (2001) Concurrent partnerships and syphilis persistence—New thoughts on an old puzzle. *Sex. Trans. Dis.* **28** 504–507.

Morris, R.S., Wilesmith, J.W., Stern, M.W., Sanson, R.L., and Stevenson, M.A. (2001) Predictive spatial modelling of alternative control strategies for the foot-and-mouth disease epidemic in Great Britain, 2001. *Vet. Rec.* **149** 137–145.

Murray, J.D. (1989) *Mathematical Biology*. New York: Springer-Verlag.

Murray, J.D. (1982) Parameter space for Turing instability in reaction diffusion mechanisms: a comparison of models. *J. Theor. Biol.* **98** 143–163.

Murray, J.D. (2003) *Mathematical Biology (3rd edition) II: Spatial Models and Biomedical Applications*. New York: Springer Verlag.

Murray, J.D., Stanley, E.A., and Brown, D.L. (1986) On the spatial spread of rabies among foxes. *Proc. Roy. Soc. Lond. B* **229** 111–150.

Narang, H.K. (2001) A critical review of atypical cerebellum-type Creutzfeldt-Jakob disease: Its relationship to "new variant" CJD and bovine spongiform encephalopathy. *Experimental Biol. and Med.* **226** 629–639.

Nasell, I. (1991) On the Quasi-Stationary Distribution of the Ross Malaria Model. *Math. BioSci.* **107** 187–207.

Nasell, I. (1996) The quasi-stationary distribution of the closed endemic SIS model. *Adv. Appl. Probab.* **28** 895–932.

Nasell, I. (1999) On the time to extinction in recurrent epidemics. *J. Roy. Stat. Soc. B* **61** 309–330.

Neal, P. (2003) SIR epidemics on a Bernoulli random graph. *J. Appl. Probab.* **40** 779–782.

Neal, P.J., and Roberts, G.O. (2004) Statistical inference and model selection for the 1861 Hagelloch measles epidemic. *Biostat.* **5** 249–261.

Nee, S. (1994) How populations persist. *Nature* **367** 123–124.

Neubert, M.G., Kot, M., and Lewis, M.A. (1995) Dispersal and pattern formation in a discrete-time predator-prey model. *Theor. Pop. Biol.* **48** 7–43.

Newman, M.E.J., and Watts, D.J. (1999) Scaling and percolation in the small-world network model. *Phys. Rev. E* **60** 7332–7342.

Newman, M.E.J., Watts, D.J., and Strogatz, S.H. (2002) Random graph models of social networks. *Proc. Natl. Acad. Sci. USA* **99** 2566–2572.

Newton-Fisher, N.E., Reynolds, V., and Plumptre, A.J. (2000) Food Supply and Chimpanzee (*Pan troglodytes schweinfurthii*) Party Size in the Budongo Forest Reserve, Uganda. *Int. J. Primatol.* **21** 613–628.

Nielen, M., Jalvingh, A.W., Meuwissen, M.P.M., Horst, S.H., and Dijkhuizen, A.A. (1999) Spatial and stochastic simulation to evaluate the impact of events and control measures on the 1997–1998 classical swine fever epidemic in The Netherlands. II. Comparison of control strategies. *Prev. Vet. Med.* **42** 297–317.

Nisbet, R.M. and Gurney, W.S.C (1982) *Modelling Fluctuating Populations*. Chichester, UK: John Wiley & Sons.

Noble, J.V. (1974) Geographic and temporal development of plagues. *Nature* **250** 726–728.

Nodelijk, G., de Jong, M.C.M., van Nes, A., Vernooy, J.C.M., van Leengoed, L.A.M.G., Pol, J.M.A, and Verheijden, J.H.M. (2000) Introduction, persistence and fade-out of porcine reproductive and respiratory syndrome virus in a Dutch breeding herd: a mathematical analysis. *Epidemiol. Infect.* **124** 173–182.

Nokes, D.J., and Swinton, J. (1997) Vaccination in pulses: A strategy for global eradication of measles and polio? *Trends Microbiol.* **5** 14–19.

Noordegraaf, A.V., Jalvingh, A.W., de Jong, M.C.M., Franken, P., and Dijkhuizen, A.A. (2000) Evaluating control strategies for outbreaks in BHV1-free areas using stochastic and spatial simulation. *Prev. Vet. Med.* **44** 21–42.

Nowak, M.A., and May, R.M. (2005) *Virus Dynamics.* Oxford, UK: Oxford University Press.

Øksendal, B. (1998) *Stochastic Differential Equations.* Berlin: Springer-Verlag.

O'Neill, P.D., Balding, D.J., Becker, N.G., Eerola, M., and Mollison, D. (2000) Analyses of infectious disease data from household outbreaks by Markov chain Monte Carlo methods. *J. Roy. Stat. Soc. C* **49** 517–542.

Olsen, L.F., and Schaffer, W.M. (1990) Chaos versus noisy periodicity: alternative hypotheses for childhood epidemics. *Science* **249** 499–504.

Onstad, D.W., and Kornkven, E.A. (1992) Persistence and Endemicity of Pathogens in Plant-Populations over Time and Space. *Phytopathol.* **82** 561–566.

Orenstein W.A., Papania M.J. and Wharton M.E. (2004) Measles elimination in the United States. *J. Infect. Dis.* **189** S1–S3.

Oxman, G.L., Smolkowski, K., and Noell, J. (1996) Mathematical modeling of epidemic syphilis transmission—Implications for syphilis control programs. *Sex. Trans. Dis.* **23** 30–39.

Park, A.W., Gubbins, S., and Gilligan, C.A. (2001) Invasion and persistence of plant parasites in a spatially structured host population. *Oikos* **94** 162–174.

Park, A.W., Gubbins, S., and Gilligan, C.A. (2002) Extinction times for closed epidemics: the effects of host spatial structure. *Ecol. Lett.* **5** 747–755.

Pascual, M., and Guichard, F. (2005) Criticality and disturbance in spatial ecological systems. *Trends Ecol. & Evol.* **20** 88–95.

Pascual, M., Mazzega, P., and Levin, S.A. (2001) Oscillatory Dynamics and Spatial Scale: The Role of Noise and Unresolved Pattern. *Ecology* **82** 2357–2369.

Pascual, M., Rodo, X., Ellner, S.P., Colwell, R., and Bouma, M.J. (2000) Cholera dynamics and El Nino-Southern Oscillation. *Science* **289** 1766–1769.

Paton, D.J., and Greiser-Wilke, I. (2003) Classical swine fever—an update. *Res. Vet. Sci.* **75** 169–178.

Petersen, L.R., and Roehrig, J.T. (2001) West Nile virus: A reemerging global pathogen. *Emerging Infect. Dis.* **7** 611–614.

Peterson, M.J., Grant, W.E., and Davis, D.S. (1991) Simulation of Host-Parasite Interactions within a Resource-Management Framework—Impact of Brucellosis on Bison Population-Dynamics. *Ecol. Model.* **54** 299–320.

Phillips, S.S. (2000) Population trends and the koala conservation debate. *Conserv. Biol.* **14** 650–659.

Potterat, J.J., Philips-Plummer, L., Muth, S.Q., Rothenberg, R.B., Woodhouse, D.E., Maldonado-Long, T.S., Zimmerman, H.P., and Muth, J.B. (2002) Risk network structure in the early epidemic phase of HIV transmission in Colorado Springs. *Sex. Transm. Infect.* **78** i159–i163.

Potterat, J.J., Rothenberg, R.B., and Muth, S.Q. (1999) Network structural dynamics and infectious disease propagation. *Int. J. STD AIDS* **10** 182–185.

Pourbohloul, B., Rekart, M.L., and Brunham, R.C. (2003) Impact of mass treatment on syphilis transmission—A mathematical modeling approach. *Sex. Trans. Dis.* **30** 297–305.

Press, W.H., Teukolsky, S.A., Vetterling, W.T., and Flannery, B.P. (1988) *Numerical Recipes in C: The Art of Scientific Computing.* Cambridge, UK: Cambridge University Press.

Rand, D.A., and Wilson, H.B. (1991) Chaotic stochasticity—a ubiquitous source of unpredictability in epidemics. *Proc. Roy. Soc. Lond. B* **246** 179–184.

Ransome, A. (1880) On Epidemic Cycles. *Proc. Manchester Lit. Phil. Soc.* **19** 75–96.

Ransome, A. (1881) On the Form of the Epidemic Wave and Some of Its Probable Causes. *Trans. Epidem. Soc. Lond.* **1** 96.

Ranta, J., Mäkelä, P.H., Takala, A., and Arjas, E. (1999) Predicting the course of meningococcal disease outbreaks in closed subpopulations. *Epidemiol. Infect.* **123** 359–371.

Read, J.M., and Keeling, M.J. (2003) Disease Evolution on Networks: the role of Contact Structure. *Proc. Roy. Soc. Lond.* **270** 699–708.

Reluga, T. (2004) A two-phase epidemic driven by diffusion. *J. Theor. Biol.* **229** 249–261.

Renshaw, E. (1991) *Modelling Biological Populations in Space and Time.* Cambridge: Cambridge University Press.

Renton, A.M., Whitaker, L., and Riddlesdell, M. (1998) Heterosexual HIV transmission and STD prevalence: predictions of a theoretical model. *Sex. Trans. Inf.* **74** 339–344.

Reynolds, J.C. (1985) Details of the geographic replacement of the red squirrel (*Sciurus vulgaris*) by the grey squirrel (*Sciurus carolinensis*) in Eastern England. *J. Anim. Ecol.* **54** 149–162.

Rhodes, C.J., and Anderson, R.M. (1996) Dynamics in a lattice epidemic model. *Phys. Letts. A* **210** 183–188.

Rhodes, C.J., and Anderson, R.M. (1997) Epidemic thresholds and vaccination in a lattice model of disease spread. *Theo. Pop. Biol.* **52** 101–118.

Rhodes, C.J., Atkinson, R.P.D., Anderson, R.M., and MacDonald, D.W. (1998) Rabies in Zimbabwe: reservoir dogs and the implications for disease control. *Phil. Trans. Roy. Soc. Lond. B* **353** 999–1010.

Rhodes, C.J., Jensen, H.J., and Anderson, R.M. (1997) On the critical behaviour of simple epidemics. *Proc. Roy. Soc. Lond. B* **264** 1639–1649.

Riley, S., Fraser, C., Donnelly, C.A., Ghani, A.C., Abu-Raddad, L.J., Hedley, A.J., Leung, G.M., Ho, L.M., Lam, T.H., Thach, T.Q., Chau, P., Chan, K.P., Leung, P.Y., Tsang, T., Ho, W., Lee, K.H., Lau, E.M.C., Ferguson, N.M., and Anderson, R.M. (2003) Transmission dynamics of the etiological agent of SARS in Hong Kong: Impact of public health interventions. *Science* **300** 1961–1966.

Roberts, M.G. (1996) The Dynamics of Bovine Tuberculosis in Possum Populations, and its Eradication or Control by Culling or Vaccination. *J. An. Ecol.* **65** 451–464.

Roberts, M.G., and Saha, A.K. (1999) The asymptotic behaviour of a Logistic Epidemic Model with Stochastic Disease Transmission. *Applied Math. Letters* **12** 37–41.

Rohani, P., Earn, D.J., Finkenstädt, B., and Grenfell, B.T. (1998) Population dynamic interference among childhood diseases. *Proc. Roy. Soc. Lond. B* **265** 2033–2041.

Rohani, P., Earn, D.J.D., and Grenfell, B.T. (1999) Opposite Patterns of Synchrony in Sympatric Disease Metapopulations. *Science* **286** 968–971.

Rohani, P., Earn, D.J.D., and Grenfell, B.T. (2000) The Impact of Immunisation on Pertussis Transmission in England & Wales. *Lancet* **355** 285–286.

Rohani, P., Green, C.J., Mantilla-Beniers, N.B., and Grenfell, B.T. (2003) Ecological interference between fatal diseases. *Nature* **422** 885–888.

Rohani, P., Keeling, M.J., and Grenfell, B.T. (2002) The interplay between determinism and stochasticity in childhood diseases. *Am. Nat.* **159** 469–481.

Rohani, P., Wearing, H.J., Vasco, D.A., and Huang, Y. (2006) Understanding Host-Multi-Pathogen Systems: The Interaction Between Ecology and Immunology. *Ecology of Infectious Diseases,* (eds: Osfeld, Keesing, and Eviner) (in press).

Rothenberg, R. (2001) Commentary—How a net works—Implications of network structure for the persistence and control of sexually transmitted diseases and HIV. *Sex. Trans. Dis.* **28** 63–68.

Rothenberg, R. (2003) STD transmission dynamics: Some current complexities—2002 Thomas Parran Award Lecture. *Sex. Trans. Dis.* **30** 478–482.

Rothenberg, R.B. (1983) The geography of gonorrhea: empirical demonstration of core group transmission. *Am. J. Epid.* **117** 688–694.

Rushton, S.P., Lurz, P.W.W., Fuller, R., and Garson, P.J. (1997) Modelling the distribution of the red and grey squirrel at the landscape scale: a combined GIS and population dynamics approach. *J. Appl. Ecol.* **34** 1137–1154.

Rushton, S.P., Lurz, P.W.W., Gurnell, J., and Fuller, R. (2000) Modelling the spatial dynamics of parapoxvirus disease in red and grey squirrels: a possible cause of the decline in the red squirrel in the UK? *J. Appl. Ecol.* **37** 997–1012.

Sabin, A.B. (1991) Measles, killer of millions in developing countries: strategy for rapid elimination and continuing control. *European J. Epidemiol.* **7** 1–22.

Sander, L.M., Warren, C.P., Sokolov, I.M., Simon, C., and Koopman, J. (2002) Percolation on heterogeneous networks as a model for epidemics. *Math. Biosci.* **180** 293–305.

Sartwell P.E. (1950) The distribution of incubation periods of infectious disease. *Am. J. Hyg.* **51** 310–318.

Sauvage, F., Langlais, M., Yoccoz, N.G., and Pontier, D. (2003) Modelling hantavirus in fluctuating populations of bank voles: the role of indirect transmission on virus persistence. *J. Anim. Ecol.* **72** 1–13.

Schenzle, D. (1984) An age-structured model of pre- and post-vaccination measles transmission. *IMA J. Math. App. Med. Biol.* **1** 169–191.

Schwartz, I.B., and Smith, H.L. (1983) Infinite Subharmonic Bifurcation in an *SEIR* Epidemic model. *J. Math. Biol.* **18** 233–253.

Seydel, R. (1994) *Practical Bifurcation and Stability Analysis.* Berlin: Springer-Verlag.

Shampine, L.F., and Reichelt, M.W. (1997) "The MATLAB ODE Suite," *SIAM Journal on Scientific Computing* **18** 1–22.

Shaw, M.W. (1995) Simulation of population expansion and spatial pattern when individual dispersal distributions do not decline exponentially with distance. *Proc. Roy. Soc. Lond. B* **259** 243–248.

Shea K. (1998) Management of populations in conservation, harvesting and control. *Trends Ecol. & Evol.* **13** 371–375.

Shirley, M.D.F., Rushton, S.P., Smith, G.C., South, A.B., and Lurz, P.W.W. (2003) Investigating the spatial dynamics of bovine tuberculosis in badger populations: evaluating an individual-based simulation model. *Ecol. Model.* **167** 139–157.

Shulgin, B., Stone, L., and Agur Z. (1998) Pulse vaccination strategy in the *SIR* epidemic model. *B. Math. Biol.* **60** 1123–1148.

Simpson, R.E.H. (1952) Infectiousness of communicable disease in the household (measles, chickenpox, and mumps). *Lancet* **263** 549–554.

Sleeman, C.K., and Mode, C.J. (1997) A methodological study on fitting a nonlinear stochastic model of the AIDS epidemic in Philadelphia. *Math. Comput. Model.* **26** 33–51.

Smith, D.L., Lucey, B., Waller, L.A., Childs, J.E., and Real, L.A. (2002) Predicting the spatial dynamics of rabies epidemics on heterogeneous landscapes. *Proc. Natl. Acad. Sci. USA* **99** 3668–3672.

Smith, G.C. (2001) Models of Mycobacterium bovis in wildlife and cattle. *Tuberculosis* **81** 51–64.

Smith, G.C., and Harris, S. (1991) Rabies in Urban Foxes (*Vulpes-Vulpes*) in Britain—The use of a spatial stochastic simulation-model to examine the pattern of spread and evaluate the efficacy of different control regimes. *Phil. Trans. Roy. Soc. Lond. B* **334** 459–479.

Smith, G.C., and Wilkinson, D. (2003) Modeling control of rabies outbreaks in red fox populations to evaluate culling, vaccination, and vaccination combined with fertility control. *J. Wildl. Dis.* **39** 278–286.

Smith, G.C., Cheeseman, C.L., Clifton-Hadley, R.S., and Wilkinson, D. (2001) A model of bovine tuberculosis in the badger Meles meles: an evaluation of control strategies. *J. Appl. Ecol.* **38** 509–519.

Sole, R.V., Manrubia, S.C., Benton, M., Kauffman, S., and Bak, P. (1999) Criticality and scaling in evolutionary ecology. *Trends Ecol.& Evol.* **14** 156–160.

Soper, H.E. (1929) The Interpretation of Periodicity in Disease Prevalence *J. Roy. Stat. Soc. A* **92** 34–61.

Sørensen, S.J., Oregaard, G., De Lipthay, J., and Kroer, N. (2004) Plasmid transfer in aquatic environments. *Microbial Ecol. Manual 2004.*

Stacey A.J., Truscott, J.E., Asher, M.J.C., and Gilligan, C.A. (2004) A model for the invasion and spread of rhizomania in the United Kingdom: Implications for disease control strategies. *Phyopathol.* **94** 209–215.

Stark, K.D.C., Pfeiffer, D.U., and Morris, R.S. (2000) Within-farm spread of classical swine fever virus—A blueprint for a stochastic simulation model. *Vet. Q.* **22** 36–43.

Stigum, H., Magnus, P., and Bakketeig, L.S. (1997) Effect of changing partnership formation rates on the spread of sexually transmitted diseases and human immunodeficiency virus. *Am. J. Epid.* **145** 644–652.

Stollenwerk, N., and Briggs, K.M. (2000) Master equation solution of a plant disease model. *Phys. Lett. A* **274** 84–91.

Stollenwerk, N., and Jansen, V.A.A. (2003) Meningitis, pathogenicity near criticality: the epidemiology of meningococcal disease as a model for accidental pathogens. *J. Theo. Biol.* **222** 347–359.

Strang, G. (1986) *Introduction to Applied Mathematics.* Wellesley, MA: Wellesley-Cambridge Press.

Strikas, R.A., Anderson, L.J., and Parker, R.A. (1986) Temporal and geographic patterns of isolates of nonpolio enterovirus in the United States, 1970–1983. *J. Infect. Dis.* **153** 346–51.

Suppo, C., Naulin, J.M., Langlais, M., and Artois, M. (2000) A modelling approach to vaccination and contraception programmes for rabies control in fox populations. *Proc. Roy. Soc. Lond. B* **267** 1575–1582.

Swinton, J., Harwood, J., Grenfell, B., and Gilligan, C. (1998) Persistence Thresholds for Phocine Distemper Virus Infection in Harbour Seal Phoca Vitulina Metapopulations. *J. Anim. Ecol.* **67** 54–68.

Swinton, J., and Gilligan, C.A. (1996) Dutch elm disease and the future of the elm in the UK: A quantitative analysis. *Phil. Trans. Roy. Soc. Lond. B* **351** 605–615.

Szendroi, B., and Csanyi, G. (2004) Polynomial epidemics and clustering in contact networks. *Proc. Roy. Soc. Lond. B* **271** S364–S366.

Tang, C., and Bak, P. (1988) Critical Exponents and Scaling Relations for Self-Organized Critical Phenomena. *Phys. Rev. Let.* **60** 2347–2350.

Terpstra, C., Wensvoort, G., and Pol, J.M.A. (1991) Experimental reproduction of porcine epidemic abortion and respiratory sydrome (mystery swine disease) by infection with Lelystad virus: Koch's postulates fulfilled. *Vet. Q.* **13** 131–136.

Thrall, P.H., Godfree, R., and Burdon, J.J. (2003) Influence of spatial structure on pathogen colonization and extinction: a test using an experimental metapopulation. *Plant Pathol.* **52** 350–361.

Tildesley, M.J., Savill, N.J., Shaw, D.J., Deardon, R., Brooks, S.P., Woolhouse, M.E.J., Grenfell, B.T., and Keeling, M.J. (2006) Optimal reactive vaccination strategies for a foot-and-mouth outbreak in Great Britain. *Nature* **440** 83–86.

Tille, A., Lefevre, C., Pastoret, P.P., and Thiry, E. (1991) A Mathematical-Model of Rinderpest Infection in Cattle Populations. *Epidemiol. Infect.* **107** 441–452.

Tischendorf, L., Thulke, H.H., Staubach, C., Muller, M.S., Jeltsch, F., Goretzki, J., Selhorst, T., Muller, T., Schluter, H., and Wissel, C. (1998) Chance and risk of controlling rabies in large-scale and long-term immunized fox populations. *Proc. Roy. Soc. Lond. B* **265** 839–846.

Tompkins, D.M., White, A.R., and Boots, M. (2003) Ecological replacement of native red squirrels by invasive greys driven by disease. *Ecol. Letters* **6** 189–196.

Tong, H. (1990) *Non-linear time series. A dynamical system approach* Oxford, UK: Oxford University Press.

Turing, A. (1952) The Chemical Basis of Morphogenesis. *Phil. Trans. Roy. Soc. Lond. B* **237** 37–72.

UNAIDS/WHO (2002) *AIDS epidemic update.* `http://www.who.int/hiv/en/`

van Baalen, M., and Rand, D.A. (1998) The unit of selection in viscous populations and the evolution of altruism. *J. Theor. Biol.* **193** 631–648.

van Buskirk, J., and Ostfeld, R.S. (1998) Habitat heterogeneity, dispersal, and local risk of exposure to Lyme disease. *Ecol. Appl.* **8** 365–378.

van den Bosch, F., Metz, J.A.J., and Diekmann, O. (1990) The Velocity of Spatial Population Expansion. *J. Math. Biol.* **28** 529–565.

van Herwaarden, O.A. (1997) Stochastic epidemics: the probability of extinction of an infectious disease at the end of a major outbreak. *J. Math. Biol.* **35** 793–813.

Vasco, D.A., Wearing, H.J., and Rohani, P. (2007) Tracking the Dynamics of Pathogen Interactions: Modeling Ecological and Immune-Mediated Processes in a Two-Pathogen Single-Host System. *J. Theor. Biol.* (in press).

Viboud, C., Bjornstad, O.N., Smith, D.L., Simonsen, L., Miller, M.A., and Grenfell, B.T. (2006) Synchrony, waves, and spatial hierarchies in the spread of influenza. *Science* **312** 447–451.

Viet, A.F., and Medley, G.F. (2006). Stochastic dynamics of immunity in populations: a general framework. *Math. Biosci.* **200** 28–43.

Wallace, R., and Wallace, D. (1993) Inner-City Disease and the Public-Health of the Suburbs—the Sociogeographic Dispersion of Point-Source Infection. *Environ. Plan. A* **25** 1707–1723.

Waltmann, P. (1974) *Deterministic Threshold Models in the Theory of Epidemics.* New York: Springer-Verlag.

Watts, D.J. (1999) *Small Worlds: The Dynamics of Networks Between Order and Randomness.* Priceton, NJ: Princeton University Press.

Watts, D.J., and Strogatz, S.H. (1998) Collective dynamics of 'small-world' networks. *Nature* **393** 440–442.

Wearing, H.J., and Rohani, P. (2006) Ecological and Immunological Determinants of Dengue Epidemics. *Proc. Natl. Acad. Sci. USA* **103** 11802–11807.

Wearing, H.J., Rohani, P., and Keeling, M.J. (2005) Appropriate Models for the Management of Infectious Diseases. *PLoS Medicine* **2** e174.

Webb, G.F., and Blaser, M.J. (2002) Mailborne transmission of anthrax: Modeling and implications. *Proc. Natl. Acad. Sci. USA* **99** 7027–7032.

Wein, L.M., Craft, D.L., and Kaplan, E.H. (2003) Emergency response to an anthrax attack. *Proc. Natl. Acad. Sci. USA* **100** 4346–4351.

Welte, R., Kretzschmar, M., Leidl, R., Van den Hoek, A., Jager, J.C., and Postma, M.J. (2000) Cost-effectiveness of screening programs for Chlamydia trachomatis—A population-based dynamic approach. *Sex. Trans. Dis.* **27** 518–529.

White, A., Begon, M., and Bowers, R.G. (1996) Host-pathogen cycles in self-regulated forest insect systems: Resolving conflicting predictions. *Am. Nat.* **148** 220–225.

White, L.J., Cox, M.J., and Medley, G.F. (1998) Cross immunity and vaccination against multiple microparasite strains. *IMA J. Math. Appl. Med.* **15** 211–233.

White, L.J., Waris, M., Cane, P.A., Nokes, D.J., and Medley, G.F. (2005). The transmission dynamics of groups A and B human respiratory syncytial virus (hRSV) in England & Wales and Finland: seasonality and cross-protection. *Epidemiol. & Infect.* **133** 279–289.

White, P.C.L., and Harris, S. (1995) Bovine tuberculosis in badger (*Meles meles*) populations in southwest England: the use of a spatial stochastic simulation model to understand the dynamics of the disease. *Phil. Trans. R. Soc. Lond. B* **349** 391–413.

White, P.J., Norman, R.A., Trout, R.C., Gould, E.A., and Hudson, P.J. (2001) The emergence of rabbit haemorrhagic disease virus: will a non-pathogenic strain protect the UK? *Phil. Trans. Roy. Soc. Lond. B* **356** 1087–1095.

Wilkinson, D., Smith, G.C., Delahay, R.J., and Cheeseman, C.L. (2001) A model of bovine tuberculosis in the badger Meles meles: an evaluation of different vaccination strategies. *J. Appl. Ecol.* **41** 492–501.

Wills, R.W., Zimmerman, J.J., Yoon, K.J., McGinley, M.J., Hill, H.T., Platt, K.B., Christopher-Hennings, J., and Nelson, E.A. (1997) Porcine reproductive and respiratory syndrome virus: a persistent infection. *Vet. Microbiol.* **55** 231–240.

Wilson, M.E. (2003) The traveller and emerging infections: sentinel, courier, transmitter. *J. Appl. Microbiol* **94** S1–S11.

Wonnacott, R., and Wonnacott, T. (1990) *Introductory Statistics.* John Wiley and Sons.

Woolhouse, M., Chase-Topping, M., Haydon, D., Friar, J., Matthews, L., Hughes, G., Shaw, D., Wilesmith, J., Donaldson, A., Cornell, S., Keeling, M., and Grenfell, B. (2001a) Epidemiology: Foot-and-mouth disease under control in the UK. *Nature* **411** 258–259.

Woolhouse, M.E.J., Taylor, L.H., and Haydon, D.T. (2001b) Population biology of multihost pathogens. *Science* **292** 1109–1112.

Woolhouse, M.E.J. (2002) Population biology of emerging and re-emerging pathogens. *Trends in Microbiol.* **10** S3–S7.

Woolhouse, M.E.J. (2003) Foot-and-mouth disease in the UK: What should we do next time? *J. Appl. Microbiol.* **94** 126S–130S.

Woolhouse, M.E.J., Haydon, D.T., and Antia, R. (2005) Emerging pathogens: the epidemiology and evolution of species jumps. *Trends Ecol. & Evol.* **20** 238–244.

Xia, Y.C., Bjørnstad, O.N., and Grenfell, B.T. (2004) Measles metapopulation dynamics: A gravity model for epidemiological coupling and dynamics. *Am. Nat.* **164** 267–281.

Xu, C., Boyce, M.S., and Daley, D.J. (2005) Harvesting in seasonal environments. *J. Math. Biol.* **50** 663–682.

Yorke, J.A., and London, W.P. (1973) Recurrent outbreaks of measles, chickenpox and mumps; II systematic difference in contact rates and stochastic effects. *Am. J. Epidemiology* **98** 469–482.

Index

Acquired immunity, **3**, **16**, 107–112, 121. *See also* Immunity: Cross or partial
Additive noise, **193–200**
Age at first infection, 31–33
  calculation, **31–33**, 40, 115, 295
  use in models, 46, 82, 161–162, 181, 302, 305
  use in parameterization, 51–52
Age structure, **77–93**
  control, 81–82, 303–306
  general, 46, 114, 159, 181–183
  measles, 84–89
  parameterization, 82–84
  rubella, 304–306
AIDS, 55, **71–74**. *See also* HIV
Annual forcing, **155–187**, 302, 327–333
  deterministic period, 177–178, 186
  dominant period, 174, 177–178, 188
  resonance, 160–164
Anthrax, **144**
Antibiotic resistance, **111–113**
Assortative mixing, 58, 60, **67–69**, 75, 82, 84, 306–308
Asymptomatic infection, 4, 148, 193, **317–320**
Attractors
  coexisting, 167–170, 179–180, 185–188
  periodic, 163–164, 177–178, 327–334
  stability, 170
  See also Equilibria

Bacterial infections, **2–3**
Basic reproductive ratio, 5, **19–21**
  for malaria strains, 115
  in multi-compartment models, 97–98
  in multi-strain models, 127
  in risk structured models, 59–62, 66–68, 75, 80
  in simple models, 19–21, 28, 35, 36, 38, 39, 42, 45, 49–52, 293
  in spatial models, 241–245, 267
  in vector-born models, 138
Basins of attractors, **167–170**, 179–180
Biennial cycles, 112–114, **162–167**, 185–187
  measles, 87–89, 112–114, 173–176, 178–180, 220
Bifurcation diagrams, **164–167**, 170
  generic for measles, 166, 169, 172, 180
  pulsed births, 186
  sinusoidal births, 185, 188
  sinusoidal forcing, 169, 172
  term-time forcing, 172, 180

Bite rate of mosquitoes, **136–138**, 150–151
Bioterrorism, 9, 144, **313–314**
Birth rate
  constant, 26–28
  density dependent, 74–75, 133, 328
  pulsed, 186, 187, 329
  sinusoidal, 185, 188
Bovine spongiform encephalopathy (BSE), 38, **89–92**, 99
Brucellosis, 144
Bubonic plague, 6, 23, 77, 105, 148

Carriers, **44–46**. *See also* Asymptomatic infection
Case reports, 23, 27, 56, 88, 106, 146, 156, 176, 179, 207, 261, 272, 324
  compared to prevalence, 50
  parameterization using, 50
Cellular automata, **257–262**
Chaos, 166–167, 172–173, 178, 180–181
Chickenpox, 5, 16, 21, 44, 156, 161, 182, 294
Chlamydia, 56, **74–76**
Cholera, 2, 124, 158
Citrus Tristeza virus, 6, 269–274
Coexisting attractors, **167–170**, 179–180, 185–188
Coexisting strains, **119**, **128**
Co-infection, **116–117**, **125–128**
Commuters, 209–211, **242–245**
Compartmental models, **16–19**
  SEIR, 41–43, 94–95
  SI, 34–39
  SIR with demography, 26–34
  SIR without demography, 19–26
  SIS, 39
Competition, 108–112, 119–122, 128
Competitive exclusion, **108–109**
Contact process, **258–259**
Contact tracing, 7, 274, 308–309, **312–321**
Control, **291–335**
  by contact tracing, 7, 308–309, 312–321
  by culling, 6, 91–92, 271, 274–276, 321–325, 327–333
  by quarantining, 6, 308–311, 313–331
  by vaccination, 5–6, 9, 63–64, 67, 81–82, 213–214, 249–250, 292–308
Coupled lattice models, **255–257**
Coupling, **235–236**, 245–246
  for commuters, 243–245, 245–246

Coupling (*cont.*)
for humans, 242–243, 245–246
lattices, 252, 256
metapopulations, 238–240, 251
networks, 280
for sessile hosts, 240–241, 245–246
for wildlife, 241–242, 245–246, 252, 253
Covariances between S and I, **195–197**, 228–229, 286
Creutzfeldt-Jacob Disease, 89
Critical community size, **205–209**
Critical vaccination threshold, **292–308**
Cross immunity, **107–109, 118–119, 125–127**.
complete, 107–112
partial, 118–125, 125–128
Culling, 6, 91–92, 271, **274–276**, 321–325, 327–333
Cyclic dynamics
due to seasonality, 155–189, 327–334
due to stochasticity, 195–196, 197–198

Damped oscillations, **29–32**, 35–36, 40–41, 42, 73
Data, 21, 23, 27, 56, 70, 71, 88, 86, 93, 114, 106, 146,
156, 176, 179, 207, 261, 272, 324
comparing model to, 23, 27, 86, 88, 93, 100, 114,
132, 171–172, 175–176, 207, 253–254
parameterization from, 21, 48–52, 64–70, 82–84,
129, 148–151, 219, 236, 253–254
Demographic stochasticity, **200–214**
Dengue fever, 34, 106, 124, 129, 135, 158
Density-dependent birth rate, 74–75, 133, 328
Density-dependent transmission, **17–18**, 35–36, 133,
241
Deterministic vs. stochastic, **190–193**, 213–214,
227–228,
Diffusion, **262–266**
Direct vs. indirect transmission, **2–3**
Discrete space, 255–262, 264, 270–271
Discrete time
approximations, 25, 204–205, 270
models, 46–48
Disease vs. infection, **3–5**, 23, 37, 50–51, 89, 100,
193, 310, 318
Disease-induced mortality, **34–39**, 72, 74–76, 109,
125–128, 133–135, 141–143, 318, 327–329
Distributions, 4, **93–102**, 208–209, 318–321
Dispersal, **240–241**, 253, **265–268**, 272–274
Drift-dynamics of influenza, **120–122**

Ebola, 129, 144, 328
Effective reproductive ratio, 63, 77, 108, 299, 306
Eigenvalues, **30–31**, 42, 47, **60–62**, 72, 170,
Eigenvectors, **60–62**
Endemic, **26–29**. *See also* Equilibria
Epidemic, **28–31, 35–39**, 43, 57–61, 116–118,
127–128, 139, 235, 245, 250–251, 267–278
Equilibria
SIS, 39
SI, 38–39

SIR, 28–29, 35, 36, 38
$SEIR$, 42
stability, 30–31, 42, 47
See also Attractor
Eradication, 63–64, 82, 91–92, 213–214, 246–250,
292–333. *See also* Control
Event-driven models, **200–221**, 321–327
Evolution, **109–112**, 120–122
Exponential distribution
age-related susceptibility, **51–52**, 80–81, 115–116,
303–304, 304–306
infectious and incubation periods, 94–96
Exposed (or latent) class, 3–4, **41–43**, 46–48, 84–87,
89–91, **94–99**, 100–102, 125–127, 149–151,
317–321, 327–329
Extinction
deterministic, 76, 82, 108, 117, 128, 133–135
stochastic, **205–214**, 236, 246–250, 295–296
See also Eradication

Fadeout, **205–209**
Faroe Islands, 260–261
Fixed points. *See* Equilibria
Fleas, **141–143**
Fluctuations
seasonal, 155–189, 327–334
stochastic, 195–196, 197–198
Fokker-Plank equations, **222–227**
Foot-and-mouth disease (FMD), 9, **131–133**,
274–276, **321–327**
Force of infection, **17–18**, 80–81, 122–124, 237–238,
243–244, 267–268, 272–273, 280
Forcing, **155–188**
pulsed birth rate, 185–186, 327–333
sinusoidal birth rate, 183–188
sinusoidal transmission, 159–170
term-time transmission, 85–89, 171–183
Forest-fire models, **259–260**
Forward Euler, 25, 194, 264
Frequency of oscillations, **29–32**, 35–36, 40–41, 42,
73, 160–162, 177–178. *See also* Cyclic dynamics
Frequency-dependent transmission, **17–18**, 36–37,
136
Fourier spectrum, 176, **177–178**, 182, 187–188

Gamma distribution, **93–102**, 208–209, 318–321
Genetic drift, 121–122
Genetics, 76–77, 102, 120–122
Gillespie's methods, **201**, 203, 204, 205
Graphs and networks, 68–70, **276–280**
Gonorrhea, 55–56

Hantavirus, 144–146
Harmonic resonance, **160–163**
Herd immunity, **294**
Hepatitis B, 44–46
Heterogeneities
age, 77–92

risk, 55–77
sexual partners, 65–67, 71–74
space, 232–290
use in control, 63–64, 67–68, 81–82, 89–92,
    246–250, 275–276, 303–306, 306–308, 321–327
HIV, 55, 70, **71–74**
High-risk groups, **57–64**
Horizontal transmission, 75, 89–90

Immigration, **210–212, 241–242**
Immunity
    acquired, 3, **16–34**
    cross or partial, **107–128**
    waning, **40–41**, 214–217, 298–301
Immunization. *See* Vaccination
Imports of infection, **209–212**, 260
Incubation period, 4, 93, 317–320
Indirect vs. direct transmission, **2–3**
Individual-based models, **217–219, 268–275**
Infected vs. infectious, **41–44**
Infection vs. disease, **3–5**, 23, 37, 50–51, 89, 100,
    193, 310, 318
Infectious class, **16–32**
Infectious period, **16–19, 93–102**, 208–209, 318–321
Influenza, 21, **26–27**, 120–122
Integration methods, **25–26**, 194, 264
Integro-differential equations, 99–100, **265–268**
Interference, **113–115**, 125–128
Invasion, **28–31, 35–39**, 43, 57–61, 116–118,
    127–128, 139, 235, 245, 250–251, 267–278
Isolation, 236, 308–321

Jacobian, **30–31**, 42, 47, 170

Koalas & Chlamydia, **74–76**
Kolmogorov-forward equations, **223–228**

Latent period, **4, 41–43**, 94–98. *See also* Exposed
    class
Lattices, **255–262**, 282–283
    cellular automata, 257–262, 269–274
    networks, 277–279
    ordinary differential equations (ODEs), 252,
    255–257
Levins-type metapopulation, **250–251**, 252–255
Likelihoods, **51–52**, 219
Livestock infections, 3, 144, 145, 210, 241, 291
    bovine spongiform encephalopathy (BSE), 38,
    89–92, 99
    foot-and-mouth disease (FMD), 9, 131–133,
    274–276, 321–327
    porcine reproductive and respiratory syndrome
    (PRRS), 214–217
Leishmania, 105, 135, 148
Low-risk groups, **57–64**

Macro- vs. micro-parasite, **2–3**
Malaria,
    age structure, 115–116
    R0 estimation, 115–116
    vector transmission, 135–141
Mass-action transmission
    formulation, 17–18
    mortality, 36–37
    See also Frequency–dependent transmission
Mass vaccination, 5–6, 9, **292–306**
    measles, 178–181, 294
    rubella, 304–306
    problems, 9, 314, 317–318
    pulsing, 249–250, 301–303
    smallpox, 5–6, 9, 314, 317–318
    stochasticity, 295–296
    thresholds, 293–294
    uses, 292, 314, 317–318
    See also Control; Vaccination
Master equations, **223–228**
Maternal antibodies/immunity, **28**, 33
Matrices for transmission, **58–61**, 76–77, 83–85, 88,
    130–131, 138, 150
Measles, **84–89**, 93, 112–115, 155–158, 159–181,
    193, 206–209, 210–211, 260–261
Meningitis, 4, 9, **106**
Metapopulation, **237–255**, 282–283
    basic structure, 237–238
    for humans, 242–245, 245–246
    Levins-type, 250–251
    phocine distemper, 252–253
    rabies, 252–255
    for sessile hosts, 240–241, 245–246
    UK communities, 243–244
    for wildlife, 241–242, 245–246
Methicillin 4, 310
Resistant Staphylicoccus Aureus (MRSA),
Micro- vs. macro-parasite, **2–3**
Migration, **241–242**, 245–246
Minimum infected ratio, **137**
MMR (measles-mumps-rubella) vaccine, **298**, 304
Models, **7–10**
    comparison to data, 23, 27, 86, 88, 93, 100, 114,
    132, 171–172, 175–176, 207, 253–254
    good practice, 10, 149, 200, 204, 269, 282–283
    parameterization, 21, 23, 48–52, 64–70, 82–84,
    148–151, 156–157, 219, 253–254, 268, 272
    simple vs. complex, 10, 42–43, 88–89, 92, 149,
    208–209, 220–221, 251, 255, 257–259, 261,
    282–283, 291–292, 318
    uses, 8–10, 63–64, 71–74, 77–78, 89–92, 100, 106,
    111, 116, 135, 144–145, 148–150, 214, 243–244,
    274–276, 298, 304–306, 313–321, 321–327,
    327–333
Moment closure,
    for networks, 283–286
    for spatial processes, 286–287
    for stochastic models, 227–229
Mortality, **34–39**, 72, 74–76, 109, 125–128, 133–135,
    141–143, 318, 327–329

Mortality (*cont.*)
    density-dependent transmission, 35–36
    frequency-dependent transmission, 36–37
    late in infection, 37–39
Mosquitoes, **136–142**, 148–152, 158, 140–142
    fast dynamics, 140–142
    minimum infected ratio, 137
    quasi-equilibrium, 140–142
    standard models, 136–138
    threshold density, 139–140
Multiple attractors, **167–170**, 179–180, 185–188
Multiple hosts, **128–151**, 274–276, 321–327
Multiple strain, **106–128**
Mumps, 156–157, 260–261, 294, 298
Mutation, 109–111, 120–122

Natural period, 29–32, 35–36, 40, 42. *See also*
    Resonance
Networks, **276–282**, 282–283
    lattices, 255–262, 277–279
    random, 277
    scale free, 279–280
    sexual, 68–70, 312–313 (*see also* Contact tracing)
    simulation, 280–282
    small world, 279
    social, 261, 276, 279–280 (*see also* Contact tracing)
    spatial, 279
Noise, **192–200**, 222–224
    demographic, 197 (*see also* Demographic
        stochasticity)
    observational, 193
    in parameters, 198–200
    process, 193–198
Number of sexual partners, **64–70**, 71–72
Numbers vs. proportions, **17–19**, 191
Numerical integration, **25–26**, 194, 264

Observational noise, **193**
Oscillations
    cross immunity, 122–125
    damped, 29–32, 35–36, 40, 42
    natural period, 29–32, 35–36, 40, 42
    seasonally forced, 155–188
    stochastic, 195–198

Pediatric vaccination, **292–296**, 298–301, 303–306
    age of first infection, 295, 304–305
    eradication threshold, 293–294
    pulsed, 301–303
Pairwise approximations, **283–287**
    networks, 283–286
    spatial, 286–287
Parameterization, 10, **48–52**, 55, 64–70, 82–84, 129,
    148–151, **219**, 236, 253–254
Parapox virus, **133–135**
Partial differential equations (PDEs), 79, 222–223,
    262–265
    from case reports, 50–51, 82–84, 87–88, 100–102,
        175–176

from serology, 51–52, 82–84, 86, 148–151, 215
    for ordinary differential equations (ODEs), 52,
        64–70, 82–84,
    for seasonally forced models, 85, 87–88, 175–176
    for stochastic models, 219, 253–254, 269, 272
Period of oscillations, **29–32**, 35–36, 40–41, 42, 73,
    160–162, 177–178. *See also* Cyclic dynamics
Period,
    exposed, 3–4, 41–43, 46–48, 84–87, 89–91,
        94–99,100–102, 125–127, 149–151, 317–321,
        327–329
    incubating, 4, 93, 317–320
    infectious, 16–19, 93–102, 208–209, 318–321
    prodromal, 4, 319
Periodicity
    determining, 177–178
    due to forcing, 155–189, 327–334
    due to stochasticity, 195–198
Persistence, **205–209**, 211–212, **212–213**
Pertussis, **112–115**, 207, 220, 260–262, 294
Plague, 6, 23, 77
Phocine distemper virus, **252–253**
Porcine reproductive and respiratory syndrome
    (PRRS), 21, **214–217**
Power-law distributions, 110–111, 235, **260–262,
    267–268**, 272–274, 279–280
Prodromal period, 4, 319
Prophylactic vaccination, **292–297, 298–308**
    eradication threshold, 293–294, 297, 299
    age of first infection, 295, 303–304
    seasonality, 301–303
Proportions vs. numbers, **17–19**, 191
Pseudo mass action
    formulation, 17–18, 133, 241
    mortality, 35–36
    See also Density-dependent transmission
Publication rates, 39271
Pulsed births, **186–187**, 329
Pulsed vaccination, **301–303**
    synchrony due to, 249–250

Quarantine, 6, 112–115, 125–128, **308–321**
Quasi-equilibrium dynamics, **140–141**

Rabbit hemorrhagic disease, **185–187**
Rabies, 21,145, 183, **251–255**
Random numbers, **190–191**
Random parameters, **198–200**
Random partnership model, **65–67**
Reaction-diffusion equations, **262–265**. *See also*
    Partial differential equations
Recovery rate, 3–4, 16, **19–24**, 26–28, 40–42, 44–45
Reproductive ratio, 63, 77, 108, 299, 306. *See also*
    Basic reproductive ratio
Resistance, 3, 16–34, 40–41. *See also* Immunity;
    Recovery
Resonance, **160–164**, 192, 196–198

Ring culling, 6, 271, **275–276**, 323
Ring vaccination, 294, 306, 314, **326–327**
Rubella, 21, **304–306**

SARS, 6, 77, **100–102**
Scale-free networks, **279–280**
School terms, **171–183**. *See also* Forcing
SEIR models, 4–5, **41–43**, 85–88, 94–95, 166–182,
    205–207
    comparison to SIR, 42–43, 94–95
    equilibrium, 42
    seasonally forced, 163, 166–168, 173–174,
       176–182
    stability, 42
Seasonality, **155–188**
    pulsed birth rate, 185–186, 327–333
    sinusoidal birth rate, 183–188
    sinusoidal transmission, 159–170
    term-time transmission, 85–89, 171–183
Serology, **51–52**, 139, 215
    age of infection, 32–33, 115–116
    parameter estimation, 51–52, 65, 80–83, 85–86,
      115–116
Sessile vectors, **141–143**
Sexually transmitted infections, 4, 40, **55–76**, 110,
    116–117, 276
    number of partners, 65–69, 70–72
    risk structure, 57–64, 72–74
SIS models, 39, 116–117, 202–203, 220–221, 223,
    312–313
Shared hosts, **130–135**. *See also* Vectors; Zoonoses
SI models, **38–39**
SIR models, **16–34**, 78–79, 107–108, 130–131, 136,
    145, 159–160, 184, 186, 194, 197, 199, 201, 237,
    241–242, 259–260, 263, 293, 309
    basic reproductive ratio, 20–21
    with demography, 26–34
    without demography, 19–24
    equilibria, 28–29
    stability, 29–31
SIS models, **39**, 116–117, 202–3, 220–221, 223,
    312–313
    basic reproductive ratio, 39
    equilibria & stability, 39
    risk–structure, 57–74
    See also Sexually transmitted infections
Simulation methods
    ordinary differential equations (ODEs), 25–26, 194,
      264
    networks, 280–282
    spatial processes, 270–271
    stochastic event-driven models, 201, 204, 217–219
    stochastic noisy models, 194
Simulation time, **203–205**
Sinusoidal forcing, **159–170, 183–188**
Slaved dynamics, **59–62**, 66, 73, 80, 139
Smallpox, 5–6, 9, 21, 292–294, **313–321**

Small-world networks, **279**
Spatial models, **232–290**
Spectral analysis, 176, **177–178**, 187–188
Squirrels, red vs. gray, **133–135**
Stability, **30–31**
    Jacobian, 30–31, 42, 47, 170
    SEIR equilibrium, 42–43
    SIS equilibrium, 39
    SIR equilibrium, 29–32
Stochasticity, **190–232**, 321–327,
    additive noise, 193–200
    demographic stochasticity, 200–221
    imports, 209–212
    need for, 190–193, 216–217
    parameterization, 219
    persistence, 205–209, 211–213, 236
    resonance, 192, 196–198
    scaled noise, 197–200
    simulation methods, 201, 204, 217–219
    and spatial models, 221, 232
    vaccination, 213–214, 295–296
    variance, 191, 195–198, 227–229, 234–235
    See also Cellular automata; Individual–based
      models; Metapopulations; Networks
Subharmonic resonance, **160–164**
Susceptibility, 90–92, 116–117, 118–120, 131–132,
    274, 308
Susceptible class, **16–22**, 26–29, 34, 38–39
Swine fever, **327–333**
Synchrony, 88, 234–235, **245–250**, 303
Syphilis, 40, **55–56**

Targeted control, 5–7, **291–292**
    culling, 6, 91–92, 274–276, 321–325, 327–333
    quarantining, 6, 112–115, 125–128, 308–321
    vaccination, 5–6, 213–214, 292–308, 314
τ–leap methods, **204–205**, 217, 270
Temporal forcing, **159–188**, 327–333
Term-time forcing, 85–89, **171–183**
Thresholds
    for eradication, 63–64, 82, 91–92, 213–214,
      246–250, **292–333**
    for invasion, 28–31, 35–39, 43, 57–61, 116–118,
      127–128 139 (*see also* Basic reproductive ratio)
    for vaccination, 213, 293–294
Ticks, 136, **141–142**
Time since infection, 5, 74, **93–95**, **98–101**, 218, 310
Toxoplasmosis, 144–145, 291
Trade-offs and evolution, **109–111**
Transmission, **16–19**
Transmission kernel, 235, **267–275**, 286–287, 322

Vaccination, 5–6, 179, 213–214, **292–308**, 314
    age-structured, 81–82, 86, 295, 303–306
    boosting, 298–301
    mass, 297–298, 325–326
    pediatric, 81–82, 86, 178–180, 292, 298–301

Vaccination (*cont.*)
  pulsed, 249–250, 301–303
  ring, 326–327
  rubella, 304–206
  targeted, 63–64, 67–68, 81–82, 306–308
  thresholds, 213, 293,294
  wildlife, 296–297
Vector notation, **64–65**
Vectors, **136–143, 148–152**
  fast dynamics, 140–142
Vertical transmission, 74–75, 89–90
Virulence, **109–111**. *See also* Disease–induced
  mortality
Virus, 1–3, 15

WAIFW (Who Acquires Infection From Whom),
  58–61, 76–77, 83–85, 88
Waves, 123, 149, 234–235, **252–255, 256–257,
  265–266**, 267–268

West Nile virus (WNV), **148–151**
Whooping cough, **112–115**, 207, 220, 260–262,
  294
Wildlife diseases, 3, 17–18, 130–131, 183–187,
  251–255, 296–297
  Chlamydia in koalas, 74–76
  metapopulation coupling, 241–242, 245–246,
    252
  parapox virus, 133–135
  phocine distemper, 252–253
  rabbit hemorrhagic disease, 38, 185–188
  rabies, 21, 145, 183, 233, 252–255, 296
  swine fever, 327–333
  transmission assumptions, 17–18, 74–75, 134
  vaccination, 296–297
West Nile virus (WNV), 148–151
  zoonoses, 143–151

Zoonoses 3, **143–151**

Parameter Glossary

N	Total population size.
X, W, Y, Z	Number of susceptible, exposed, infectious, and recovered individuals.
X_A, Y_A etc	Number of susceptible, infectious individuals who also belong to class/species/population A. (Chapters 3 and 4)
S, E, I, R	Proportion of population that are susceptible, exposed, infectious, and recovered. ($S = X/N$, etc).
n_A	Proportion of entire population that belong to class A. (Chapter 3)
S_A, I_A	S_A is the proportion of the entire population that belongs to both the susceptible class and to class A. I_A is similarly defined (Chapters 3)
β	Transmission rate of infection.
β_{AB}	Transmission rate of infection to class/species A from class/species B. (Chapters 3 and 4)
γ	Recovery rate. $1/\gamma$ is the infectious period.
σ	Rate of moving from exposed to infectious class. $1/\sigma$ is the latent period.
μ	Natural per capita death rate.
ν	Birth rate; often we assume $\nu = \mu$.
R_0	Basic reproductive ratio; average number of secondary cases produced by an average infectious individual in a totally susceptible population.
R_∞	Final epidemic size; expected proportion of the population infected in a simple epidemic.
A	Average age of first infection.
p	Proportion of the population vaccinated at birth.
υ	Rate of vaccination of susceptibles.
ρ	Mortality probability; probablity of dying due to infection before recovery.
m	Disease-induced mortality rate; increased rate of mortality due to infection.
w	Rate of waning immunity; rate of moving from recovered to susceptible classes.
l_A	Rate at which individuals mature and leave class A for the subsequent class; $1/l_A$ is the average amount of time spent in class A. (Chapter 3)
N_{SI} etc	The proportion of the population that are in state S with respect to disease 1 and state I with respect to disease 2. (Chapter 4)

b	Bite rate of mosquitoes or other vectors. (Chapter 4)
T_{AB}	Transmission probability to species A from species B per vector bite. (Chapter 4)
β_1	Relative amplitude of sinusoidal transmission forcing. (Chapter 5)
b_1	Relative amplitude of term-time transmission forcing. (Chapter 5)
δ	Stochastic rate at which infectious imports join a population. (Chapter 6)
ε	Stochastic rate at which susceptibles are infected due to imports from an external population. (Chapter 6)
ρ_{ij}	Degree of spatial interaction to population i from population j. (Chapter 7)
$K(d)$	Transmission kernel; measures how the risk of infection declines with distance d between an infectious and susceptible individual. (Chapter 7)
ξ	Normally distributed error rate (Chapter 6).